1420

A Practical Guide to
GAS CONTRACTING

A Practical Guide to GAS CONTRACTING

PennWell

Ann O'Hara

A Practical Guide to Gas Contracting

PennWell Books
PennWell Publishing Company
Tulsa, Oklahoma

Copyright 1999 by
PennWell
1421 S. Sheridan Road/P.O. Box 1260
Tulsa, Oklahoma 74101

Library of Congress Cataloging-in-Publication Data
O'Hara, Ann
    Natural Gas Contracts: A Practical Guide / Ann O'Hara
        cm.
    Includes index
    ISBN 0-87814-764-0

All rights reserved. No part of this book may be reproduced, stored in a retrieval system, or transcribed in any form or by any means, electronic or mechanical including photocopying or recording, without the prior permission of the publisher.

Printed in the United States of America.

Cover design by Morgan Paulus
Book layout by Brian Firth

Table of Contents

List of Figures vi

List of Acronyms vii

Preface ix

Acknowledgements x

Chapter 1 Introduction 1

Chapter 2 Gas Industry Players (and why they play that way) 51

Chapter 3 Natural Gas Transportation Laws and Regulations 91

Chapter 4 U.S. Gas Sales Laws 109

Chapter 5 North American Sale of Goods 151

Chapter 6 Essentials of Contracting 185

Chapter 7 The Gas Contract 241

Chapter 8   The Gas Contract—Part 2 289

Chapter 9 New Issues: Over-the-Counter (OTC) Derivatives
Documentation and Electronic Contracting 339

Chapter 10 Practical Tips for Gas Contractors 377

Glossary 403

Index 431

List of Figures

1-1	Hedging to Achieve Price Goals	26
1-2	A Legal Futures Contract	29
1-3	Goal Achievement through Futures Contracts Sale and Purchase	33
1-4	Using a Swap to Establish a Fixed Price	36
1-5	Using INDEX-based Prices	37
1-6	Transferring Risk to a Derivatives Dealer	38
1-7	Using Options to Hedge	39
2-1	Producer's Force Majeure Clause	58
2-2	Marketer's Force Majeure Clause	64
2-3	Example of Force Majeure Pass-on	65
2-4	Result of Prudency Review	69
2-5	LDC Force Majeure Clause	72
2-6	Industrial Consumer Force Majeure Clause	76
2-7	Example of Two-Tiered Gas Supply Contract	85
3-1	Capacity Release Transaction—Privately Negotiated	102
3-2	Capacity Release Transaction—Publicly Bid	103
4-1	Basic Layout of Gas Gathering and Trunk System	117
6-1	Situation Establishing a Breach of Warranty Action	232
8-1	Transactions Under Base Contract	299
9-1	Early Termination Process	353

List of Acronyms

A/P	Accounts payable
A/R	Accounts receivable
ABA	American Bar Association
ADR	Alternate dispute resolution
ANSI	American National Standards Institute
ALI	American Practicing Law Institute
Btu	British thermal unit
CD	Contract demand
CFTC	Commodity Futures Trading Commission
CISG	Convention for the international sale of goods
CODE	Uniform Commercial Code (UCC)
COLA	Cost of Living Index
CPUC	California Public Utility Commission
CRE	Commission de Regularia Energia
DUNS	Dun and Bradstreet Number
EBB	Electronic bulletin boards
EDI	Electronic data interchange
EDM	Electronic delivery mechanism
EIA	Energy Information Administration
EPAct	Energy Policy Act
ERMA	Energy Risk Management Association
FERC	Federal Energy Regulatory Commission or Commissioners
FPC	Federal Power Commission
FLEX	Fictitious name used as an example
GASCO	Fictitious name used as an example
Gas EDI	Canadian standards-setting organization
GASFLOW	FERC work group
GIC	Gas inventory charges
GISB	Gas Industry Standards Board
ICC	International Chamber of Commerce
IFGMR	Inside FERC's Gas Market Report
ISDA	International Swaps and Dealers (Derivatives) Association
JOA	Joint operating agreement
KCBT	Kansas City Board of Trade

L3D	Three trading days prior to settlement
LDC	Local distribution company
LNG	Liquified natural gas
LOI	Letters of intent
MDQ	Maximum daily quantity
MMS	Mineral management service
NAFTA	North American Free Trade Agreement
NCCUSL	National Conference of Commissioners on Uniform State Laws or Uniform Law Commissioners
NEA	National Energy Act
NEB	National Energy Board (Canada)
NGA	Natural Gas Act
NGPA	Natural Gas Policy Act
NGWDA	Natural Gas Welhead Decontrol Act
NOPR	Notice of Proposed Rulemaking
NYCE	New York Cotton Exchange
NYMEX	New York Mercantile Exchange
OCC	Office of the Comptroller of Currency
OCS	Outer continental shelf
OCSLA	Outer Continental Shelf Lands Act
OTC	Over-the-counter
PEMEX	Petroleos Mexicanos, S.A.
PIPE	Fictitious name used as example
PUC	Public Utility Commission
PUMP	Fictitious name used as example
REFLEX	Fictitious name used as example
RFP	Request for proposal
SFV	Straight fixed variable
SGA	Sale of Goods Act
T&E	Transportation and Exchange
TSA	Transportation service agreement
TSP	Transportation service provider
UCC	Uniform Commercial Code (CODE)
UNCID	Uniform rules of conduct for interchange of trade data by teletransmission
WIDGET	Fictitious name used as example

Preface

The United States natural gas industry is undergoing dramatic changes that began when the government responded to the energy crises of the 1970's. From total regulation of nearly every aspect of natural gas sales and transportation to deregulation in nearly every arena, the industry today is one in which competition is indeed alive and well. In 1988, I first walked into the offices of PSI, Inc., a small gas marketing company in Omaha, to begin my on-the-job training as a contract attorney. I began that job on a dead run, and I've been running ever since—just to keep pace with changes to the industry, changes in contracting practices, and changes in the law.

I wrote this book to help others who might be going through the same type of experience that I had—trying to learn a job while keeping up with the industry. That's why the guide is filled with illustrations of everyday contracting problems, but all such illustrations are hypothetical only and not based upon any actual cases or facts.

Many legal aspects are discussed in this guide as well. But this should just be a starting point for you when analyzing any particular set of facts. This guide is not intended to provide legal advice, but is being provided as a learning tool. Any individual situation in your own company should be examined in light of its own special circumstances, and legal, tax and financial counsel should be sought in appropriate circumstances. The contract language examples found in this guide are included for educational purposes only, and no recommendations are made as to form, style, or legal consequences.

Acknowledgements:

Were it not for my wonderful clients through the years (and all the fascinating legal issues I encountered because of those relationships), this book could not have been written. For their kindness and generosity with time and talents, I especially want to thank my many friends at Aquila Energy Corporation and at Tenaska Marketing Ventures. My husband, Paul O'Hara, edited this book for me. I cannot express sufficiently my thanks to him for his support, his kind words of encouragement, and mostly for his valuable editing suggestions. Finally, I want to thank my parents, whose many sacrifices made it possible for their children to have a better life. I am a blessed person.

Chapter 1 Introduction

This is a book about natural gas sale and purchase contracts. It is intended to be a practical tool for anyone involved in contract negotiation, drafting, or revision. The fundamental premise of this guide is that successful gas contractors are those who understand the natural gas industry in particular, as well as the laws governing sales of goods in general, and who succeed because of those understandings. Grounded in those two assumptions, the following chapters will highlight the areas within which successful gas contracts may be crafted.

The guide has been formulated for both attorneys and non-attorneys alike. My experience has revealed that the gas industry has far more non-attorney employees working in the contracts area than attorneys, so when I first decided to write this book, it was primarily with the intention of giving some guidance to those who haven't had formal training in contracts. But, attorneys may find the material to be a helpful starting point for specific research on any gas contracting topic as extensive endnotes provide more in-depth information for issue research. The language and terminology used will typically be that of the gas industry, legal terms will only be used when the context requires, and practical applications of contracting principals as well as specific contract language examples will be generously scattered throughout the book.

Where do you start when you've been handed responsibility for negotiating, drafting, or revising a gas contract? How do you know which contract language to accept and which to reject? Which risks can your company absorb and which risks must be minimized? How do you weigh benefits for your organization against potential costs? The most logical starting point is that you gain a general understanding of the physical natural gas industry. Whether gas is sold at all depends on whether it is available for delivery, and in many instances, some event involving production, transportation, or other physical activity beyond the control of the seller or the buyer, interferes with ready availability of the gas.

To negotiate a natural gas contract, you don't need to know the history of the industry, but you're going to learn what you should know. You do need to know how gas moves from its geological formations to the burner-tip and to know the history of gas regulation. Today's gas contracts are effected both by what nature constrains and allows in moving gas from one place to another, and what regulators allow in the same flow of gas.

One of the continuing challenges of the gas industry is the tension between things that never change, and things that change all the time—physical laws of nature don't change, regulations and marketing opportunities do. Immutable laws of science govern whether, and to what extent natural gas is removed from the ground and transported through pipelines—gases move only from areas of higher pressure to areas of lower pressure. While scientific and technological advances in natural gas exploration and production have opened vast new gas reserves to commercial usage, one fact remains unchanged that scientific gas laws cannot be coaxed to operate other than the way nature dictates. Gas contracts must reflect the limitations caused by natural laws.

On the other hand, the manner in which gas is bought and sold is subject to nearly continuous change, as organizations seek out new and better ways to improve margins in times of low profitability on sales of the commodity itself. You are going to learn about the transportation of natural gas, which is regulated, the wholesale sales, which are not. You will also learn that gas contracts must provide flexibility to buyers and sellers who have long-term relationships, and want their contracts to be dynamic and adaptable to regulatory changes, as well as changes in the market. This constant tension between rigid laws and regulations and entrepreneurship at its best can be hard on some gas industry participants, and even harder on third-party consumers. The bottom line for people employed in the gas industry actually boils down to this: If you require certainty and constancy in your business or profession, work on the science of gas movement. If your work is on the business side, stay nimble and alert because it is always an exciting ride. Let's take a look at what this unique industry actually is.

GAS INDUSTRY HISTORY AND DEVELOPMENT

The natural gas business has been a vigorous part of the North American commercial landscape for most of the Twentieth Century, but the businesses referred to collectively as the *natural gas industry* are really a compilation of several discreet business sectors bound together by common interests and interdependencies, as well as common ownership in some cases. The major business sectors comprising the industry include *exploration, drilling, production, processing, transportation, storage, distribution,* and *marketing*. The exploration industry is responsible for determining where to drill wells in the hope of finding underground natural gas in commercial quantities. Once a site has been identified and other necessary prerequisites satisfied, *drilling* begins. Drilling will either result in a well completion, if tests indicate a winner, or a plugged and abandoned hole in the ground if it is not economically feasible to extract the gas in commercial quantities.

Production is the process of physically removing the natural gas from completed wells, after which the *transportation* sector takes over. Any movement of natural gas through pipelines is referred to as transportation, yet even this sector is further subdivided. First, gas is typically moved from the wellhead[1] through a feeder line to a common, relatively small-diameter pipeline gathering system.[2] A gathering system's function is to collect gas from multiple wells and deliver that commingled gas to a larger diameter, more highly pressurized transmission line.

Before gas can be transported long distances, however, it generally will be *processed* to remove both unwanted and valuable constituent elements of the native gas stream. The unwanted non-combustible products such as carbon dioxide, nitrogen, and hydrogen sulfide—if left in the gas stream—could corrode pipelines, damage fittings, and valves, in addition to interfering with measurement and combustion. The same holds true for water and sediment produced with the native gas stream. Heavier hydrocarbons such as ethane, butane, and propane are commercially valuable and may be removed from the gas, usually through the fractionation[3] process. Gas that has been processed to levels satisfactory to meet the transporting pipeline's tariff[4] requirements is termed *pipeline quality.*

Long distance transportation of gas, typically subject to regulation by either the state or federal government is called *transmission.* Transmission may involve thousands of miles of movement from supply to consumption, and the gas must be repressurized by artificial compression at regular intervals along the pipeline system at compressor stations. Gas may be physically stored—either in above-ground tanks, if the gas has been pressurized and cooled to become liquefied natural gas (LNG) or, more commonly, the gas is stored underground. This *storage* may occur close to areas of production or close to market areas.

The final phase of gas transportation, treatment, and measurement is referred to as *distribution.* At the *city gate* gas is measured, perhaps depressurized, cleaned and mercaptans[5] are added. Then the gas is distributed through local distribution company (LDC) pipelines with ultimate delivery to residential, commercial, and industrial end-users. All of the previously listed sectors together comprise what is known as the *physical* gas industry.

The final segment—sometimes totally separate and apart from the physical industry—is the *marketing* sector. Marketing entities are the organizations that typically *aggregate* gas, i.e., they contract with multiple gas suppliers and multiple markets. They purchase and sell natural gas, and profit when gas sale prices are greater than their gas purchase costs.

Exploration, drilling, production, transportation, processing, storage, distribution, marketing—each of these sectors is a unique industry completely separate from the others, yet each is also a part of the natural gas industry as a whole and dependent on the others to survive. For example,

Chapter 1 Introduction

the natural gas transportation network in the U.S. alone includes more than one million miles of pipeline, but if natural gas were not produced, the need for this intricate system of both discreet and interconnected physical pipelines would no longer exist. If industrial consumers of natural gas didn't require huge quantities of gas to fuel their manufacturing processes, if homes and businesses weren't heated with natural gas, and if the electric industry were not using large quantities of natural gas to generate electric power, both the production and transportation segments would have no purpose.

Incidentally, each of the services provided by the physical industry requires a contract—gathering service agreements, storage service agreements, transportation service agreements and other service-related agreements are separate from, but related to, the gas sale contract. The interrelationships between gas gathering, storage, sales, and transportation agreements result in multiple agreement failures from time to time when a failure under one contract causes a domino effect of failures under other related contracts.

Separate and unequal levels of governmental regulation further differentiate the natural gas industry. Federal, state, or local governments may impose regulation, and inconsistent layers of regulation aren't always compatible with a national or North American gas industry. Some U.S. transportation pipeline companies are subject to federal regulation that is evenly applied to all other similarly situated pipelines regarding pipeline services, operations, and rates. Other transportation pipeline companies are subject to regulation by state, county, or municipal governments and this regulation isn't uniform from jurisdiction to jurisdiction—certainly not from county to county in North America. Since any given natural gas sale may involve both federally regulated transportation and state or provincial regulation, the regulatory framework confronting organizations that ship gas through more than one pipeline company's facilities may at times be rather complicated. The potential regulatory incompatibilities must be considered in contract drafting as well.

Finally, to make the natural gas business even more interesting is its involvement with the oil sector. Natural gas is being produced virtually everywhere that commercially producing reserves have been discovered. In some underground rock formations (reservoirs), both natural gas and oil may be recoverable in the same wells. Although some wells produce only natural gas, others produce oil as well as gas, so both industries, in many respects, share a common history. In the U.S., natural gas and oil are abundantly produced in certain areas of the country—Texas, Alaska, Oklahoma, Louisiana, Wyoming, New Mexico, Kansas, and Colorado. In Canada, the Western Canadian Sedimentary Basin, located in British Columbia and Alberta, produces most of that country's natural gas, although vast new reserves are located near the Sable Islands, 150 miles offshore of Nova Scotia. Most of these major production areas also produce oil. Mexico has large natural gas reserves, ranking thirteenth in the world in natural gas production and seventh in oil production.

Colorful stories of oil discoveries abound to this day. Picture, if you can, the earliest wildcatter[7] days in the U.S.—let's go back about 100 years. The first oil wildcatter in Oklahoma City arrived in 1890, convinced that the city floated on a pool of black gold. He dug a test well in the city, but at 600 feet he gave up. Although he was correct in his assessment, it wasn't until 1928 that oil was discovered in Oklahoma City, a city built among oil wells. Today the Oklahoma State capitol grounds are virtually covered with oil wells still producing measurable amounts of black gold. In California, the town of Mentryville was established in 1876 when Pico No. 4, the first commercially successful oil well in the western U.S., began production. That well holds the world longevity record as production continued until 1990, long after the town of Mentryville had died. John Galey, the greatest Nineteenth Century American wildcatter, was so good at finding oil that it was said he could smell it. Galey was the one who said, "drill here," in Beaumont, Texas, for the great well that became Spindletop and opened the Texas oil industry. Galey said, "only Dr. Drill knows for sure"—a statement as valid today as it was then.

A gunsmith named William Aaron Hart discovered the first recoverable amounts of U.S. natural gas near Fredonia, New York in 1821. While drilling for water, he heard a hissing sound coming out of the hole he had dug. Experimentation with the properties of the gas proved to Hart the combustibility of natural gas, which is comprised essentially of methane.[8] He fashioned a crude pipeline out of tree trunks, captured the gas coming out of his "water well," and transported enough gas to provide light to a number of buildings in Fredonia. Even though other areas of the country are generally more closely associated with natural gas supplies and production, Fredonia, New York is where natural gas production was born in the U.S.

Natural gas was discovered in plentiful quantities in the 1800s, but it was much more difficult and costly to transport natural gas than oil, so while oil became a valuable energy source almost immediately, natural gas did not. Natural gas wasn't used commercially until it became feasible to transport gas to consumers for lighting and other purposes. Prior to the industrial age, the process of manufacturing and laying the cast iron pipeline necessary for natural gas transportation was difficult at best. In the late 1800s, horsepower was just that—horses and mules and men doing the backbreaking labor required to manufacture and lay the pipes necessary for this new industry. After the process for making steel was formulated and welding techniques were developed, the gas industry began to grow. Not until the 1950s, however, did the U.S. natural gas industry begin to flourish, and then primarily because of the many metallurgical advances made during World War II.

Once pipelines were laid and distribution systems developed, the nascent structure involving production, transportation, and distribution of gas was formulated. Practices were very different in the southwestern and southern oil fields of the U.S. from those in the northeast where coal methane gas was being used in Baltimore. Methane gas removed from coal, also called *manufactured gas*, was first used for lighting purposes in highly populated cities. Customers who purchased methane gas from distribution companies in Baltimore typically paid a flat monthly fee for those services, based upon the number of lights in the home or commercial establishment. Those sales and purchases were made in the tradition of the day and contractual relationships were dictated by the distribution company.

Contrast that urban system of final gas sales with the oil fields of Texas and Oklahoma where initial oil sales were in large part consummated only by a handshake—if at all. Because of the hardship in transporting natural gas in the early years, the gas was most frequently flared or vented into the atmosphere. Early production companies produced oil and looked at natural gas as more of a nuisance than a product with value. Many of the traditional ways of selling gas at the wellhead in the natural gas industry were born in the oil fields of Texas and Oklahoma. A hundred years ago, a "man's word was as good as his bond" and many industry pioneers would have been insulted if anything beyond a handshake had been required to memorialize the deal. Today a man's word may still be as good as his bond, but if gas buyers and sellers want to enforce their contractual rights and obligations in a court of law, that court will undoubtedly require some written evidence of the deal.

From diverse industry beginnings in the oil fields where strict contract structures were avoided, and in urban distribution systems where the merchants' rules were strictly followed, different traditions and practices formed the soul of this dynamic industry. Those original traditions and practices still bedevil the industry to an extent, particularly in the contracting realm. Producers still tend to view gas contracting differently than do LDCs and other consumers. One of the unfortunate results of differing approaches to gas contracting is it may take an inordinate amount of time to negotiate a contract if the seller and buyer come from industry cultures that view contracting differently.

In today's wholesale gas environment, many transactions are valued at hundreds of thousands—or millions—of dollars; a gas sale or purchase transaction is typically for large quantities of gas. It would seem prudent to negotiate and sign contracts evidencing the agreements between buyers and sellers of gas when the stakes are so high. This isn't always the case. Millions of dollars of natural gas are bought and sold daily between organizations that have not signed a gas sale contract. It is not unusual for contract negotiations today to involve weeks and sometimes months of intense bargaining. For organizations that have active physical trading functions, two painkillers—the fax confirmation and telephone tape recording of trader activities—have alleviated the administrative discomforts related to contract negotiations. Both of these will be discussed in chapter 4, where you will also find guidance for resolving the issues that occur when gas flows without a contract in place.

SIX ISSUES CRITICAL TO GAS CONTRACTING

Gas contracting needs a *roadmap*. In the evolutionary continuum, ours is a new industry. It is suffering the growing pains of its youth. As you have already seen, contracting for gas transactions is a developing art. Unlike your income tax forms, there are no standardized instructions available. For those of us whose lives are spent perfecting this art, there are standards, principles, clauses, and phrases that enable the contract negotiator and writer to construct a durable document. That is our goal. Let's look at what is happening in the gas industry that will effect gas contracting into the future.

There are six focal points of change, and each will have an effect on how natural gas contracts are negotiated, written, and revised. The six areas of focus are:

- regulatory changes on the federal and state level
- a changing role for regulators as competition increases
- creation of an *energy* marketing industry
- natural gas futures contracts and other hedging products
- electronic contracting and communications enhancements
- Gas Industry Standards Board and further standards development

Later chapters will provide more detailed information about these topics, particularly regulation and electronic contracting. For now, you need some historical context for each of the subjects.

Regulatory changes at the federal and state level

The history of regulation for interstate[9] pipelines in the U.S. began toward the end of the Great Depression. Prior to that time, the U.S. government didn't regulate interstate transportation of natural gas because no law had been passed to effect such a regulatory framework. Typically, the state of origin of a pipeline system would provide whatever regulation may have been applicable. When gas pipelines began crossing state lines in the 1920s and 1930s, individual states lost the ability to regulate natural gas activity. The U.S. Constitution reserves the pre-emptive right to regulate interstate commerce[10] to the federal government. States, even those with extensive pipeline systems located within their borders, could not regulate the commercial activities on those pipelines after the contiguous system originating in that state physically crossed into another state. Congress acted for two primary reasons. First, pipelines are natural monopolies. Two pipelines cannot be in the same place at the same time so if one pipeline can provide adequate service to a particular market, all others will be excluded from that market. This creates the monopoly and since no competition exists where one entity maintains an exclusive presence in the area, at that time Congress felt the need to provide the leveling influence of regulatory oversight to compensate for the lack of competition. Second, the pipelines were crossing state lines and the states lost the ability to regulate transportation activities or rates charged by the pipeline companies for transportation and sales services.

Congress passed the Natural Gas Act[11] in 1938, and the regulation of all interstate gas sale and transportation activities began. Regulation was imposed with a heavy hand on the pipeline companies themselves, on their rate structure and rates, and on the transportation services they provided. The industry at that time had relatively few participants. Pipeline companies either owned their own gas production or negotiated with producers to purchase gas. If necessary, the pipeline companies contracted with other pipeline companies for upstream transportation and transported gas through their own proprietary systems for delivery and sale to LDCs, who were their customers. The pipelines provided a *bundled service*[12] as LDCs paid a flat rate for the gas itself, for transportation, gathering, processing, and any other service necessary to deliver the gas to the customer. The price paid by the pipelines to producers for gas at the wellhead was not regulated.

In 1954, the U.S. Supreme Court, in the now-famous "Phillips" case[13] decided prices paid to producers at the wellhead were also subject to regulation. The totally regulated interstate gas market now began to see increased competition from the less costly non-regulated intrastate gas pipelines. When faced with a choice of using a regulated interstate system with relatively high bundled rates or a nonregulated intrastate system, customers who had a choice chose the intrastate pipeline services. Even before the two energy crises of the 1970s, the interstate natural gas industry struggled with the laws of supply and demand. Over the years, as demand dropped on the interstate system, so did the need for new interstate production. Low demand for natural gas, coupled with a realization of the danger of dependence on foreign oil brought on by the oil embargo of the 1970s, prompted

Congress to spur natural gas development by passing the Natural Gas Policy Act[14] (NGPA) in 1978.

The purpose of the NGPA was to jump-start exploration and production of gas by setting maximum wellhead prices for gas produced from different types of wells. For example, wells drilled after a certain date were classified as "new" wells and had a relatively high maximum lawful price attached to production. "Old" wells, those already in production in 1978, had relatively low maximum lawful prices. (For a further description of the price categories, see chapter 3.) The plan worked. New wells were drilled and began production. Absent any economic incentives to sell *old* gas, producers began to *shut in*[15] their old wells, so pipelines were competing for access to high-priced *new gas*. The competition between pipelines led to *take-or-pay* contract clauses in which the pipelines agreed to contract for an amount of gas and pay for it even if they didn't take it.

Predictably, the laws of economics exerted their superiority over the laws of regulators and lawmakers. High gas cost drove consumers to alternative fuels such as oil, which were now cheaper. This produced a *gas bubble*, a surplus of production that existed for several years. By the 1980s, the industry was again facing a new crisis—too much available gas and a chokehold of inflexible laws caused by the solution to the previous crisis. In 1984, the Federal Energy Regulatory Commission (FERC or Commission), the successor to the FPC, took action to correct the economic imbalance and stimulate a more competitive atmosphere in the natural gas industry.

In 1985, FERC issued Order No. 436,[16] authorizing third-party transportation on the interstate pipelines. It was a radical change. Prior to Order No. 436, interstate pipeline companies owned all the gas moved through their pipes. Producers sold gas to the pipelines and LDCs bought gas from them. Order No. 436 opened access to the transportation grid so that a third party could buy or sell gas and pay the pipeline to transport it. The pipelines weren't required to provide open-access transportation. The order contained sufficiently attractive incentives[17] as well as some hefty disincentives[18] for any interstate pipeline company planning to reject open-access transportation. Eventually all interstate pipelines did "go open access" and negotiated or litigated their way out of those long-term take-or-pay contracts with producers.

This FERC order spawned the concept of gas marketing companies.

Individuals and small groups of people working for producers, pipelines, and LDCs began figuring out ways to buy, sell, and transport gas on the *spot* [19] market. More than one nationally known energy marketing company began as a couple of LDC employees with some vision and existing contacts within the gas industry.

The order was momentous in its impact, but not perfect. Order No. 436 did not prohibit pipeline companies from buying and selling gas through their own marketing affiliates. Competing marketers, producers, and other shippers soon complained to the FERC about discriminatory treatment the pipelines could give to their own affiliates as shippers. FERC Order No. 497[20] addressed that issue and required separation of the pipeline's transportation and marketing functions—sometimes to the point of requiring the pipeline operations personnel and marketing personnel to be in separate buildings. Pipeline companies were in both the bundled gas sale business and the third-party transportation business after Order No. 436 was issued. Even with its problems, Order No. 436 heralded the era of *self-implementing*[21] transportation. Except for some adjustments along the way, culminating in the virtual elimination of interstate pipeline companies as gas merchants, the U.S. interstate gas transportation industry today continues to operate under the format of original Order No. 436.

With Order No. 436, the FERC was attempting to provide equal access to markets for gas producers and equal access to producers for gas consumers. The statements expressed by FERC Commissioners during open-access proceedings indicated interest in providing an atmosphere for full and fair competition. As the gas industry performed under the open-access order, though, it became apparent to the Commission that more work needed to be done. As long as the pipeline companies could provide a bundled sale and transportation service to their own customers, the transportation component of that service would always be superior to the firm[22] transportation provided to third-party shippers. In the early 1990s, the FERC concluded the only answer to this continuing competitive dilemma was to change the structure of the industry again.

Order No. 436 was not finally implemented until 1991[23] when the U.S. Supreme Court refused to hear a case threatening the legality of portions of the order.

That same year the Commission issued a "Notice of Proposed Rulemaking" (the *MegaNOPR*) regarding the issues eventually embodied in Order No. 636[24]—primarily unbundling sales and transportation services on the interstate pipeline systems. During its attempts to deregulate the wellhead market, the FERC's goal had been to ensure that consumers could realize the full benefits of competition right at the wellhead. Order No. 436 did indeed open the pipeline to producers, new marketers, consumers and distribution companies. But the commissioners believed that the interstate pipeline companies gave better transportation service to its own sales customers than to its third-party shippers. It was a reliability issue.

The major request of third-party shippers to pipeline companies as described by the FERC in Order No. 636 was, "Give us the same level of firm transportation service that you give yourselves." Pipeline companies countered that they couldn't improve third-party transportation service because doing so might put their obligations to LDC customers at risk.

The LDCs had yet another viewpoint. They knew they would get all the gas they needed to fill their customers' gas needs on the coldest day of the year with their current level of service. They were unwilling to settle for anything less. The LDCs had obligations as utilities to serve customers within their own franchised service areas and they needed the reliability of a bundled service. Everyone wanted Grade AAA service and there just wasn't enough Grade AAA service to go around.

When the FERC examined the concerns of LDCs and other industry participants, the commissioners concluded that the only way to provide for full access to gas markets, while at the same time improving the competitive structure of the natural gas industry, was to eliminate the pipeline companies' ability to sell gas at the city gate. Their future was to be like truckers—transporters only. This regulatory framework is a description of today's interstate gas transportation industry to which all participants, both regulated and non-regulated, must adhere.

The result of the reliability issue so hotly debated in 1992 is that now all firm shippers receive the same level of transportation service. The high-peak usage LDCs, concerned about incurring penalties for taking gas they had not nominated on peak days, may obtain *no-notice* firm service from the pipelines that were providing them with essentially the same no-notice service prior to Order No. 636. This level of service continues to give the LDCs the reliability they had before Order No. 636 was issued.

The Changing Role of Regulators

Regulators have shifted from a passive role of reacting to change to *causing* change, not just in the U.S., but in Canada and Mexico as well. Add the leveling effect of the North American Free Trade Agreement (NAFTA), and the result is an increasingly dynamic multi-national business environment. The most active regulators have been the FERC and Public Utility Commissions (PUCs) in the U.S.; the National Energy Board (NEB) in Canada, and the Comision de Regularia Energia (CRE) in Mexico.

REGULATION IN THE U.S.

For anyone who has experienced regulation from the outside—*i.e.*, outside the Washington D.C. beltway—it isn't uncommon to have a rather myopic view of the FERC. Change is just as traumatic inside the beltway in Washington D.C. as it is in a Texas or Alberta gas field—and sometimes more so because of the political climate within which any governmental agency must operate.

The FERC is an independent agency within the Department of Energy. The president, with required senate confirmation, appoints the five commission members. No more than three members may be from one political party, and the members serve staggered five-year terms. This agency is charged with oversight of the Natural Gas Act (NGA) of 1938, the NGPA of 1978, the Outer Continental Shelf Lands Act (OCSLA), the Natural Gas Wellhead Decontrol Act (NGWDA) of 1989, and the Energy Policy Act (EPAct) of 1992. The FERC's control extends to the following activities:

- transportation of natural gas in interstate commerce
- transportation of oil by pipelines in interstate commerce
- transmission and wholesale sales of electricity in interstate commerce
- municipal and state hydroelectric projects; and
- related environmental matters

Since Order No. 436 was issued, the Commission has generally favored increased competition in the gas industry. The FERC does not have a role in direct industry competition, but ensures that the rules of engagement do not favor one player over another. Although the Commission has, since the mid 1980s, exhibited this preference for some competition, when the EPAct of 1992 was passed by Congress in 1992,[25] the business of energy industry restructuring or re-regulation was placed squarely on the FERC's front burner, and the commissioners have acted in accordance with the mandate.

The Commission's role is to protect the consumer in an industry in which monopolistic practices are often the most practical and economical, and at the same time, assure access to supplies and markets for all industry participants. At the end of 1999, Order No. 636 had not been fully implemented by all interstate pipeline companies.

The FERC will ultimately assess the success of this mandate after full compliance and may take additional action if necessary. It would be very surprising if the Commission would not continue to make public pronouncements on developing issues, such as further electronic communication mandates, appropriateness of different rules and regulations for different types of transportation, differing rate treatment, and other issues. One thing will remain constant through all current and future FERC activities— the U.S. natural gas industry will not return to the former days of total regulation from wellhead to burnertip.[26]

Generally, all interstate pipelines are bound by the same rules, which has provided comfort to those utilizing the interstate gas transportation system, and heartburn for pipeline executives who have had to recast their operations to fit the new regulatory mold. An order of the FERC has the same force and effect as does a law passed by congress. Those who are comforted by uniformity of regulation, however, need to be aware that while the major FERC orders made the rules uniform, they did not make the pipelines uniform. Pipeline companies may implement the same FERC orders in a variety of acceptable ways. Each pipeline is unique both in location and in configuration. Some have gathering or storage facilities. Others do not. Some pipelines transport gas produced in Louisiana, others in West Virginia, or the Rocky Mountains, or from offshore. Some pipelines are considered to be trunklines,[27] others are grid systems,[28] and some are a combination of trunkline and grid.[29]

Each individual pipeline system has its own unique ownership structure, and load profiles[30] are different from pipeline to pipeline. On a case-by-case basis, the FERC will, within its mandate, massage and push to a certain extent, if one pipeline company's situation doesn't quite fit into a provision of a major order. One example of the FERC's willingness to listen to those it regulates, is seen in its position on rate development. Interstate pipelines, as public utilities, have traditionally used a cost-based rate structure. In Order No. 636, the FERC required all interstate pipelines to utilize the straight fixed-variable (SFV) rate-setting methodology. Under the SFV methodology, all of the pipeline's fixed costs are allocated to the demand[31] component of its rates, and the variable costs are allocated to the commodity[32] portion of its rates. In some circumstances, the SFV methodology would not produce an equitable result for a particular pipeline company or its customers. In those cases, the commission may allow the pipeline to utilize a variation of the SFV methodology. The ultimate transportation rates charged to shippers would, of course, be effected by any such change.

Regulation in the U.S.—intrastate pipelines and LDCs

Pipelines located entirely within one state are regulated by that state. Unlike federal regulation of interstate pipeline rates and services, each state has its own regulatory system and those systems can range from distant oversight to heavy-handed regulation similar to that of the FERC. Because intrastate pipelines and LDCs are public utilities, the state assumes the obligation of balancing the interests of the consuming public with those of the companies providing necessary services to the consumers.

The agency or commission given regulatory authority over intrastate transportation rates and services is typically called the Public Utility Commission or the Public Service Commission (collectively called PUCs, or, among some industry ruffians, "pukes"). Some states have adopted other names for their utility commissions.[33] A search of any state's statutes would reveal statutory authority for the PUC as well as any limitations of that authority. The PUC in turn establishes rules and regulations for public utilities and maintains oversight authority.

At the distribution level, either the state agency or the local municipality or both, may have regulatory jurisdiction over distribution facilities and rates. It is difficult to provide any general guidelines or common definitions for state regulation for either an intrastate pipeline company or an LDC. Each state is different. For example, the California Public Utility Commission (CPUC) is an extremely powerful agency having exclusive jurisdiction over the transportation and distribution systems of LDCs serving millions of customers in California. It isn't uncommon to see disputes between the FERC and the CPUC because of the aggressive positions both entities routinely take. On the other hand, some states have little gas production or transportation activities and a low population, and the PUCs in those states may attend to the other industries subject to their jurisdiction, most recently telecommunications and cable activities.

Because of the vast differences in level and type of regulation of any given state-regulated utility, the watchword for contract negotiators unfamiliar with a particular state's regulatory scheme is carefulness. There are no rules of thumb for regulation of state utilities. Research into the relevant PUC, its policies and its directives, may reap information that could well prevent an inadvertent contracting mistake for a particular transaction. These same cautions apply if the transportation being provided is in Canada or Mexico.

Regulations in Canada—interprovincial

While a virtual global energy industry may be possible given today's technology and industry economics, physical and regulatory barriers still remain even to creation of a North American gas grid. The regulatory barriers include some differences in the industry that make a seamless grid difficult. One of those differences between the U.S. and Canada is in the open market philosophy utilized by the National Energy Board (NEB) that regulates Canadian interprovincial[34] pipeline companies and their rates and services as compared to sometimes burdensome regulation by the FERC.

The Canadian pipeline system itself is quite different from the U.S. system. If it were a matter of which country has the longest pipeline, Canada would be a hands-down winner with the TransCanada pipeline system that traverses nearly all of the country from west to east. Canada covers far more landmass than does the U.S. but Canada has fewer pipelines and a much less complicated physical interconnection system than does the U.S. The total number of Canadian interprovincial pipeline companies (both gas and oil) is fewer than 50, while in the U.S. there are 110 natural gas pipeline companies. Canadian pipeline companies have fewer customers, but those customers are long-term large-capacity holders while in the U.S. many pipelines have a large number of short-term small-capacity shippers.

The physical and regulatory history of natural gas in Canada began after World War II when natural gas and oil resources were explored, discovered, and drilled in the western Canadian provinces. Because pipeline construction contemplated both interprovincial and international deliveries, the federal government was faced with immediate international trade considerations. In November 1959, legislation was passed by the Parliament in Ottawa creating the National Energy Board Act. Under the Act, the NEB undertook the following:

- responsibility for pipelines from the Board of Transport Commissioners
- responsibility for oil, gas and electricity exports from the Minister of Trade and Commerce
- authority to regulate tolls and tariffs
- later was declared to be an independent court of record

In 1994, legislation expanded the board's jurisdiction to include decision-making authority for Frontier Lands not administered through provincial/federal management agreements. The Act has not been changed much since it was originally passed in 1959, but like the U.S., market conditions in Canada have changed dramatically since that time. Natural gas exports have quadrupled, estimates of the Western Canada Sedimentary Basin resource potential have increased by nearly 50%, and more short-term transactions are being negotiated, both domestically and internationally. Throughout this expansion, the NEB has encouraged competition with their regulatory design.

Up to nine permanent members may be appointed by the governor in council. Reappointments may be for any period up to seven years until the member reaches the age of 70. An additional six temporary board members may also be appointed. Three members constitute a quorum and the board may meet at any location it chooses. All proceedings before the board are formal and the board is a court of record. Its decisions are public documents. Once a proceeding is filed, the board has discretion to determine whether an oral or written public hearing is warranted in that case. Natural gas pipelines are classified as *contract carriers* and are not obligated by statute to accept all gas offered to them for transportation. The board does have authority under the Act to require a pipeline to transport gas for a shipper.

Though it may appear that they control their pipelines, gas transporters do operate on an open-access basis. The tariff may contain provisions for queuing and access procedures, minimum duration of contracts, renewal requirements and the bidding mechanism used to secure interruptible service. In Canada, at least at the federal level, interprovincial pipelines have been given the opportunity to operate with very light-handed regulation.

Regulation in Mexico

According to its own estimates, Mexico ranks 7th in world oil production and 13th in natural gas reserves. However, the country and its energy industry face a two-fold challenge, as described by the (CRE)[36]. First, energy demand is currently growing faster than Mexico's economy. Given the importance of energy as a crucial input for the rest of the economy, the gas industry infrastructure will require large capital investments.

Second, Mexico, like the energy sectors of many countries, is going through profound processes of structural reform aimed at introducing a greater degree of private participation and competition. Mexico is studying this new global framework to ensure its competitiveness. A feature of the Mexican gas transportation system unique to North America is government ownership of the major pipeline system. Petroleos Mexicanos, S.A. (Pemex) is the state-owned national pipeline system and, until the end of 1999, transported all interstate gas in Mexico. The first privatized gas pipeline in Mexico, the Yucatan Peninsula Pipeline, was completed before the end of 1999.

The CRE regulates the construction, operation, and ownership of power generation and natural gas transportation, storage, and distribution systems. This entity commenced operations in January 1994 as a technical and consultative body of the Ministry of Energy. Nonetheless, 1995 reforms brought about fundamental changes in the energy sector that called for the redefinition of regulatory authorities in the industry.

The CRE's mission is to foster productive investment and efficient markets for the benefit of consumers by regulating natural and legal monopolies in the natural gas and power industries in Mexico. It is committed to applying a regulatory framework that is stable and clear in a transparent, impartial, uniform, flexible, and agile manner. The Commission is headed by five commissioners, including a chairman, who are supported by five areas: the executive secretary, three operational areas (natural gas, power, and law), and economics.[37]

The Mexican government has only recently authorized open-access on its state-owned natural gas pipeline and is still developing its regulatory philosophy and trying to contribute to the solution of a country with vast natural resources and an infrastructure currently incapable of sustaining much expansion.

Gas contractors now operate on a landscape of international dimension with three non-complimentary regulatory frameworks: one is intrusive, one is passive, and one is yet to be defined. As the lines between continental gas sales, distribution, and transportation begin to blur, all three national regulatory agencies are going to have to look to a supra-national framework to most efficiently operate on the North American gas grid.

CREATION OF THE *ENERGY* MARKETING INDUSTRY

Remember that prior to 1985 when Order No. 436 was issued, gas sellers were producers that sold to pipelines, natural gas pipeline companies that sold to LDCs, and LDCs that sold to ultimate consumers. After Order No. 436, virtually anyone or any organization could sell natural gas and many newly formed marketing entities took advantage of the non-regulated wholesale gas sale industry. As a practical matter, most wholesale gas marketing companies were started as affiliates of LDCs, producers, and pipeline companies—all three accustomed to heavy doses of state and federal regulation and more than willing to share that burden.

Gas sales had traditionally been made by producers at the wellhead, at the end of a gathering system, or from a lateral line directly into the mainline of a long-distance transportation system, and those sales were made only to pipeline companies. The dawn of open access meant that producers could still make sales at the same locations, but to a larger customer base. After 1985, producers were making sales to pipeline companies, marketers, and directly to end users and LDCs. In addition to the direct market access producers gained, new marketing companies began springing up everywhere. Small niche companies[38] and marketing divisions of established companies quickly gained the national presence they coveted.

With spectacular growth, the gas marketing segment has become one of the largest players in the gas industry. Characterized by "war room" atmospheres of youthful phone salesmen, these companies required minimal capital and virtually no assets. The original marketing entities were not individual subsidiaries or affiliate companies—they were divisions, groups, or departments of the asset-based production, pipeline, or distribution companies. However, with the large potential liability from a gas sale gone bad (*most often because of poorly crafted contracts or inattention to contract terms*), and for other business reasons, most asset-based gas industry companies ultimately formed subsidiary marketing affiliates. During the decade of the 1990s, independent and non-affiliated[39] marketing companies flourished.

Now, at the beginning of a new decade, the organizational pendulum is beginning its return swing—but with a bit of a twist. Large energy organizations producing, selling, and distributing various forms of energy are acquiring, merging with, and incorporating individual natural gas marketing companies into larger, more powerful energy service companies. With their regulatory shackles loosened, previous natural gas companies have begun to diversify into many non-traditional service areas such as real estate development and telecommunications. Of the many gas marketing companies in existence at the beginning of a new century, how many will survive to 2005—five, ten, or twenty? Those surviving companies will have a completely different look and perhaps a different mission, whatever their number. Those organizations that don't become part of the mega-energy company complex must hold a niche in which the mega-marketers cannot economically compete.

Another trend in gas marketing is movement from selling the commodity to diversifying into related services. With the advent of natural gas futures trading in 1990,[40] natural gas pricing became more transparent, and margins narrower and narrower. With the field now saturated, gas marketing companies have evolved into gas sales and energy services. Sometimes the service is the only thing sold. Services can include gas balancing,[41] nominations[42] and accounting, gas procurement, fuel portfolio management, asset management, billing, and embedding hedging products in a purchased or delivered price. Bundling sales with services raises a number of contractual issues addressed in a later chapter.

Electronic Contracting and Communications Enhancements

From fax machines to bicycle couriers careening through city streets, to a dependence on overnight delivery services, activities over the last 15 years have shown what happens when competitive businesses lose patience with the postal system. Now the limitless Internet is available to provide impatient merchants their ultimate tool—instantaneous communication.

The first mass intrusion of electronic communications into the natural gas world came in 1993 when the FERC required all interstate transporters to provide the same information to all shippers on their system at the same time. This was mandated through electronic bulletin boards, or EBBs. EBBs were first used to provide information—they were a form of electronic publishing—but now nominations, transportation service agreements, and other information is being transmitted electronically. The FERC required that all interstate pipeline companies provide direct Internet access and communication capabilities by June 2000.[43]

In addition to e-mail and interactive Internet communications, electronic data interchange, or EDI is being used extensively as a method of electronic business communications. EDI has become a standard mode for conducting business in other industries, but it is still new to the gas industry—used primarily for shipper communications with pipelines.

Where EDI is concerned, a gambling analogy might be appropriate. Industry players are like the first-time player at a craps table. Anyone who has watched a craps game has observed that everyone already in the game seems to understand the rules. The problem for the new player who sees potential for success at the table is this: what are the rules, and how can I learn them and participate without looking like a fool?

Electronic contracting is the exchange of messages between buyers and sellers, structured according to an agreed format. The contents are processed by computers and automatically give rise to contractual obligations. Electronic contracting may include an electronic solicitation of bids, an electronic bid or proposal, or an electronic purchase order, invoice, or payment order. The implications for gas marketing are nearly boundless as more and more organizations are offering on-line gas trading. The world isn't very far from the day when electronic contracting will be the usual course of business dealing for all major industries, but there are substantial hurdles to overcome before everyone *wants* to conduct business on-line. There are many lessons yet to be learned about the impact of pushing that little *Enter* button—particularly with regard to contractual obligations.

NATURAL GAS FUTURES CONTRACTS AND OTHER HEDGING PRODUCTS

The basics of gas pricing

There are two basic spheres of natural gas prices: cash prices (current market) and future prices (gas sold for delivery in the future). Current prices and future prices aren't the same, so the mechanisms of price development for both the cash and the futures market should be differentiated. First, let's deal with the issue of current gas prices. How does your supply or marketing representative know what a fair price should be for gas being purchased or sold by your organization?

Current prices. Gas prices are first set at the wellhead, but most gas is not consumed at the wellhead, so the gas must be processed, transported, stored, and distributed for final consumption. Each of those functions increases the delivered cost of gas, so your representative could theoretically determine a delivered price for gas tomorrow by knowing all the component costs. The daily price for most gas sold occurs in a liquid and transparent market, but other factors including production and storage levels, weather, and industrial output make it somewhat difficult to predict even tomorrow's gas prices.

Price indicators. Price indicators available to gas buyers and sellers take many forms today, but when open-access became a reality in 1985, and the spot market for gas began its explosive growth, things were vastly different. Future pricing mechanisms for long-term sales were determined according to various pre-established price escalation formulas.

In one sense, the price of gas was linked to nomination procedures mandated by the interstate pipeline companies. These pipelines, used to doing business the same way for 50 years, were reluctant to make changes other than those required by Order No. 436. All pipeline companies required shippers and customers to provide nomination information prior to the beginning of the month of actual gas flow. Once made, nominations for firm transportation were effective for the entire month. These nomination requirements gave rise to what we still call *bid week*, a concept that began with pipeline nomination deadlines and that continues in another form today tied to the futures contract settlement date.

Bid week. Here is how bid week began. The pipelines needed to assure that all preparations for first-of-the-month gas flow were completed during the previous month, so each pipeline established its own nomination deadline toward the end of the month. If Panhandle Eastern had a nomination deadline of four days before the end of the month, most of the buying and selling of gas to be transported on Panhandle would be done on the 5th day before the end of the month. If Tennessee Gas Pipeline had a nomination deadline of the 25th day of the month prior to nominated gas flow, the gas that would be transported on the Tennessee system would be bought and sold on the 24th. In all, the pipeline nomination deadlines spanned about a week—bid week, the busiest time of the month for all industry players.

Billing, payment, measurement, nominations—all aspects of the business, in fact—were geared toward monthly performance. Wholesale transactions lasting a year or more would still be nominated on a monthly basis. When all spot business was done this way, the prices for firm gas were established for a month as well.

Monthly pricing. How did gas buyers and sellers after Order No. 436 manage to competitively price gas they bought and sold? Fixed prices have always been used extensively, but because of the difficulties in determining exactly what the price of gas at a particular delivery point should be, many organizations began to use an approach called index pricing. An index, in this context, refers to the published result of price information gathered by various industry publications. The most commonly used index price in the late 1980s was *Inside FERC's Gas Market Report* (IFGMR),[45] which published prices for gas at the most liquid delivery points on the major pipeline systems. By canvassing gas buyers and sellers for price information, IFGMR would collect a variety of price indicators at the same delivery points. After data collection, IFGMR would calculate the results and publish. Index prices were published during the first week of the relevant month, so February gas prices wouldn't be reported until the first week in February, although gas to be delivered in February was purchased, sold, and nominated to the pipeline during the last part of January.

Today there are many publications that publish index prices on daily, weekly, and monthly bases and each publication has its own calculation and reporting methods. At that time, IFGMR published only monthly prices, and no publication either then or now publishes future index prices. Only regulated contract markets such as the New York Mercantile Exchange (NYMEX) publish future natural gas prices, depending on the prices being paid and received by futures contracts buyers and sellers at the time.

Future gas prices. Regulatory changes have driven most industry innovations from the beginning. That was true with the FERC and open-access and became true with the Commodity Futures Trading Commission (CFTC) when the NYMEX, one of the CFTCs regulated futures markets, started trading natural gas futures contracts in April 1990. For the first time, instead of calculating esoteric future price mechanisms in gas contracts, gas buyers and sellers were able to harness future prices in advance of actual gas deliveries and either incorporate a NYMEX-based price in forward contracts[46] or as a stand-alone hedge transaction. For hedgers,[47] this was achieved by buying or selling futures contracts and later liquidating those positions before settlement of the futures contract. When hedged according to the book, the profit or loss on buying and selling futures contracts will offset the loss or profit on the physical transaction, and the hedger is able to contain its risk of future adverse price movement. I'll discuss the mechanics of futures contracts trading with some examples later in this section.

Hedging. To understand how the futures markets help buyers and sell-

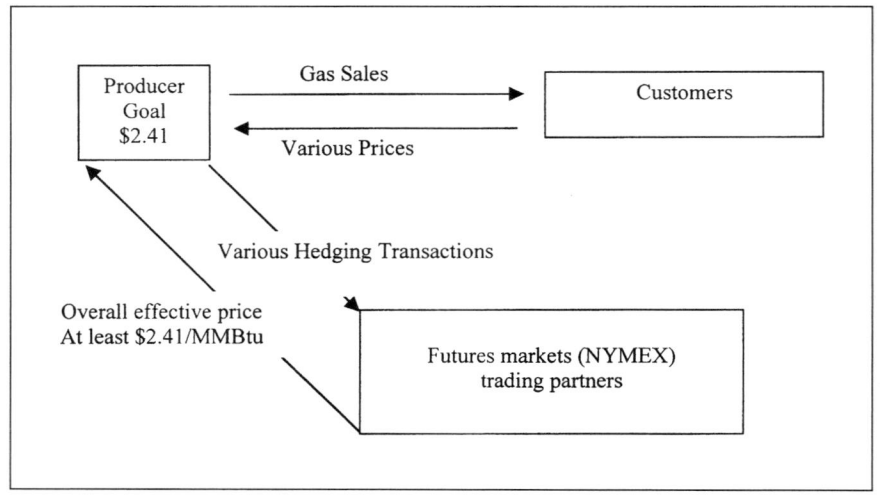

Fig. 1-1 Hedging to Achieve Price Goals

ers one must understand the concept of risk transfer. Let's say that a producer's gas sale representative sells 2,000 MMBtu/day for two years at a downstream delivery point, beginning the following July, at a published index price, plus a premium of 2 cents (Fig. 1-1). This producer must have at least $2.41 per MMBtu for all gas it sells, so when the index price is high, the producer will benefit and when the price is below $2.41/MMBtu, it will lose money. Now the producer may decide to keep the falling market risk, thinking that it will profit when the market is high. Another option, however, is to transfer all price risk below and above a price range with $2.41 as the low point of the range. Hedging tools like those discussed in this chapter provide the opportunity to transfer any unwanted price risk to a third party. One thing you must keep in mind throughout this is that hedging, in the traditional sense, is not used to increase profits but rather to transfer unwanted future price risk. One risk is essentially traded for another, so that if a hedger's goal is in setting a fixed price during an identified term, it will transfer both the potential floating risk and reward. The most accurate statement of a hedger's goals may be to *quantify* a future price risk in accordance with the organization's price goals.

Relationship between cash prices and future prices. As noted earlier in this chapter, the laws of supply and demand, while the most important indicators of price, don't operate in a vacuum, and many other elements can effect gas prices at any given location on any day. No one, including the producer mentioned above, knows what the market price for gas will be tomorrow at a particular delivery point. We do know, however, that there are certain relationships between the price of gas today and price of gas at some time in the future. For this discussion, the two most relevant relationships are *parallelism* and *convergence*. Both are related to the relationship between cash and futures prices, and understanding both can assist the producer's supply representative in making hedging decisions.

Parallelism. Parallelism occurs because the same factors that effect cash prices tend also to effect the price of the commodity for future delivery, so prices in the cash market and prices in the future market tend to move in the same direction at any given time. There is likely to be a high correlation between cash and futures prices because of two factors. First, both financial and physical traders have access to the same pricing information. Second, storage. The major component of carrying costs, the difference between the cash price and future price of a commodity, is cost to store that commodity for a stated period.

Convergence. Convergence is the second principle of futures prices. Cash and futures prices tend to be the same or to converge at the expiration of the futures contract. As the futures contract approaches expiration it becomes a close substitute for the cash commodity because it is a contract for delivery of the cash commodity. At expiration, owning the futures contract is essentially the same as owning a contract for immediate delivery of the cash commodity.

Futures contract markets. Until April of 1990, the only way to account for future adverse price changes in a forward contract was through some type of price adjustment provision in the contract. The most common types of such clauses were automatic price escalation or reduction, price renegotiations upon the happening of an event, *e.g.*, passage of time, change in the cost of living index (COLA) or other economic indicator, and increase or reduction in a referenced index price by a certain percentage. This type of pricing tool is still being used today, but in recent years, forward contract pricing has undergone a change of huge proportions brought on by natural gas futures contracts trading through established commodity exchanges.

U.S. futures markets in their present form came into existence about 100 years ago, but the principles involved have been used for hundreds of years. The first recorded instance of contracting for goods in the future at a fixed price occurred during the time of the Phoenicians around the Sixth or Seventh Century BC. The practice of speculating on the future price of tulip bulbs in Holland in 1635-36 nearly caused the economic downfall of the country. Standardized futures contracts on rice were traded in Japan in the middle of the Eighteenth Century.

In the U.S., the very first energy futures contract was traded on a commodity exchange in the mid 1960s. This was the New York Cotton Exchange (NYCE) propane contract. But due to lack of hedge participation and low price volatility, the futures contract was not initially successful and trading soon ceased. The first crude oil futures contract through the NYCE in the early 1970s required Rotterdam delivery. This concept might have made sense to the exchange but not to business—the remote delivery coupled with other problems, including the 1974 oil embargo, forced the contract into dormancy.

In 1978, the NYMEX opened a No. 6 oil contract with New York delivery and the NYCE converted its dormant heating oil contract to a No. 2 oil contract with New York delivery as well. The NYMEX contract specified a very low sulfur content (.3%) consistent with New York City pollution laws. Most of the country allowed a higher sulfur content than was required by the No. 6 oil contract, so the NYMEX contract suffered from the same flaw that Rotterdam delivery had produced for the NYCE. In the spring of 1983, the NYMEX started trading a futures contract on crude oil calling for delivery of sweet crude in Cushing, OK. It was a success, mostly because of the realistic contract specifications not present on other failed contracts. Since then, energy-related futures contracts and markets have achieved steady growth. Among all energy futures contracts, crude oil has still been the most successful.

Natural gas futures contracts. On April 3, 1990, the NYMEX began trading Gulf Coast natural gas futures contracts for a 12-month forward period. Every futures contract is comprised of various specifications regarding such points as quantity, grade, quality, point of delivery and trading limitations. Each natural gas futures contract was for a quantity of 10,000

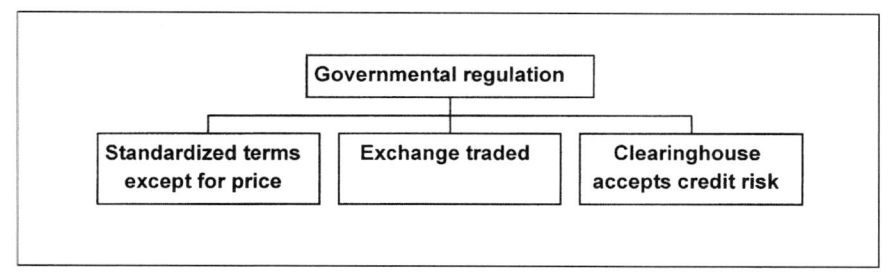

```
                    ┌─────────────────────────┐
                    │ Governmental regulation │
                    └─────────────────────────┘
         ┌──────────────────┼──────────────────┐
┌──────────────────┐ ┌──────────────────┐ ┌──────────────────────┐
│ Standardized terms│ │ Exchange traded  │ │    Clearinghouse     │
│  except for price │ │                  │ │ accepts credit risk  │
└──────────────────┘ └──────────────────┘ └──────────────────────┘
```

Fig. 1-2 *A Legal Futures Contract*

MMBtu.[48] The delivery point for contract positions not liquidated[49] prior to the settlement date was specified as the Henry Hub, an area near Erath, LA, where 13 major pipelines have receipt/delivery points. If a position is not liquidated prior to the settlement date specified by the NYMEX, and the futures contracts transaction results in a short[50] or long[51] position, the holder of those contracts will be required to make or take delivery of the gas at the Henry Hub.

Natural gas trading business wasn't very good on the first day as fewer than 1,000 contracts were traded. That number has grown exponentially through the years to thousands of contracts per day as more and more companies have learned to take advantage of the opportunity to reduce risk and, in some cases, increase profitability through use of the futures market. Now thousands of contracts are traded each day and natural gas futures on the NYMEX are traded for up to 36 months in the future. The Kansas City Board of Trade (KCBT) experimented with a West Texas contract but later converted the "Waha Contract" to a "basis contract."

Hedging with energy futures has become a way for natural gas buyers and sellers to focus on business concerns rather than expending time and energy attempting to forecast prices.

Futures contracts aren't secured in a fireproof file like other contracts. The major distinguishing factor of futures contracts is that they are regulated by the federal government.[52] They are completely standardized by the specifications and regulations established by the relevant exchange (Fig. 1-2). The terms of each contract on a particular exchange are the same in all respects except price. Futures contracts are readily transferable with only a small transaction cost. The commodity exchange provides a mechanism whereby contracts may be purchased or sold. In futures contracts there is no concern about the creditworthiness of the other party (a critical element of gas contracts) because the futures exchange places itself between the buyer and seller of each futures contract. Once the agents for the buyer and seller meet on the floor of the exchange and negotiate the price, all connection between the buyer and seller is severed. The exchange becomes the buyer's seller and the seller's buyer so that the solvency of a particular transaction as far as the parties are concerned is dependent only upon the solvency of the exchange involved.

What makes a successful futures contract?

In the above discussion of various unsuccessful attempts to jump-start energy contracts, there are some common threads running through the failures. Each lacked one of the three essential elements for a successful futures contract listed below.

Homogenous commodity. The contract must be for a fungible commodity, a concept discussed earlier in this chapter. Grains are fungible, as is natural gas, gold, silver, and other commodities traded on organized exchanges. Natural gas delivered pursuant to the NYMEX Henry Hub futures contract must meet the pipeline quality specifications of the Sabine Pipeline Company, operator of the hub that is the delivery point for this futures contract. Every futures contract has standardized specifications for the commodity being traded.

Substantial trading activity. The second element of a successful futures contract is substantial trading activity. Although hedgers—those transferring risk of an underlying physical transaction—are a substantial presence in trading activities, another type of futures trader adds necessary trading liquidity to the futures contract market as well. These are the speculators. Speculators do not have an underlying physical risk they are trying to transfer. They are taking on risk. Speculators predict where a particular market is going by the use of fundamental analysis and how the market is going to get there by using technical analysis. Based upon their calculations and analysis, they attempt to take advantage of price changes.

Volatility. Anyone familiar with natural gas markets in recent years can attest to the fact that natural gas prices certainly meet the third elemental requirement—price volatility. This issue is quite a paradox in the natural gas business because one would ordinarily suppose that the natural supply/demand theory would prevail to the point that in times of cold weather and high demand, prices would be higher than when the weather is warm and demand low. During formative years of the natural gas spot market, gas pricing did predominately work that way and it is still true to an extent—particularly for peak cold periods—but demand is not the only determinative of natural gas prices. Otherwise, how could we account for price spikes in the relatively warm spring months and low prices during late fall months? A number of additional factors—from wellhead production statistics to long-range weather forecasts and storage predictions, to name just a few—join together to form pricing models and ultimately the price.

At the time of copyright, only the NYMEX offered natural gas futures contracts, but with probable development of more hubs or market centers[53] in North America, little doubt exists that additional natural gas contracts will be offered for trading.

Using futures contracts for gas delivery

More than 90% of futures contracts positions are liquidated prior to the contracts' settlement date or closing date. But buyers and sellers who plan to make or take delivery instead of liquidating a position may also use futures contracts to accomplish their delivery goal. Futures contracts may be purchased or sold instead of trying to find a willing seller or buyer particularly if gas is being delivered at a futures contract delivery point. Let's say WIDGET—an industrial consumer located in Louisiana—typically purchases gas produced in southern Louisiana to fuel its manufacturing operations. The NYMEX natural gas futures contract specifies delivery at the Henry Hub delivery point located in Louisiana. Instead of phoning around to find gas for its manufacturing needs, WIDGET may simply place an order for futures contracts calling for Henry Hub delivery and take physical receipt of the gas instead of liquidating its futures contracts positions.

Using futures contracts to transfer price risk

Futures contracts may also be used to transfer price risk to a third party. For the illustration given below, you'll need to make a few assumptions, so

the "givens" are as follows:
1. *Given*: FLEX (a marketer) needs a price of at least index plus 2 cents for all sales made. Even though index-plus-2 is always FLEX's target, it can't always find a buyer willing to pay that price. Absent a buyer willing to pay its price, FLEX can still make the sale at index, but, in order to meet organizational goals, it will need to find another source for the 2 cents. This is exactly how buying and selling futures contracts can pass along the 2-cent risk to another party
2. *Given*: FLEX observes that futures contracts prices have begun a gentle, but real decline. This is just the kind of opportunity FLEX needs to capture the 2 cents it won't be receiving on the gas sale to WIDGET
3. *Given*: The index-based price is an unknown at the time the deal is made, but once the publication provides necessary price information, the price for gas at the futures contract delivery point is $2.28/MMBtu

Example of Hedging with Futures Contracts

REFLEX, another marketer, is willing to purchase gas from FLEX at a straight index price. Since REFLEX's purchase terms other than price are agreeable to FLEX, it agrees to the sale for a fixed quantity per day of 5,000 MMBtu for the *future* period of April through October.

FLEX may utilize financial hedging tools to transfer the risk of receiving less than index plus 2 cents on that particular sale. One way FLEX can accomplish this goal in a falling futures market is to sell futures contracts when the price is high and buy futures contracts when the price is lower to realize that additional 2 cents (Fig. 1-3).

FLEX first sells futures contracts (5,000 x No. of days April-October = 107 contracts @ 10,000 MMBtu each). The price in this example is $2.31/MMBtu for the futures contracts on the day they were purchased.

Then, if the prices for the relevant futures contracts fall at least 2 cents, FLEX can buy futures contracts. In this example, FLEX was able to purchase 107 futures contracts at a $2.28 price, thus capturing 3 cents in addition to the index price it is receiving from the physical sale, and making the effective price for the physical gas *Index-plus-3-cents*.

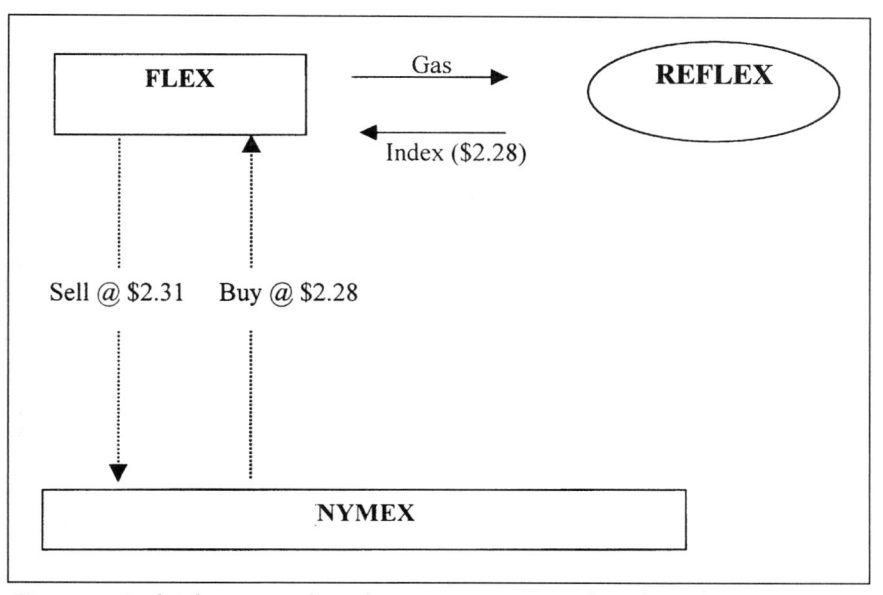

Fig. 1-3 Goal Achievement through Futures Contracts Sale and Purchase

Over-the-counter derivatives as a hedge option

With the advent of natural gas futures trading, another type of hedging product neither standardized nor traded through an exchange also was available to parties wishing to hedge physical transactions. The transactions comprising this product are known generically as over-the-counter (OTC) derivatives. Privately negotiated OTC derivative products involve a combination of elements, each transaction is tied to a unique "underlying"—the asset or index from which the value of the derivative product is derived. For example, in a natural gas OTC transaction, the natural gas itself is the underlying *asset*. In an interest rate transaction, the underlying *index* would be that provided in the index publication for the relevant interest rate(s). Because the transaction derives its value from this underlying asset or index, it is known as a *derivative*. For natural gas hedging purposes, the three most common OTC derivative products are commodity swaps, commodity options and commodity swaptions.

Swaps

A swap is a financial transaction between two parties who each agree to make future payments to the other based upon an agreed asset or commodity basis called the *notional quantity*. One party makes payments based on a fixed price (as agreed by traders) multiplied by the notional quantity (also agreed to by the traders) and the other makes payments based on an agreed floating amount multiplied by the notional quantity. This calculation results in a *notional amount* for each counter party that is owed to the other party. For a *plain vanilla*[54] swap, the notional quantity is only a basis for measuring payments and does not create an obligation to deliver the product. Actual delivery may occur with currency derivatives where the counterparties exchange different currencies, rather than making payments based on fixed and floating prices.

Delivery of the underlying does not typically occur with interest rates, commodities or energy. A natural gas commodity swap will usually contain a daily notional quantity. For example, assume that a swap is being used to hedge a gas sale of 5,000 MMBtu/day for a month. To match the hedge to the physical sales and purchases being made—*i.e.*, to transfer the total unwanted risk, the notional quantity would, in an offsetting financial transaction, be 5,000 MMBtu per day for the month.

Now, physical swaps or gas exchanges are also possible, so the term "swap" may refer to either a financial transaction or a physical sale or purchase. A financial swap is not a physical sale transaction. It is rather a financial tool used to transfer risks incurred in the physical transaction. Take a look at the three swap examples that follow of how a producer, marketer or consumer of natural gas might hedge sales and purchases with this tool. Several "givens" are also necessary to fully appreciate any swap example.

1. *Given*: When viewing a hedging transaction using this tool, do not presume that one party always wins and the other loses. Profiting on hedging activities is a nice plus, but the real purpose of the swap from an energy industry participant's perspective is to transfer risk to another party
2. *Given*: In the producer example, PUMP (the producer) has no economical way to halt production when prices are low for short periods of time. Shutting-in a well can be both complex and costly. From a con-

tractual point of view there may be additional problems. If PUMP is operating a well system for other third party gas owners, restrictions in the joint operating agreement (JOA) may limit the producer's ability to shut-in wells
3. *Given*: Each swap counterparty meets all credit requirements of the other party, and is otherwise eligible to enter into this derivative swap transaction
4. *Given*: The examples in this book are totally focused on the future price risk and do not speak to other types of risk or the hedging tools that might be used to account for price variations based upon transportation costs differentials

SWAP EXAMPLE — PRODUCER (SUPPLIER)

PUMP must sell all of its production at a fixed price of $1.80 or it will lose money. PUMP has not been able to find a fixed-price buyer for 2,000 MMBtu/day for the upcoming November and December, but it still must sell and deliver the gas. FLEX offers to purchase the gas for an index-based price. Since all of FLEX's terms, other than price, are acceptable, PUMP agrees to the sale.

One way FLEX can achieve a fixed price of at least $1.80/MMBtu for its gas is to find a swap counterparty willing to take the index price risk (Fig. 1-4). In this case, FLEX contracts with an OTC derivatives dealer—an entity that develops and sells derivatives products and is not hedging.

In this example, PUMP received an index-based price that averaged $1.74, 6 cents below its required fixed-price goal. However, because PUMP was able to find a swap counterparty willing to pay a fixed price of $1.82, the calculation looks like this:

Payment to PUMP for physical	= $1.74 x 5,000 x 61 days = $530,700
Swap payment from PUMP	= ($1.74) x 5,000 x 61 days = $(530,700)
Swap payment to PUMP	= $1.82 x 5,000 x 61 days = $555,100
Effective result	$1.82 x 5,000 x 61 days = $555,100

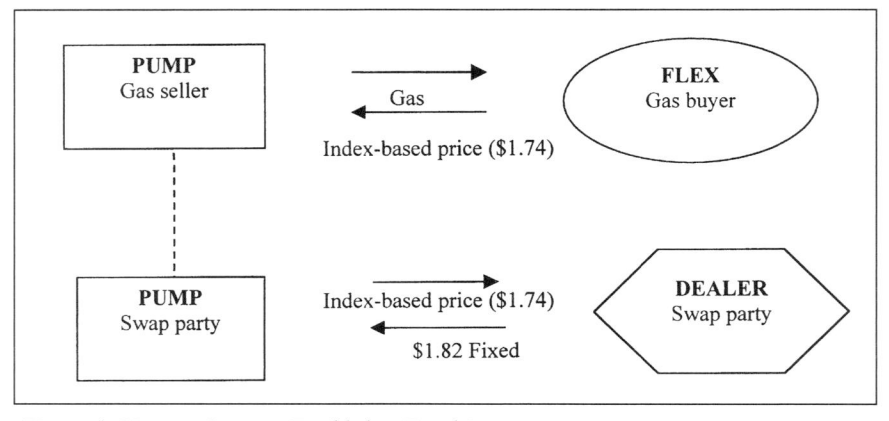

Fig. 1-4 Using a Swap to Establish a Fixed Price

SWAP EXAMPLE — MARKETER

Let's say that FLEX has its overall portfolio goal (both purchases and sales) set at index plus 1 1/2 cents. Setting that goal means that at the end of the designated period, the total received for gas sold minus the total paid to suppliers should at least equal index plus 1 1/2 cents. The indexes being used to assess whether goals have been met are the major points where FLEX buys and sells the majority of its gas, including the NYMEX Henry Hub prices and Permian Basin delivery points.

Assume that FLEX is receiving a fixed price of $2.00/MMBtu/day for some of the gas it is selling for future delivery. At the time the sale was made, but before delivery, the fixed price was greater than index plus 1 1/2 cents. Let's also assume that during the period of time between making the deal and actual delivery dates, the index price experiences an upward trend and that this increase is tied to rising futures prices. FLEX may decide to use price as a tool for a capturing profit from hedging (Fig. 1-5).

If the INDEX price is greater than $2.00/MMBtu, FLEX will profit from the difference and achieve more than its goal of index plus 1 1/2 cents. The futures market may also move against FLEX if futures prices fall after FLEX has entered into the swap deal. In that case, FLEX will not have profited from hedging, but it would still have achieved its goal of INDEX pricing.

Chapter 1 Introduction

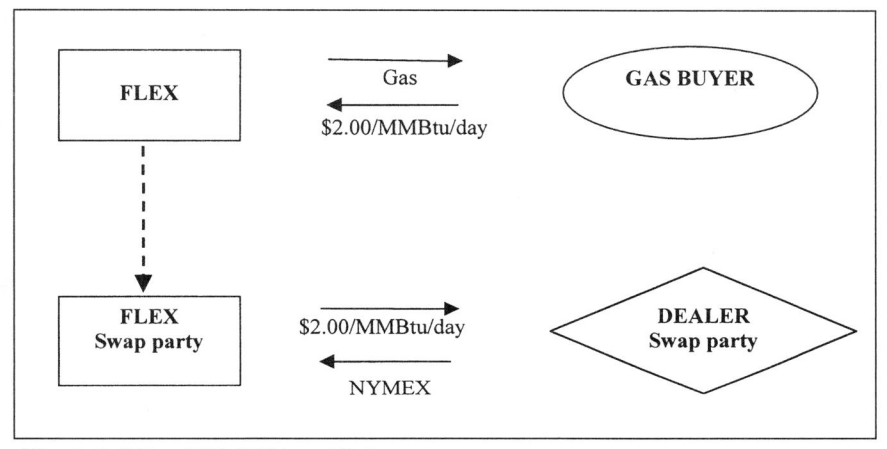

Fig. 1-5 *Using NYMEX-based Prices*

HEDGING EXAMPLE — CONSUMER

Assume that WIDGET is buying its natural gas fuel at fixed prices. This way WIDGET knows exactly what fuel procurement costs will be both at the time the purchase is made and when natural gas is delivered. If WIDGET's high manufacturing times occur during Spring months, it may have no problem finding low-cost gas and inexpensive transportation to move the gas to its facility, but what if prices on the spot market go down? In that case, WIDGET may be paying more for its long-term gas supply than it would have to pay if it were buying gas on a monthly basis on the spot market. Wishing to transfer the fixed price risk to another party, WIDGET contracts with an OTC derivatives dealer to pay an index-based price and to receive a fixed price with the goal of paying less than its fixed price for natural gas fuel (Fig. 1-6).

WIDGET may be correct or incorrect in its assumptions about gas prices in the future. If the Index price is less than $2.00/MMBtu, WIDGET will have achieved its goal of realizing a fuel cost of less than $2.00/MMBtu. If WIDGET is wrong, it will pay more than its fixed-price goal of $2.00/MMBtu. But, remember also that once a hedge is put on, markets must be monitored to see if adjustments to the total hedging position are necessary.

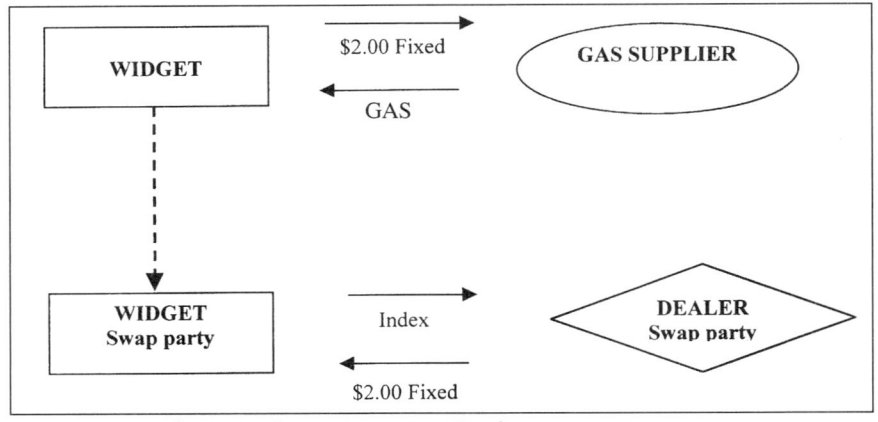

Fig. 1-6 Transferring Risk to a Derivatives Dealer

Options

An option contract gives the buyer (or holder) of the option the right, but not the obligation, to purchase or to sell a specific commodity at the specified price at any time the option contract is in effect. Options on futures contracts may be traded through a regulated exchange. But in the OTC context, the option writer (often referred to as the seller) will, in return for a premium payment (usually made up-front), provide the option buyer with the ability to exercise that option at the specified time(s) during the specified term. Upon exercise, either actual sale/purchase of the gas or a payment bond upon the difference between a strike price and another price is made by the swap writer. The buyer's risk is limited by the premium paid for the option, but the seller's risk is theoretically not quantifiable because of potential market volatility. The two types of options are calls (sometimes called *caps*) and puts (sometimes called *floors*).

A *call* option gives the buyer the right to *buy* the specified commodity at the specified price at any time the option contract is in effect. A *put* option gives the buyer the right to *sell* the specified commodity at the specified price at any time the option contract is in effect. Exercise rights vary, depending on the terms negotiated by the option seller and buyer.

Chapter 1 Introduction

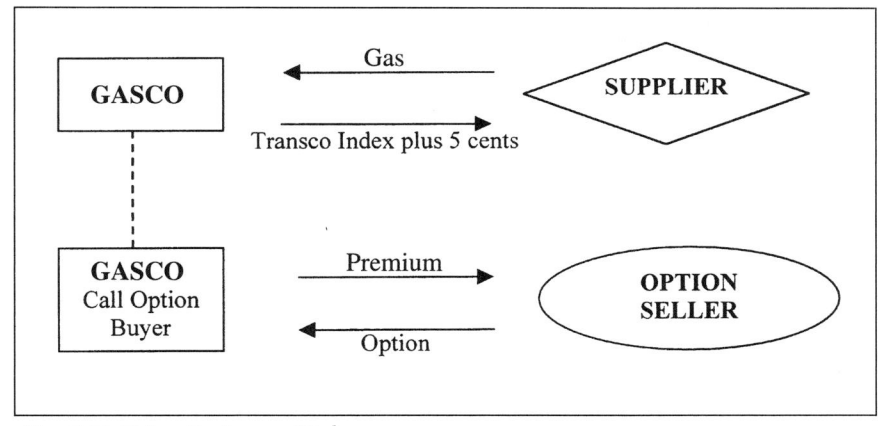

Fig. 1-7 *Using Options to Hedge*

OPTION EXERCISE

GASCO, a bundled-service utility located in the northeast U.S., has received permission from its state regulators (PUC) to hedge the cost of gas it purchases for its customers. GASCO is limited in the types of hedging transactions it may use and the PUC has established certain pricing benchmarks as goals for its hedging activities (Fig. 1-7).

GASCO's supply portfolio includes fixed-price purchases—including both specifically stated fixed-prices (*e.g.*, $2.00/MMBtu) and exchange-based prices (e.g., NYMEX L3D)—and index-based purchases. GASCO is most concerned about prices it will pay for gas during its customers' peak-usage times. It has contracted for firm peaking supplies during the upcoming winter months at a price of Transco Station 60 index plus 5 cents delivered at a point downstream. GASCO has determined that winter index plus 5 cents will probably be higher than its benchmark price mandated by the PUC.

In this example, GASCO has purchased a call. One of the elements of any option is the "strike" price—a pre-agreed price that designates the point at which an option is exercised. If, during the term of the option, the agreed index price is above the strike price, the option will be in the money to the buyer and it will be exercised.

Once the option is *in the money* (strike price is less than the market price) to the GASCO option buyer, the option seller is required to sell the product at the lower price. Typically, instead of going to physical delivery, the seller will instead pay to the option buyer the difference between the index price and the strike price multiplied by the notional quantity.

SWAPTIONS

A swaption is a combination of a swap and an option and may take several forms. The two basic swaption forms are:

- performance of the swap is underway and the buyer's option is to stop the swap
- upon occurrence of a stated event (usually price-related) the already negotiated (but dormant) swap will become effective and performance will begin

The combined effect of regulatory changes and natural gas futures trading mean that adjustments to standard gas contract language are necessary in contracts being negotiated today and will be even more important in the future as newer, more sophisticated and complex transactions become commonplace. Succeeding chapters will provide illustrations, examples and suggestions for new contract language.

GAS INDUSTRY STANDARDS BOARD AND FURTHER STANDARDS DEVELOPMENT

Prior to 1996, every U.S. interstate gas transporter was free to set its own policies regarding measurement, invoicing, nominations, and other operational requirements subject to the FERC. Today, natural gas shippers typically contract to ship gas on a number of different pipeline systems, including any or all of interstate, intrastate, intraprovincial, interprovincial, and LDC systems. Many times the path of the gas from production to a downstream[55] market includes a variety of different pipeline systems. To fully understand the permeating effect of pipeline regulations, one now must also understand the relationship between the FERC and the gas industry standards board (GISB). As will be discussed later in this section, standards developed by GISB are totally voluntary upon industry participants. However, the FERC has incorporated a number of GISB standards into Order No. 587,[56] and once put into a FERC order, the interstate pipelines must adopt those standards.

GISB has developed and adopted several categories of standards many of which are very broad in scope. This has led to requests for interpretations of current standards. GISB also processes requests for change to current standards and requests for new standards. Some examples of standardization include common deadlines for interstate transportation nominations, establishment of allocation methodologies, standards for measurement of gas, requirements for similarity in invoices from transporters to their customers, deadlines for resolution of invoices and balancing provisions, electronic standardization of communications between transporters and customers, and several contracts.

GISB standards may be developed for business practices as indicated in the preceding paragraph, or they may be electronic delivery mechanism (EDM) standards, designated as the electronic rules of the road for transacting business through EDI. To illustrate the difference, business practice standards (nomination deadlines, definition of a gas day, definition of MMBtu or Mcf, and many others) relate to substantive matters. In order to communicate information necessary to comply with a business practice standard such as nominations by EDI, every participant (interstate pipelines and their shippers) must adhere to the same requirements for computer communications. As business practice standards relate to *substance* the EDM standards relate to *communication procedures* to implement standards electronically.

One important point to keep in mind regarding the GISB standards is that both the business practice standards and the EDM standards developed through the GISB processes—even after approval of GISB members—are not binding on any segment of the natural gas industry. The only time any GISB standards become mandatory is when the FERC incorporates any of the standards in FERC Order No. 587. Since Order No. 587 is issued as installments, the FERC may issue new orders under Order No. 587 for many years.

Each reiteration incorporates a new requirement, but all orders bearing the number 587 will relate to implementation of standards developed by GISB. As an example, Order No. 587-I[57] changed a deadline and Order No. 587-K[58] adopted Version 1.3 of the GISB Business Practice Standards.

There is a practical business reason for industry participants (other than interstate pipelines who are bound to comply) to voluntarily adopt the GISB standards. The reason is ease of doing business with all pipelines. Many intrastate pipelines and LDCs have voluntarily adopted the GISB standards in the U.S. The gas industry in Canada has developed an organization similar to and compatible with GISB. Called GasEDI,[60] the organization operates much as GISB and has undertaken a number of important tasks with the goal of standardizing business practices and electronic data communications for the Canadian natural gas industry.

Producers are not directly required by any law to accept GISB standards. Neither are marketers, distribution companies, or consumers. The interstate pipelines through which these entities ship gas set the rules and shippers must follow those rules. To the extent the North American gas industry as a whole can agree to some basic rules, every individual organization will, in turn, benefit as well. Just keep in mind that the standardization process is a dynamic exercise. Standards are added, amended and interpreted by various GISB working groups, so it is important to monitor activities of this organization. The GISB standards, available through the GISB Internet home page, may be purchased for a small fee.

The foregoing regulatory issues only underscore the importance of thoroughness for anyone reviewing a gas contract. Regulations change. Interstate pipelines have adopted FERC orders in different ways that can be found in only the pipelines' tariffs. Standards development in an industry as diverse as the combined "natural gas industry" is a controversial topic. The five representative GISB groups—producers, transporters, end-users, LDCs, and services—suffered from the lack of standardization prior to 1996 when the first GISB standards were adopted and they also suffered from a lack of cooperation between and among themselves. Standardization, while still controversial, has probably done more to bring the industry together than has anything before.

The Six Issues and Their Relationship to Gas Contracting

We began with the premise that gas contracting traditions and practices originated in the oil fields and with early, unregulated distribution systems. How did we move from handshakes to signatures?

After 1938, all interstate gas contracts were regulated cumbersome documents without a hint of free negotiations, while contracts for intrastate gas sales and transportation varied from pipeline to pipeline and state to state. Our contracting journey begins after Order No. 436 because the modern industry was born with that order. The new gas contracts remained surprisingly like the pre-Order No. 436 regulated sale and purchase contracts for several years. Those regulated contracts contained many standard clauses filled with language regarding physical structures necessary to move, measure, price and deliver the gas. Buying and selling gas on the new spot market was very different from buying and selling under traditional long-term (typically 20-year) regulated contracts. Gas marketers, companies that rarely owned hard assets such as wells, pipelines, or other necessary equipment, and who consequently had no ability to control, check or maintain any such equipment, drafted their own gas contracts to remove much of the operational language so important in gas contracts pre-Order No. 436.

Although the pre-Order No. 436 language became outdated and unnecessary for spot sales in many instances, years passed before spot market contracts completely shed the regulated look and feel of long-term regulated gas contracts. Today the only gas contracts still containing operational clauses are for sales of gas production at or close to the wellhead and some end-use or distribution gas sale contracts. For most sales, that language just isn't necessary. With all the contracting changes seen from 1985 until the present though, gas contract language still has not kept pace with actual industry practice. This is partly because commercial law hasn't changed to accommodate today's commercial realities, but also because current contract language may not be sufficiently versatile to accommodate new and different transaction types or, for that matter, electronic contracting.

The description and numbers of gas industry participants has changed dramatically since the late 1970s as well. In early times, industry participants were either producers, transporters, or distributors, and many times one company provided all three services as integrated natural gas companies. The modern gas industry has many more participants, both in numbers and in services provided—many of the industry players are multifaceted service providers. No more do the interstate pipelines provide a bundled transportation and sales service to their LDC customers. Today's industry participants who buy and sell gas utilize the service of third party providers to transport, store, gather and process the gas.

Transportation and sales are two discreet functions. Sellers and buyers contract for gas sales and purchases with other buyers and sellers. Transportation service providers contract with gas sellers and buyers to transport the gas being sold and purchased. Even though natural gas is becoming commoditized[61] to greater degrees, most natural gas sales still involve some necessary transportation, so modifications to transportation regulation necessarily have an impact on gas sales, hence on the contracts memorializing those sales. Today's gas sale contract is in many ways evidence of an attempt to make the sale itself somehow work within a prescribed transportation structure, both physical and regulatory.

Now we'll move to a fresh look at gas contracting, not with a reflective examination of where the industry has been, but with a view to future possibilities. For typical gas sale and purchase contracts, the same general contract language should be adaptable for use by producers, marketers, LDCs and other end-users. The ensuing chapters will propose a variety of approaches to contracting that may be beneficial to any spot market sale or purchase. The days of marketing only natural gas are coming to an end—*service marketing* is now the watchword. Organizations are broadening their energy service portfolio to provide a variety of energy products for their customers. Gas sale contracts must be adaptable enough to provide a vehicle for marketing other forms of energy commodities in addition to natural gas.

The gas sales contract, today so intertwined with financial hedging transactions, must be flexible enough to accommodate new concepts of pricing, embedding of financial products, and additional service provisions.

Today's gas sale contracts have many faces, but whether the contract itself is one page or 50 pages in length, written on paper or executed electronically, each has certain characteristics in common with every other gas sale contract. The beginning chapters of this book will focus on common contract characteristics. As the industry changes to incorporate the latest commercial and regulatory trends, gas contracts must be adaptable to incorporate the realities of gas trading. Incorporating the six critical gas contracting issues discussed in this chapter, the newest trends in energy marketing—and, therefore, the newest trends in energy contracting—will be discussed in later chapters. Recommendations for the future of energy contracting will complete this book. In chapter 4, we'll take a look at the first FERC order that paved the way for a multitude of industry participants and opened the door to a variety of innovative gas sale and purchase options.

---

1. The physical point at which gas is severed from the ground. It is the furthest upstream point.

2. A series of small diameter, low pressure pipes collecting gas from individual wells, commingling the gas in a common pipeline, and delivering the gas into a larger pipeline to carry it into commerce.

3. Removal of liquid hydrocarbons by utilizing the different boiling points of the hydrocarbons.

4. All interstate pipelines must receive FERC approval for services offered and rates charged for those services. The result of this approval received after formal proceedings at the FERC is evidenced in tariffs that contain general terms and conditions relevant to all services, rate schedules for each type of service offered and rates—both minimum and maximum—for each of the services.

5. The distinctive odorant in gas that moves through populated areas.

6. Local Distribution Company or LDC is the public utility responsible for distribution of natural gas to customers within its franchised service area.

7. One that drills a well in an area not previously known to contain oil or gas deposits.

8. Methane (CH4) is the primary component of natural gas. Other natural components, in varying amounts, include the methane chain (ethane, propane, butane, pentane,

hexane, heptane, octane, nonane and decane), saltwater, sediment, and various noncombustible gases (carbon dioxide, nitrogen and hydrogen sulfide). Native natural gas must be processed to remove many of the impurities prior to transmission in a pipeline.

9. Pipelines that cross state boundaries.

10. Commerce subject to regulation only by the federal government. U.S. Constitution Article VIII (Commerce Clause).

11. 15 U.S.C. §717-717w.

12. A "bundled" service is one that includes different service components but does not separately price each service. A pipeline company's bundled sales service could include costs for transportation, gas procurement, storage and gathering, among others.

13. Phillips Petroleum Co. v. Wisconsin, 647 U.S. 672 (1954).

14. Natural Gas Policy Act of 1978, (15 U.S.C. § 3301-3432).

15. Refers to the process of stopping production in a well, usually on a temporary basis.

16. Regulation of Natural Gas Pipelines After Partial Wellhead Decontrol, Order 436, 50 FR 42408 (Oct. 18, 1985), vacated and remanded, Associated Gas Distributors v. FERC, 824 F.2d 981 (D.C. Cir. 1987), cert. Denied, 485 U.S. Association, v. FERC, 888 F.2d 136 (D.C. Cir. 1989), readopted, Order No. 500-H, 54 FR 52344 (Dec. 21, 1989), reh'g granted in part and denied in part, Order No. 500-I, 55 FR 6605 (Feb. 26, 1990), aff'd in part and remanded in part, American Gas Association v. FERC, 912 F.2d 1496 (D.C. Cir. 1990), cert. Denied, 111 S. Ct. 957 (1991. This order offered incentives for interstate pipeline companies to open their system to third party shippers for the first time. Because of a successful court challenge to this order, the D.C. Circuit Court remanded Order 436 back to the FERC for changes regarding take-or-pay requirements. The resulting order is No. 500, so Order 436 is frequently cited as Order 436/500.

17. One incentive for participation in the open access structure was issuance of individual blanket transportation certificates so the transportation service provider would not be required to go through separate certification and abandonment proceedings to provide transportation services. Other incentives included the ability to pass through a portion of the pipeline's stranded contract take-or-pay costs.

Chapter 1 Introduction

18. Disincentives included competitive disadvantages and potentially disgruntled distribution company customers not given the opportunity to choose gas merchants, among others.

19. Sales other than sales made to pipeline companies in the long-term market were originally considered to be sales on the spot market—in other words, the leftovers every month. Many gas marketing companies made fortunes on these leftovers.

20. Inquiry into Alleged Anticompetitive Practices Related to Marketing Affiliates of Interstate Pipelines, Order No. 497 53 FR 22139 (June 14, 1988), order on hrh'g, Order No. 497-A, 54 FR 52781 (Dec. 22, 1989), order extending sunset date, Order 497-B, 55 FR 53291 (Dec. 28, 1990), order extending sunset date and amending final rule, Order No. 497-C, 57 FR 9 (Jan. 2, 1992), reh'g denied, 57 FR 5815 (Feb. 18, 1992), 58 FERC Sec. 61, 139 (1992), aff'd in part and remanded in part, Tenneco Gas v. FERC, No. 89-1768 (D.C. Cir. July 21, 1992).

21. Self-implementing transportation does not require a separate FERC approval process. The transportation is provided under a blanket transportation certificate issued to the interstate pipeline.

22. A level of transportation service where the pipeline company has an obligation to deliver the quantity of gas that was scheduled by the pipeline company.

23. See endnote 15 supra.

24. Pipeline Service Obligations and Revisions to Regulations Governing Self-Implementing Transportation Under Part 284 of the Commission's Regulations; and Regulation of Natural Gas Pipelines After Partial Wellhead Decontrol, Order No. 636, 57 FR 13267 (April 16, 1992), order on reh'g, Order No. 636-A, order on reh'g. Order No. 636-B, reh'g denied, 62 FERC Sec. 61,007 (1993).

25. Energy Policy Act (EPAct) Public Law 102-486, 1992. The EPAct was a broad-based attempt to clean up the environment, and a part of the multifaceted structure involved deregulation of the energy industry.

26. "Burnertip" is the name given to the point of gas consumption. It is the furthest downstream point.

27. Trunkline pipelines are long-distance transporters that carry gas from areas of production to areas of consumption. Examples of U.S. trunkline pipelines are Transcontinental Pipe Line Company (Transco), Tennessee Gas Pipeline Company (Tennessee) ANR Pipeline Company, Texas Gas Pipeline Company (Texas Gas),

Transwestern Pipeline Company (Transwestern), Texas Eastern Pipeline Company (TETCO) and Panhandle Eastern Pipe Line Company (PEPL).

28. Grid systems are usually a network of many interconnection and delivery points that operate in and serve major market areas. Two grid systems are Columbia Gas Transmission System (Columbia Gas) and CNG Transmission (CNG).

29. Combination trunk/grid system pipelines include Williams Central (Williams), Northern Natural Pipeline Company (NNG or Northern), Koch Gateway Pipeline Company (Koch) and Southern Natural Gas Company(Southern).

30. The total services provided to the pipeline's customers, how and when those services are distributed. For example, a pipeline that delivers gas to the northern U.S. may have a high seasonal profile since the winter load would be higher than the summer load.

31. The demand component of firm service (sometimes called a "reservation" rate) is paid to the transporter in advance to reserve capacity in the pipeline.

32. The commodity charge is a per-unit rate charged for each Dekatherm or MMBtu of gas actually transported by the pipeline company.

33. e.g., Virginia State Corporation Commission, Oklahoma Corporation Commission, New Jersey Board of Public Utilities and Tennessee Regulatory Authority.

34. The interprovincial gas pipeline industry in Canada is comprised of the pipelines that cross provincial boundaries.

35. The name given to holders of pipeline capacity will typically be either "shipper" or "customer."

36. The Comission de Regularia Energia has its own Internet Website at: http://www.cre.gob.mx.

37. See generally the CRE's Website.

38. Niche marketing companies may sell gas within a set geographical region and aggregate supplies in specific regions. They may provide services only to a LDC or group of distribution companies. They may provide specialized services for customers or groups of customers. In fact, any marketing company not national in scope may have developed its own successful niche.

Chapter 1 Introduction

39. Regulated entities such as pipelines and LDCs in nearly all circumstances have been required to separate themselves from their heretofore-affiliated marketing function. For this reason, energy marketing companies associated with regulated entities are always non-affiliated.

40. The first natural gas futures contract for Henry Hub delivery was traded on the NYMEX in April of 1990. It is still the most successful commodity futures contract on that exchange. Since that time, other gas futures contracts have been introduced with varying degrees of success.

41. Gas balancing services may include a number of activities—all of which combine to assist a customer in staying within permitted transporter tolerance levels between the quantity of gas nominated and the quantity actually measured. Shippers outside these "tolerance" levels could incur various imbalance charges assessed by the transporter.

42. The communications from a pipeline customer to the pipeline regarding anticipated gas flow for a prospective time, e.g., tomorrow.

43. 87 FERC ¶ 61,021, [Docket No. RM 96-1-011; Order No. 587-K], Standards For Business Practices Of Interstate Natural Gas Pipelines (Issued April 2, 1999).

44. Gas buyers and sellers now use bid week primarily as a time to settle any outstanding futures positions prior to the settlement date of the next month's futures contract prices.

45. Inside FERC Gas Market Report, a McGraw-Hill Publication.

46. A legal agreement to make or take delivery in the future. The most commonly used definition in the natural gas industry is that found in the Bankruptcy Code, 11 U.S.C. § 101(25).

47. Assuming an opposite position in a futures market or derivative transaction to that in the cash market.

48. One million British thermal units. One Btu equals the amount of heat required to raise the temperature of water from 60 degrees Fahrenheit to 61 degrees Fahrenheit and a constant pressure of 14.73 psia and on a dry basis.

49. To enter into an offsetting transaction so that if futures contracts had first been sold, liquidation would involve buying the same amount of contracts at a later time. Similarly, if the first futures contracts were bought, the liquidating position would be to sell the same number of contracts.

50. The futures contract(s) holder has an obligation to deliver gas.

51. The futures contract(s) holder has an obligation to take delivery of gas.

52. The CFTC regulates commodity futures contracts through exchanges like the NYMEX.

53. Geographic areas with sufficient market activity (buying, selling, storage, transportation, etc.) to provide the liquidity needed for hub pricing.

54. Any basic, uncomplicated swap. Swaps that are more complicated are referred to as "exotic."

55. Gas moves from an "upstream" supply source to downstream markets.

56. Standards For Business Practices Of Interstate Natural Gas Pipelines, Order No. 587, Final Rule (Issued July 17, 1996).

57. Standards for Business Practices of Interstate Natural Gas Pipelines, 63 Fed. Reg. 53565, October, 1998.

58. Standards for Business Practices of Interstate Natural Gas Pipelines, 68 Fed. Reg. 17276, April 9, 1999.

59. The Internet web site for GISB is http://www.gisb.org.

60. The Internet web site for GasEDI is http:/www.gasedi.ca/.

61. Commoditization refers to the practice of buying and selling natural gas that may or may not involve delivery or receipt of the product. At any given delivery point, the same gas may be purchased and sole multiple times. These multiple purchases and sales add liquidity to the market.

Chapter 2: Gas Industry Players (and why they play that way)

We began this guide with an overview of the eight business sectors that together comprise the natural gas industry:

- Exploration
- Drilling
- Production
- Processing
- Transportation
- Storage
- Distribution
- Marketing

Of the eight major players, though, only four comprise the bulk of the gas sales industry—producers, marketers, LDCs, and consumers—and the individual business organizations that form each of the four sectors fulfill different needs to fulfill through gas contract language. What this means to you as a gas contractor is that four different types of needs result in four different ways to approach the same contract language.

For example, gas *production* embraces a unique type of operations, regulations, and assets ownership. Producers need a certain level of control to protect operations and assets, and to comply with myriad governmental regulations. Marketers are not restrained by either the controls or regulations imposed on producers.

Marketing organizations aren't usually asset-based, hence no gas field operations. Competition is keen in the marketing industry, so gas sales and purchases are neither regulated nor subject to third party review. Since marketers don't have anything they can directly control (such as pipeline valves, measurement, and quality determination), they need flexibility in contracting to avoid losses they cannot control. LDCs have neither the flexibility nor the competitive environment within which to operate.

LDCs own their distribution systems, customer services are regulated by the state PUC, and prices paid for gas supply are subject to review and redress by those same regulators. With an obligation to serve customers in its designated service area, an LDC must have reliable sources of gas. Of the four major players, consumers—particularly large volume commercial users—have the most in common with LDCs, but almost nothing in common with either producers or marketers.

Industrial consumers may be subject to various levels of regulation. Physical asset protection and consequences of lost markets are constant concerns of the manufacturing consumer. Industrial consumers sell products for which they must ascertain a price; therefore, the cost of gas supply as fuel is a critical element to these consumers.

Standard contract language doesn't always fill the disparate needs of the four industry players, so you'll see how each sector leaves its own identifying fingerprints all over the contracts it creates. On the other hand, cross-sector buyers or sellers share a number of common interests with all other buyers or sellers. In this chapter, you'll learn:

- who the players are and their unique business interests
- how to identify the needs behind gas contract language from a third party
- ways to respond to unacceptable contract language
- the common interests that join all gas sellers together and all gas buyers together

*Chapter 2 Gas Industry Players
(and why they play that way)*

GAS PLAYERS AND INDUSTRY STANDARDS

Because we're beginning to talk about real contract language in this chapter, we need to differentiate between contract language drafted for an individual producer, marketer, LDC, or consumer and the standardized gas contracts that have been adopted by GISB.[1] The most famous of the GISB contract offerings to date has been informally dubbed the "GISB Contract." The contracts developed through GISB and adopted by its members reflect consensus of the five GISB membership categories—producers, transporters, LDCs, end users, and services (including marketers)—for use in routine gas transactions.

The primary purpose for development of the GISB Contract was to streamline the gas contracting process for organizations that process multiple transactions on a daily basis. Because there are no industry-wide contract standards, time spent in negotiating, reviewing and processing gas contracts has become a major burden to the gas sale industry. The resulting administrative backlog is a plague effecting all gas contractors. The drafting committee members' overwhelming joint interests to more efficiently negotiate and process routine transactions resulted in accommodations made by all participants in furtherance of that shared interest. None of the drafting committee was willing to go too far in concessions, but all were willing to go far enough when the contract was limited to short-term transactions.

Representatives of the five GISB membership segments at the drafting table did not agree on all contract terms for high-risk, long-term contracts—the type of contracts generally discussed in this guide, but they did find some areas of commonality for routine monthly gas transactions. In chapter 7, we'll address in detail the differences between so-called short-term and long-term contracts. For now, and for the purpose of this chapter, keep in mind that not all industry segments agree on every contract issue. There are few, if any, industry standards unofficially adopted by producers, consumers, marketers, and LDCs that would be acceptable in every contracting situation. Many industry participants who have elected not to use the GISB Contract have done so because they need greater protections than those offered in that form.

Are there industry standards for non-standard gas contracts? Have you ever encountered a negotiation counterparty that insisted on its language proposal because the language was the *industry standard*? If you were to ask a producer what the industry standard is for force majeure,[2] for gas quality requirements, for delivery obligations, or for other contract clauses, you would probably receive a different answer than if you asked the same question of an industrial consumer. Now if you asked two producers that same question, chances are fairly good that the replies would be similar, because there is some standardization within each business sector. Few of these similarities result in across-the-board agreements on contract issues.

Any new gas contractor could easily be confused when encountering multiple ways of crafting the same contract clause, each such clause being represented as *standard in the industry* by the negotiation counterparty providing the contract. Without the assistance of some fallback standard contract language, how then can one successfully review and negotiate another party's gas contract when the two organizations represent different parts of the industry, and therefore have different needs and interests to protect?

Luckily some fairly distinctive clues are available in the contract language itself. Some of these clues alone could tell most experienced gas contractors whether a producer, marketer, LDC, or industrial consumer developed the contract. The first part of this chapter will reveal the clues in three ways:

- by the types of business needs common to each sector
- by the manner in which contract language is structured to those needs
- with examples showing how each organization leaves its own distinctive fingerprints on its own contract language

THE PLAYERS IN GAS SALES AND THEIR BUSINESS INTERESTS

Given all the different perceptions of ideal contract language, in my nearly 20 years of practice, and notwithstanding common interests between sellers and between buyers, I've noticed some additional similarities as well—similarities that cross industry sector boundaries in the form of contracting themes. There also appear to be five *thematic* motivations, each a part of every contract to greater or lesser degrees. In later chapters the legal motivations driving each will be discussed.

Disparate needs and common contract terms

Organizations using the GISB contract are attracted to its standardization of terms, a worthy goal in itself. Companies using their own form contract want the same thing—to have the same or *similar terms in all gas contracts*. This motivation is the primary reason that most organizations develop their own form of gas contract and will press to use that form. Since most gas sale organizations have developed their own "house" contract, the first matter for negotiation is usually the decision of which party's form contract to use as the basis for negotiations. If all your gas supply contracts had identical terms, wouldn't life be simpler? You would always know the payment date, force majeure inclusions and exclusions, governing law, and so forth. Unless you can wield hypnotic powers of persuasion, or your counterparty hasn't read this guide, you will both make concessions and you will have to return to the document to recall what was agreed. Don't be intimidated if your company is small, though. If your "house" contract is well crafted, even the gas titans sometimes will agree to use the other party's contract.

Of the four remaining motivations, *control, flexibility, reliability,* and *price,* one predominates in a producer's gas contract, another in a marketer's contract, yet another in an LDC gas contract, and the final one in industrial consumer gas contracts. These motivations are reflected in contract language proposed by each of the four sectors.

How to Identify the Needs of an Organization through Contract Language

As will be discussed in chapter 7, nearly all gas contracts have the same general structure and contain the same types of clauses. This chapter's focus is one of nuance—how does a producer take the same language being used by a marketer and massage it to meet its unique needs? How does a marketer, LDC, or industrial consumer do the same thing when presented with a contract from another organization? Let's take a closer look at each of the four major players and identify both their particular needs and how those needs are typically filled through contract language. By using the same basic contract clause, you'll learn how each of the sectors might address the issues covered by that clause.

Control

A natural gas *producer* may be as large as a national or international energy company like Exxon-Mobil or Shell, or as small as one individual owning and operating one well on the family homestead. All producers—whether gas units sold are measured in billions or hundreds—need to control production, facilities, and sales. Three types of limitations face producers—laws of nature, governments and regulators, and economics.

The physics of natural gas determines whether it flows from underground to the surface and whether it moves from the wellhead into a pipeline. A pressure differential must exist. Wells may be engineered to produce at certain rates in order to extend production life from months in some cases, to years in others. Wells and related facilities must be routinely monitored, maintained, and repaired if necessary.

Producers are also subject to an abundance of laws and regulations concerning everything from environmental safety to production itself. Production laws may limit the amount of associated[3] gas produced in some states. Producers must comply with laws regulating unitization[4] and well spacing.[5] They generally must account to royalty owners and make extensive tax and production filings. Like pipelines, producers have fixed costs and variable costs, and ideally they sell their gas at a price sufficient to cover those costs and make a profit. While this is happening, gas is being produced on a 24-hour basis and sold on a *gas day*[6] basis, sometimes under many different gas contracts with varying contract terms.

The production sector is unlike all the others, and a producer's gas sale contract language will be distinctive as well. Some producers sell only their own production and others purchase additional supplies to meet their contractual sale obligations. When a producer is selling only its own produc-

*Chapter 2 Gas Industry Players
(and why they play that way)*

tion, it will probably develop a *gas sale contract* that will typically tie the producer's delivery obligations to actual production, rather than an established daily quantity. Multiple contract clauses will reflect the producer's right to control production, to maintain flows at its discretion, and to operate the well free of any interference from the buyer.

A producer would rarely agree to contract language in a customer's gas contract that could force it to shut-in wells when gas prices are high or keep a well operating when a loss would result. Yet that same producer needs the contractual right to shut-in wells for operational reasons. Maintenance and repair of wells and connecting pipelines are critical to a producer's physical operation, and language reflecting the producer's need to preserve its assets will nearly always be contained in a producer's form contract. Gas sales "off the lease"[7] may contain detailed measurement, quality and pressure requirements too, depending upon the blend of gases, gas liquids, and impurities coming out of the ground.

One way to address control issues is through the *force majeure* clause, a form of which is found in virtually every contract for firm or absolute delivery requirements. These clauses list the reasons and circumstances under which one party may be excused from its obligation to make or take delivery of the gas. Force majeure clauses will typically be of the following structure:

- a definition of force majeure
- a list of included events
- a list of excluded events
- notice requirements
- strike settlement provisions that allow the effected party to settle labor disputes at its discretion
- obligation to resolve the problem

While a thorough discussion of force majeure will be found in chapter 8, this ubiquitous clause is an ideal vehicle to illustrate different contracting approaches (Fig. 2-1).

Producer's Force majeure clause discussion

In paragraph 1.2 (i), the producer reserved for itself the right to main-

57

A Practical Guide to Gas Contracting

> "1.1 Except with regard to a buyer's obligation to make payment due, neither party shall be liable to the other for failure to perform a firm obligation delivery, to the extent such failure was caused by force majeure. The term *force majeure* as used herein means any event or occurrence not within the parties' contemplation at the time this Contract was executed, that is not reasonably within the control of the party claiming excuse to prevent or overcome, and which prevents performance partially or entirely. Any valid force majeure will excuse the affected party's delivery obligations to the extent and for the duration of such force majeure.
>
> "1.2 Force Majeure shall include but not be limited to the following: (i) physical events such as acts of God, landslides, earthquakes, blowouts, fires, floods, washouts, explosions, breakage or accident or necessity of repairs to machinery or equipment or lines of pipe; (ii) weather related events such as storms or storm warnings, hurricanes, lightning, blizzards, tornadoes, low temperatures that cause freezing or failure of wells or lines of pipe; (iii) interruption of firm transportation and/or storage by transporters; (iv) acts of others such as strikes, lockouts or other industrial disturbances, riots, sabotage, insurrections or wars; and (v) governmental actions such as necessity for compliance with any court order, law, statute, ordinance, or regulation promulgated by a governmental authority having jurisdiction.
>
> "1.3 Neither party will be entitled to the benefit of force majeure excuse to the extent performance is affected by any or all of the following circumstances: (i) the curtailment of interruptible or secondary firm transportation unless primary, in-path, firm transportation is also curtailed; (ii) the loss of released capacity transportation through recall by or other reservation of the releasing shipper; (iii) loss of market; or (iv) economic hardship.
>
> "1.4 Notwithstanding anything to the contrary herein, the parties agree that the settlement of strikes, lockouts or other industrial disturbances will be entirely within the sole discretion of the party experiencing such disturbance.
>
> "1.5 The party whose performance is prevented by force majeure must provide notice to the other party as soon as practicable under the circumstances. Upon providing written notification of force majeure to the other party, the affected party will be relieved of its obligation to make or accept delivery of gas as applicable to the extent and for the duration of force majeure."[8]

Fig. 2-1 Producer's Force Majeure Clause

tain and repair its equipment. Most clauses of this type will clearly identify the producer's right to claim force majeure for both scheduled and unscheduled maintenance, allowing it the control necessary to operate its well system free from interference from a gas buyer.

Chapter 2 Gas Industry Players
(and why they play that way)

Now let's focus on subpart (ii), weather related events. Hurricane season in the late summer and fall months usually means that gas production in the Gulf of Mexico will be affected. Production onshore may be slowed or halted by hurricanes, but the greater impact is felt in offshore production, all of which is owned either by the shoreline state[9] or by the federal government.[10] Outer Continental Shelf (OCS) platforms are subject to evacuation and shutdown rules enforced by the Mineral Management Service (MMS),[11] and platform abandonment of state-government leased blocks are subject to the policies of individual producers and state regulation. During hurricanes affecting OCS platforms, when the platform is abandoned, operations must be suspended as well, although platforms in state-controlled waters may be subject to different rules.

Cold temperatures have affected gas production in most regions of the U.S. at one time or another, so this is nearly always included as an event of force majeure. If even one well is affected by cold weather such that the well does not produce (or produces less than the usual amount), no producer, and specifically a producer with fixed delivery requirements under a gas contract (such as 5,000 MMBtu/day) will want to be faced with the prospect of paying its customer for failure to perform for a reason beyond its control.

Parts of Texas, most of Oklahoma and Kansas, and parts of Nebraska and Iowa are in what is known as "tornado alley." These violent storms are so common in spring and summer months that in most years, it isn't a question of *if*, but rather of *when* tornadoes will strike. Vast mid-continent production is affected by this phenomenon, and because of the widespread damage tornadoes can cause, producers in the affected areas will frequently include this specific event in force majeure. Recent changes in weather patterns in fact, may warrant adding tornadoes as an excused event in all producer contract force majeure clauses. Any producer that has had production affected by the long-term consequences of tornadoes may wish to add the *aftermath* of storms as a force majeure event as well.

Blizzards and cold weather are different, and perhaps they should be treated separately in force majeure clauses. Sub-freezing temperatures frequently occur on the sunniest days, so cold weather may or may not be the result of a storm. Blizzards, on the other hand, are unique in that the storm itself is ferocious, but the aftermath may be even worse. Transportation to and from gas fields to repair or monitor wells, feeder lines or gathering lines may be impossible even after the blizzard has passed, so where blizzards are included as force majeure events, a specific addition of the aftermath may be well-advised.

The above examples are ways a producer might view weather in its force majeure clause. Nearly every contract clause in the producer's form contract, however, will evidence the same sort of self-protective language. There are only so many things producers, at the mercy of the weather, can control, but they can control contract language.

Flexibility

The oldest marketing company in the U.S. was organized only after the advent of Order No. 436 in 1985. Today marketing companies are spread across the country and across the gas industry. These companies may be directly affiliated with production companies, or indirectly affiliates of regulated pipelines or LDCs. They may also be independent—*i.e.*, not affiliated with a natural gas production, transportation, or LDC. Sometimes, marketing independents are subsidiaries of holding companies, but the independent marketing company itself will usually have few, if any, assets other than its office equipment, perhaps its office building, and other non-gas property.

Marketing companies, or marketers, buy and sell gas according to the plans, policies, and goals of their own company. Some are regional—*i.e.*, they focus their marketing activity:

- in a certain geographical region
- on specific pipeline or LDC systems
- to selected customers

One marketing company may devote its time and energy to short-term transactions, including day trading where gas may be bought and sold on a daily (or partial day) basis. Another marketing company may focus its efforts only on long-term deals, or only on sales to industrial end users, or to sell another company's production, or on any of a number of other variables.

Sometimes a marketing company with a nationally known name will

*Chapter 2 Gas Industry Players
(and why they play that way)*

actually buy and sell gas only in a particular region. Some large marketers will perform a variety of services involving many different types of transactions. As noted in chapter 1, it wouldn't be unusual for a marketer to provide gas procurement services as well as nominations and billing services for a customer.

The marketer buys and sells gas under its own name.[12] Marketers hold title to the gas while it is in their possession. They may enter into transportation arrangements to enhance their flexibility to ship gas where customers want the gas or to transport reliable supplies to their customers. No authorizations are required of independent marketers in order to buy and sell gas in the wholesale market. Any legal entity with good credit, a good business reputation, and good connections can become a natural gas marketer. Sufficient financial backing distinguishes those in business from those with hopes. As states are deregulating gas sales to consumers, however, there may be accompanying registration or licensure requirements for marketers making retail sales "behind the city gate."[13]

After Order No. 436, many shippers voiced concerns that any pipeline-affiliated marketing company would have an unfair advantage as a shipper on its affiliated transportation system. The FERC responded[14] by requiring a *Chinese wall* between the pipeline company and its marketing affiliate, necessitated by the open access order. LDCs have similar restrictions on their marketing function, to greater or lesser degrees, depending on the PUC of the state in which the LDC is located. Every state has its own form of consumer protection administered through the PUC.

Many marketing companies purchase gas from a number of suppliers to provide more flexibility in deliveries to their multiple customers. Since marketing companies usually don't own equipment or operations affecting the flow of gas, they cannot directly control gas movement. For all of the reasons given in the preceding paragraphs, the operative word for marketers and other aggregators of natural gas is "flexibility."

An independent marketer doesn't have the ability to physically control the gas, its movement, its measurement, or in fact, any physical aspect of the product it is buying and selling. So the marketer must pass along any liability for failure to control gas transportation to either its supplier or its customer. To accomplish this, the marketer's house contract will typically give both parties a great deal of maneuverability to make the deal work. Back in the late 1980s, certain organizations were suppliers to the marketer and other organizations were customers. So, most marketing companies developed two base contracts—one a gas sale contract and the other a gas purchase contract. The contracts, while essentially the same in appearance, were different. The gas sale contract gave the marketer certain advantages as the seller and the gas purchase contract gave the marketer the advantage as the buyer.

One of the advantages was the payment date, typically around the 20th day of the month following deliveries in the gas sale contract, and the last business day of the month in the gas purchase contract. Another positive advantage was in the tax clause. When the marketer was selling gas, the buyer would automatically be required to provide *tax exemption certificates*,[15] and when the marketer was buying gas, the seller would have to request the certificates from the buyer. The marketer as seller or buyer would not have liability for any taxes imposed at the point of delivery, as this obligation would always be placed upon the other party. A marketer as seller would give only a warranty of title and disclaim all other warranties. As buyer, the marketer would receive extensive warranties[16] from the seller along with a broad indemnification[17] relating to those warranties, and no warranty disclaimer.[18]

As marketing began to blossom, the entire nature of the gas industry changed. To expand the potential list of suppliers and customers, the original bifurcation of suppliers and customers was discarded. Everyone started doing business with everyone else, so organizations were buying gas from and selling gas to the same companies. This presented a contract language problem since companies using the other party's contract were understandably reluctant to agree to contract terms that always favored the party, that had produced its house contract. Consequently, nearly every marketing organization went through the process of developing contracts suitable for both purchases and sales. Most marketers now buy and sell gas for routine business under those newly developed buy/sell contracts[19] that no longer continually favor the marketer.

As noted in chapter 1, a marketer aggregating supplies and markets usually doesn't contractually match a particular supply with a particular market. It will buy gas from multiple suppliers and sell gas to multiple markets. Even though the marketer will probably match a long-term supply with a long-term market, it is usually not contractually bound to use a particular supply source, and is consequently free to use other supplies for that market. This type of freedom is essential to the aggregator's success and contract language will reflect that freedom. Let's take a look at the same force majeure weather language that was fashioned for a producer's needs as converted to marketer language in a buy/sell contract (Fig. 2-2).

Marketer's Force Majeure Clause Discussion

From the opening line of this clause, marketers approach force majeure contract language differently. First is the recognition that either party could owe the other money under the contract, so neither party can be excused from an obligation to pay money due the other. The most common application is when one party causes the other to incur transportation imbalance charges.[20] Most gas contracts require the party causing any transportation imbalance charges to pay or reimburse the other party for those charges. Trouble, like gas, flows downstream. Sellers more frequently cause transportation problems for their buyers rather than the reverse. A seller with multiple sources of supply may have an easier time finding an alternate market or transportation than would an industrial consumer with limited suppliers and transportation capacity.

Notice also that the language in this force majeure clause excludes many of the specific weather events that had been necessary for the producer. This raises the logical question of why a marketer who typically uses the same contract to buy and sell gas would not give itself the advantage of multiple force majeure excuses when it is the gas supplier. The answer is based on the practical nature of gas marketing.

Usually, independent marketers who aren't selling their own organization's production aren't directly affected by events at one wellhead that would stop gas flow. They are most directly affected when they are buying production within a large area experiencing bad weather. If the inclement weather results in stopped or slowed production in several or many wells, or if transportation from the affected production area is limited as a result, the

"1.1 Except with regard to either party's obligation to make payments when due, neither party shall be liable to the other for failure to perform a firm delivery obligation, to the extent such failure was caused by force majeure. The term *force majeure* as used herein means any event or occurrence not within the parties' contemplation at the time this Contract was executed, that was not reasonably within the control of the party claiming excuse to prevent or overcome, and which prevents performance partially or entirely. Any valid force majeure will excuse the affected party's delivery obligations to the extent and for the duration of such force majeure.

"1.2 Force majeure shall include but not be limited to the following: (i) physical events such as acts of God, landslides, earthquakes, fires, floods, washouts, explosions, breakage or accident or necessity of repairs to machinery or equipment or lines of pipe; (ii) weather related events such as storms, storm warnings that require abandonment of offshore platforms, low temperatures that cause freezing or failure of *multiple* wells or lines of pipe; (iii) interruption of firm transportation and/or storage by transporters; (iv) acts of others such as strikes, lockouts or other industrial disturbances, riots, sabotage, insurrections or wars; and (v) governmental actions such as necessity for compliance with any court order, law, statute, ordinance, or regulation promulgated by a governmental authority having jurisdiction.

"1.3 Neither party will be entitled to the benefit of force majeure excuse to the extent performance is affected by any or all of the following circumstances: (i) the curtailment of interruptible or secondary firm transportation unless primary, in-path, firm transportation is also curtailed; (ii) the loss of released capacity transportation through recall or other reservation by the releasing shipper; (iii) the failure of individual wells; (iv) loss of supply, (v) loss of market; (vi) economic hardship or (vii) disallowance of passthrough of gas costs by any regulatory body. Any interruption in service under a third-party gas contract may qualify as an event of force majeure under this Contract if such event or occurrence fulfills the requirements of this Article.

"1.4 Notwithstanding anything to the contrary herein, the parties agree that the settlement of strikes, lockouts or other industrial disturbances shall be entirely within the sole discretion of the party experiencing such disturbance.

"1.5 The party whose performance is prevented by force majeure must provide notice to the other party. Initial notice may be given orally; however, written notification with reasonably full particulars of the event or occurrence is required as soon as practicable. Upon providing written notification of force majeure to the other party, the affected party will be relieved of its obligation to make or accept delivery of gas as applicable to the extent and for the duration of force majeure."

Fig. 2-2 *Marketer's Force Majeure Clause*

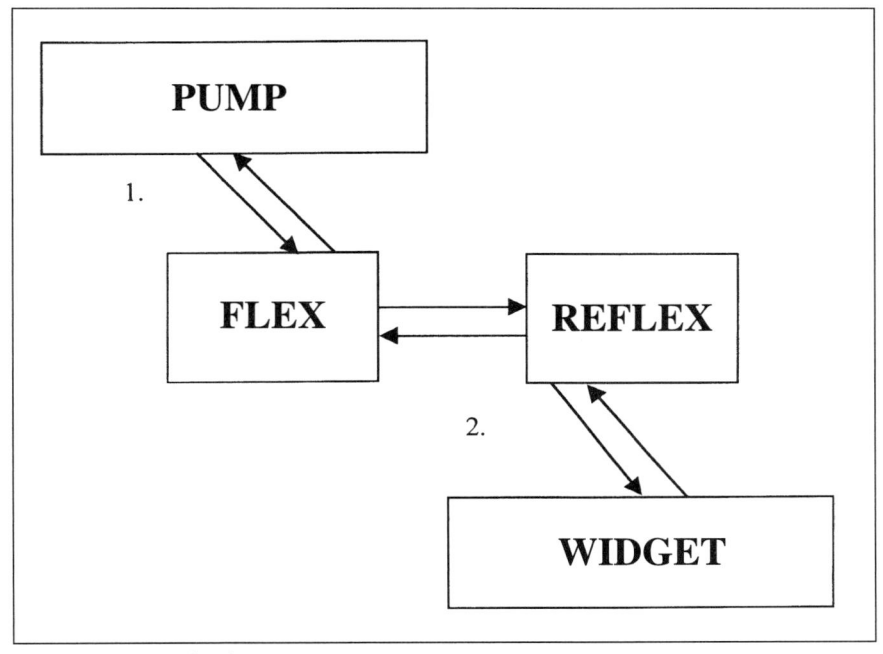

*Chapter 2 Gas Industry Players
(and why they play that way)*

Fig. 2-3 *Example of Force Majeure Pass-on*

marketer will have more difficulty finding gas supplies to feed its markets.

Since most marketers are gas aggregators, when one source of supply is lost, the marketer will find a new one. As a general rule, marketers expect the same performance level of *their suppliers*—even if those suppliers are producers. Would contract language reflecting common delivery obligations be changed sometimes? Absolutely. But a marketer will usually expect the same level of performance that it gives. Weather-related events in a marketer's contract are usually limited to those that effect a broad, general area and not a single well or line of pipe. Also note the final sentence in paragraph 1.3. The parties to this contract may pass along any valid force majeure received under another gas contract. An example of this phenomenon in action is found in Figure 2-3.

1. If FLEX receives a valid force majeure notice from PUMP (the pipeline transporting gas), it can accept the force majeure and then use that

same force majeure event to excuse its own performance to REFLEX if the definition of force majeure in the FLEX-REFLEX contract includes the specific event that prevented PUMP's performance
2. If REFLEX receives a valid force majeure notice from WIDGET, and the event would be a valid force majeure excuse in the FLEX- REFLEX contract, REFLEX can excuse its performance under the REFLEX-FLEX contract as well

Not all marketers use this force majeure pass-on clause. They may wish to limit the ability of both parties to excuse nonperformance. On the other hand, as will be discussed in chapter 8, it may facilitate settlement of otherwise protracted disputes if the domino effect of upstream or downstream force majeure is allowed to occur.

The final marketer exclusions from force majeure typically include:

- regulatory disallowance of gas cost passthrough (for reasons discussed in the following section)
- loss of supply (for the same reasons discussed earlier in this section)

This is a good time to talk about the psyche of gas marketing. Since many independent marketers don't own gas reserves to back them up, the most valuable asset of any marketer is its reputation—does it perform its obligations? Many companies are willing to sell gas, prices are not terribly competitive, and margins are thin. It is in the performance of a deal that a marketer is measured, and a marketer's gas contract frequently contains purposefully flexible language. Lawyers representing marketer clients generally disapprove of using language that may be seen as too ambiguous or vague in the contract, an argument they sometimes lose.

But I've heard from many clients that from a business perspective, contract language capable of more than one interpretation will probably tend to facilitate settlement of a dispute rather than prolong the dispute. The result of using ambiguous contract language could well work to the advantage of either the buyer or the seller in any given situation. A valid argument can be made that narrowly restrictive contract language may cause as many problems as overly broad language. At any rate, the marketer using any ambiguous contract language to its benefit and to the other party's detriment must always be mindful that without a good reputation for performance, marketing companies can not survive. Integrity earned by a sound performance record still really means something in the gas industry.

Reliability. The third prime motivation in gas contracting is reliability. Every gas buyer wants a reliable supply, but LDCs that are still providing a merchant function to their customers have a greater need. As a public utility granted an exclusive franchised service area, an LDC has an obligation to provide gas to customers in its service area. The LDC must have sufficient gas to meet the needs of its ratepayers.

Sales to LDCs are sales to the private or public utilities in every community that, in turn, distribute the gas to ultimate consumers. LDCs that for many years operated much like interstate pipeline companies underwent dramatic changes in the 1990s because of the deregulation song being sung by the federal government and by many states. An LDC that has not gone through natural gas deregulation or "restructuring behind the city gate[21]"purchases gas, typically from a variety of suppliers, to fill the needs of its customers. It may take delivery of its gas supplies at the city gate or further upstream, in which case it would ship the gas through a third-party pipeline system. Once the gas owned by the LDC enters its distribution system at the city gate, the LDC is in total control of sales and distribution. Ultimately the gas will be sold to the distribution company's residential, commercial, and industrial ratepayers. Only in recent years have third parties such as producers and marketers been able to make sales to industrial, commercial, and retail customers behind the city gate, as LDC systems had a virtual stronghold on and total control of both sales and transportation through their proprietary distribution systems with exceptions for certain industrial and commercial customers.

Except in the relatively rare case of an LDC *bypass*, where a large-volume consumer (usually industrial) can tie-in directly with the transmission pipeline, all gas in commerce ultimately moves through a local distribution system.

Regulation of LDCs by the state PUC serves essentially the same function on a state level as that of the FERC for interstate pipelines. Simply stated, LDCs are allowed a rate of return on their transportation (distribution) function, but gas supply costs are passed on to ratepayers.

Profits come to the LDC through rate of return on gas distribution, and not through gas sales. The cost of gas is passed on to ratepayers, based upon usage. The result of all this is that the LDC has a burden unlike that of any other gas buyer or seller, oversight of gas procurement costs by the PUC. In some states, the regulatory overseer gives careful attention to gas supply costs of the LDCs it regulates. In other states, a lighter approach is taken, but in either event, gas supply contracts are subject to review by the PUC. This review may be quite formal and for some LDCs may occur before a major gas purchase occurs. At other times, a review may take place after the fact, during performance of the gas contract. This process is usually called a *reasonableness* review or a *prudency* review. The distribution company can pass on its cost to procure supplies as long as that purchase is "prudent and reasonable." Any portion of gas supply costs not found to be prudent and reasonable may not be passed through to ratepayers. If the LDC is investor-owned, the shareholders will bear the burden of any imprudent excess gas costs. If the LDC is publicly owned, the ratepayers will ultimately bear that cost.

Gas contractors working for organizations that sell gas to multiple distribution companies in multiple states know well that in some states (this can vary from LDC to LDC in the same state) the PUC takes a highly intrusive approach to gas supply contracting, while in others a more relaxed attitude prevails. In most states, the LDC gas procurement employees will have pricing guidelines to assist them, such as former opinions and policy statements of the PUC. Even if the employees have followed the guidelines, gas costs may be very high and may not pass a reasonableness test. In that case, the LDC would want the opportunity to declare force majeure because of price limitations imposed by the PUC. The LDC's other alternative would be to attempt gas price renegotiations with its gas supplier. For an example of prudency at work, consider this situation (Fig. 2-4).

GASCO buys its gas supplies from a variety of suppliers, producers, marketers, and brokers. GASCO purchased gas from FLEX at a flat index price. Now, ordinarily we would say that an index-based price would appear to be quite reasonable since the index price is supposed to reflect the "average[22]" price for gas at a particular delivery point. What if a cold spell lingers (such as the first week in February 1996) and gas prices skyrocket? If GASCO is paying an effective price of $60.00/MMBtu for gas that ordinar-

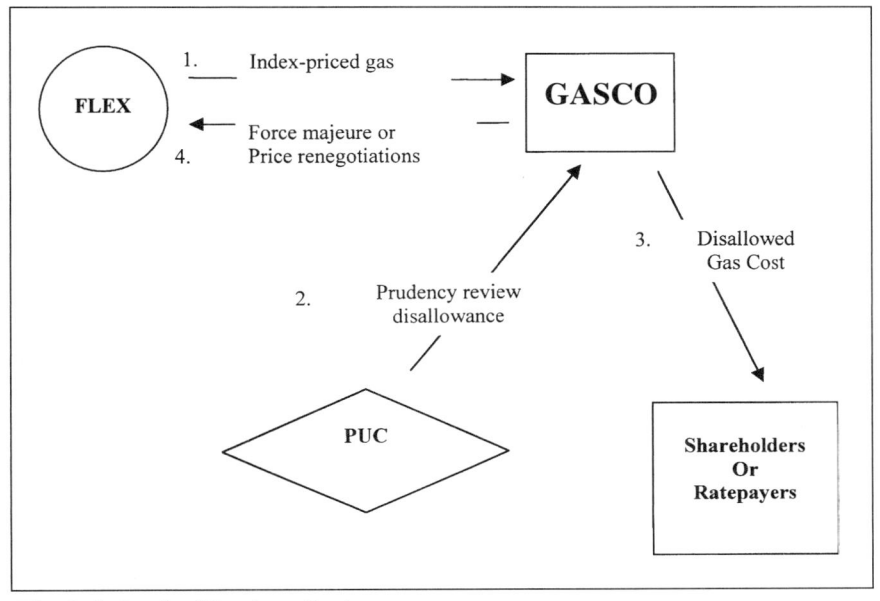

*Chapter 2 Gas Industry Players
(and why they play that way)*

Fig. 2-4 *Result of Prudency Review*

ily sells for a price around $2.50/MMBtu, the review panel may well find that some portion of the $60.00 unit price was unreasonably incurred. What if GASCO's employees wisely anticipated the possibility of price spikes during the winter months? If GASCO had negotiated contract language with FLEX that capped the price it would pay for gas in the event of a price spike—say $15.00/MMBtu—the review panel might well find a unit cost of $15.00/MMBtu to be reasonable in a $60.00 market.

LDCs gas supply load profile. LDCs are unique within the gas industry and so is the way they buy gas for their customers.

Baseload

LDCs typically purchase gas for their customer needs (and their own, to balance the distribution system) on a tiered basis (and, in many respects, industrial consumers will follow similar fuel procurement planning patterns). The gas may be purchased as *baseload*—which to the LDC means the part of its gas supply portfolio that is constant every day of the year. It knows that it will require at least this amount of gas with no seasonal fluctuations. It establishes the base of the portfolio. This part of the total gas portfolio will typically be from extremely reliable suppliers and may be for a fixed price.

Intermediate load

Usage beyond baseload needs occurs frequently during the year, even on a daily basis. During the summer, gas utilities in cold climates use less gas. During the winter months they use more. Loads may fluctuate during the spring and autumn months, so the second tier comes into play to fill those somewhat predictable, but still unknown future gas needs on a daily basis. This next tier is commonly referred to as the *intermediate* load.

LDCs use load profile projections to help them determine gas needs into the future, so the intermediate load of each LDC will depend on the weather in that region, the peak manufacturing times of industrial consumers, transportation availability, and many other variables. Intermediate load gas needs may be determined well in advance of actual needs.

Peak load

Utilities try to forecast customer gas needs as precisely as possible, but in recent years, this has been a difficult task with unusually warm winters in much of the country and unusually cold and snowy winters in other areas. Because gas supply forecasting is about as difficult as accurate weather forecasting, LDCs also contract for a third tier of gas for *peaking* needs.

Peak usage times occur most frequently when the temperature plummets, but can also occur in southern states when the temperature skyrockets and gas is needed for electricity generation. If Buffalo, NY has sub-zero temperatures on January 9th through the 12th, it will be in a peak usage period. If the LDC has contracted for sufficient quantities of gas under its base load and intermediate load contracts, no shortages will occur. More likely, the LDC has contracted with a number of suppliers for peaking gas. Because this tier is called upon only during high demand times, this is typically the most costly gas in the LDC's portfolio.

If the cold weather is localized in Buffalo, and other areas in the north-

Chapter 2 Gas Industry Players
(and why they play that way)

east U.S. aren't affected, the peaking gas will probably be available in sufficient quantities and at a relatively low price. Unfortunately, that isn't usually the case. More commonly, other large population areas will be effected by inclement weather as well, demand for gas supplies and transportation capacity will increase, and so will prices. Because of high needs and finite pipeline capacity to transport gas to the affected area, the immediate problem of the Buffalo LDC is worsened by similar situations facing other LDCs in the northeast. If the production zone is affected as well, problems can intensify exponentially.

LDCs try to protect the reliability of their gas supply through a variety of contractual commitments made by their gas suppliers in the LDC's own form gas purchase contract. Unlike marketers that typically use so-called buy/sell contracts with identical obligations for either party as buyer or seller, LDCs and industrial consumers both use contracts that do not contain many of the small concessions made in marketer's contracts as discussed earlier. The most common reliability language in a LDC gas purchase contract is found in warranties given by the seller.

Warranty Commitment. (i) "Seller confirms and warrants that it has and will maintain supplies of gas in sufficient quantities from its production or under firm contract or both to service this contract. Seller further warrants that it will supply gas as nominated by buyer under this Contract prior to supplying gas to any other of its customers."
(ii) "Seller warrants and agrees that there will be provided under the terms of this Contract a quantity of gas sufficient to enable seller to have available for delivery hereunder on any day or days a quantity of gas not less than the maximum daily quantity."

When producers were selling gas to pipelines, it was not uncommon for a producer to dedicate all of its production to one pipeline company under a *dedication of reserves* contract. The sale commitment was a warranty—*i.e.,* a promise to sell the gas only to that purchaser and under the terms of the contract. The concept of a warranty sale has survived, particularly in LDC contracts, although very little legal evidence exists as to the consequences for breaching a "delivery" warranty. In the clause above, the seller is being asked to warrant that it has sufficient gas supplies to deliver gas under the contracts, so the warranty isn't directly related to delivery itself, but rather to the ability to deliver. In chapter 6, you will learn the remedies available for

A Practical Guide to Gas Contracting

"1.1 Except with regard to either party's obligation to make payments when due, neither party shall be liable to the other for failure to perform a firm delivery obligation, to the extent such failure was caused by *force majeure*. The term *force majeure* as used herein means any event or occurrence not within the parties' contemplation at the time this Contract was executed, that was not reasonably within the control of the party claiming excuse to prevent or overcome, and which prevents performance partially or entirely. Any valid force majeure will excuse the affected party's delivery obligations to the extent and for the duration of such force majeure.

"1.2 Force Majeure shall include but not be limited to the following: (i) physical events such as acts of God, landslides, earthquakes, fires, floods, washouts, explosions, breakage or accident or necessity of repairs to machinery or equipment or lines of pipe; (ii) weather related events such as storms, storm warnings that require abandonment of offshore platforms, hurricanes, lightning, tornadoes, low temperatures that cause freezing or failure of multiple wells or lines of pipe; (iii) interruption of firm transportation and/or storage by transporters; (iv) acts of others such as strikes, lockouts or other industrial disturbances, riots, sabotage, insurrections or wars; (v) governmental actions such as necessity for compliance with any court order, law, statute, ordinance, or regulation promulgated by a governmental authority having jurisdiction *including disallowance of any or all passthrough of gas costs by any regulatory authority; and (vi) loss of buyers' sales market or territory due to regulatory changes or deregulation, system bypass, or failure of equipment.*

"1.3 Neither party will be entitled to the benefit of force majeure excuse to the extent performance is affected by any or all of the following circumstances: (i) the curtailment of interruptible or secondary firm transportation unless primary, in-path, firm transportation is also curtailed; (ii) the loss of released capacity transportation through recall or other reservation by the releasing shipper; (iii) the failure of individual wells; (iv) loss of supply; or (v) economic hardship.

"1.4 Notwithstanding anything to the contrary herein, the parties agree that the settlement of strikes, lockouts or other industrial disturbances shall be entirely within the sole discretion of the party experiencing such disturbance.

"1.5 The party whose performance is prevented by force majeure must provide notice to the other party. Initial notice may be given orally; however, written notification with reasonably full particulars of the event or occurrence is required as soon as practicable. Upon providing written notification of force majeure to the other party, the affected party will be relieved of its obligation to make or accept delivery of gas as applicable to the extent and for the duration of force majeure."

Fig. 2-5 *LDC Force Majeure Clause*

breach of warranty. There may be a problem in reconciling warranty language created to address the *condition* of the goods actually delivered with the concept of not delivering because of insufficient gas supplies.

LDCs can also seek many of the same protections through a force majeure clause crafted to protect the reliability of supply it must have. Here is how a typical LDC might adapt the force majeure language illustrated above for its own use (Fig. 2-5).

LDC Force majure clause discussion. If reliability is what the LDC needs, it has found a variety of ways to close the loopholes of force majeure excuse from its gas supplier in this clause. In paragraph 1.2, the LDC limits the ability of its supplier to claim force majeure if gas is being sourced from an offshore operation during a hurricane, while at the same time giving itself the opportunity to declare force majeure in an emergency caused by lightning or tornadoes. The seasonal load LDC needs gas during blizzards and cold weather more than at any other time of the year, so it may be reluctant to specifically list "blizzards" as an event of force majeure.

In the final sentences of paragraph 1.2, the LDC has granted itself the right to claim force majeure for disallowance of gas cost passthrough to customers and has also allowed force majeure if, for any reason, it loses part of its service obligation. For example, a large industrial consumer may receive the right to bypass the LDC, thus lowering the LDC's gas purchase needs. Some LDCs might view the excuse for loss of market on a different basis, depending on the ratio of firm to interruptible gas supply obligations and its unwillingness to resell that excess gas to another market.

LDCs must carefully balance the need for a reliable supply with potential market losses that would lessen their need for gas. This is an imperfect science and requires a major effort of those who do gas supply analysis.

Price. The most prized customers of the gas sales industry are the industrial consumers that routinely burn large quantities of gas to fuel their manufacturing operations. Cogeneration facilities,[23] steel plants, paper and pulp mills, cement plants, chemical plants—these are just a few examples of the industrial uses for natural gas. Most gas suppliers would prefer selling 100,000 MMBtu/day to one customer over a long term rather than 1,000 MMBtu/day to 100 customers. Economies of scale are tripped. Curiously, the relationship between industrials and gas suppliers has historically been a bit strained. As I've traveled the country speaking at various conferences, I've learned two things about the relationship between suppliers (particularly marketers) and their industrial customers.

First, industrial consumers want honesty, reliability, and gas at a fair price from their suppliers. The large consumers don't want to be a part of the gas industry, nor do they want to know every facet of the gas industry in order to effectively purchase fuel. They just want to deal with someone they can trust. The second thing I've learned is that many suppliers trying to sell gas to industrials feel that these customers, because they don't always understand the gas industry, don't necessarily appreciate that not all things are possible, and there are limits on what any organization in the gas industry can do.

Many industrial consumers don't develop their own standard gas purchase contract, but the largest do. Many of the industrials that have developed their own house gas *purchase* contract have found ways to contractually get what they feel they cannot get through the relationship alone. Since the contract is one for purchase by the consumer only, the contract will typically contain protective contract language similar to that in LDC contracts.

Industrial consumer contracts reflect a combination of the three needs—flexibility, reliability, and price. If a large paper company loses one of its major customers, thus reducing the need for gas supply to fuel its manufacturing process, that company will want the flexibility in its contract to reduce the quantity of gas purchased without penalty. This protection will usually be evidenced in force majeure language that includes loss of market as an excusable event. Other times the parties will agree to a formula that effectively reduces the consumer's obligations upon the occurrence of certain specified events such as loss of a major market.

Industrial consumers have production deadlines, so they will demand a reliable supply of gas to fuel operations. A consumer that does not have the ability to switch fuels for its manufacturing process will likely place the highest emphasis on reliability of supply. If a plant's equipment requires maintenance or repair, the consumer needs the flexibility to lower its gas takes according to the needs of the plant, so flexibility and supply reliability are prime motivators in an industrial consumer's contract.

The cost of gas supply is a component of a manufacturing business's profit projections, and since the selling price for manufactured items incorporates all costs incurred to produce those items, these gas users must economize with fuel costs. The operative word in gas purchase contracts provided by industrial consumers, therefore, is "price."

Price calculations for gas sold to industrial consumers are typically the most complicated in the gas industry. Since the advent of natural gas futures trading[24] and other non-regulated hedging activities, the price determination in long-term sales to this sector has become much less complicated in a way, but also more complicated in another. Prior to 1990, when the NYMEX began trading natural gas futures contracts, gas buyers and sellers only had to understand one industry. Now they must understand both the natural gas industry and the complicated financial hedging business to succeed.

Now, let's see how an industrial consumer might use the force majeure clause to protect its needs for reasonable price, supply reliability, and flexibility. (Fig. 2-6)

Industrial Consumer Force Majeure Clause Successor. Through its own force majeure clause, an industrial consumer may address the issues of flexibility and reliability much like the LDC approach. The one issue that is not addressed in force majeure clauses is price.

If the industrial consumer is located in California and purchasing gas from West Texas, it may not be concerned about hurricanes. But it will include blizzards and earthquakes as accepted events.

Many manufacturing companies are unionized, so industrial consumers always want to ability to settle strikes without affecting gas purchase contract obligations. Since many consumers take delivery of the gas right at the plant, most will include stringent limitations on transportation-related force majeure events.

Responding to Unacceptable Contract Language

The preceding section touched on a few of the major contract issues facing gas contractors exposed to a gas contract from another industry sector, but once you encounter any contract language that isn't acceptable, how can you avoid negotiation gridlock, and at the same time protect the interests of

"1.1 Except with regard to either party's obligation to make payments when due, neither party shall be liable to the other for failure to perform a firm delivery obligation, to the extent such failure was caused by force majeure. The term *force majeure* as used herein means any event or occurrence not within the parties' contemplation at the time this Contract was executed, was not reasonably within the control of the party claiming excuse to prevent or overcome, and which prevents performance partially or entirely. Any valid force majeure will excuse the effected party's delivery obligations to the extent and for the duration of such force majeure."

"1.2 Force Majeure shall include but not be limited to the following: (i) physical events such as acts of God, landslides, earthquakes, fires, floods, washouts, explosions, breakage or accident or necessity of repairs to machinery or equipment or lines of pipe; (ii) weather related events such as storms, storm warnings that require abandonment of offshore platforms, hurricanes, lightning, blizzards, tornadoes, low temperatures that cause freezing or failure of multiple wells or lines of pipe; (iii) interruption of firm transportation and/or storage by transporters; (iv) acts of others such as strikes, lockouts or other industrial disturbances, riots, sabotage, insurrections or wars; and (v) governmental actions such as necessity for compliance with any court order, law, statute, ordinance, or regulation promulgated by a governmental authority having jurisdiction and, (vi) as to buyer, any plant shutdown or slowdown caused by a change in buyer's fuel needs prompted by a decrease in buyer's obligations to a customer or customers.

"1.3 Neither party will be entitled to the benefit of force majeure excuse to the extent performance is affected by any or all of the following circumstances: (i) the curtailment of interruptible or secondary firm transportation unless primary, in-path, firm transportation is also curtailed; (ii) the loss of released capacity transportation through recall or other reservation by the releasing shipper; (iii) the failure of individual wells; (iv) loss of supply; (v) economic hardship or (vi) disallowance of passthrough of gas costs by any regulatory body.

"1.4 Notwithstanding anything to the contrary herein, the parties agree that the settlement of strikes, lockouts or other industrial disturbances shall be entirely within the sole discretion of the party experiencing such disturbance.

"1.5 The party whose performance is prevented by force majeure must provide notice to the other party. Initial notice may be given orally; however, written notification with reasonably full particulars of the event or occurrence is required as soon as practicable. Upon providing written notification of force majeure to the other party, the affected party will be relieved of its obligation to make or accept delivery of gas as applicable to the extent and for the duration of force majeure."

Fig. 2-6 Industrial Consumer Force Majeure Clause

*Chapter 2 Gas Industry Players
(and why they play that way)*

your employer or client? The answer lies in learning how to read the contract for clues. Let's assume that a consumer's gas purchase contract contains the following clause:

> *"Warranty and Covenants of Deliverability and Security of Supply. Seller warrants and covenants the delivery of the quantities of gas to the Delivery Point. Seller further warrants and covenants that it has sufficient quantities of gas which it has contractual rights to, or which it owns or controls, and which it will have contractual rights to or which it will own or control, sufficient for the timely performance of Seller's obligations hereunder. Seller further warrants and covenants that it will use all reasonable efforts to maintain for the full term of this Agreement all necessary agreements or arrangements for the transportation of gas, for the full Contract Quantity, to the Delivery Point. Seller acknowledges that the warranties and covenants made herein are material conditions under the Agreement."*

This is fairly strong language. As noted in the previous section, clauses of this type are usually reserved to gas buyers that absolutely must have the gas. It is obvious from this contract language that the buyer is extremely concerned with the seller's deliverability, but if your company (the sellers) is an independent gas aggregator relying on short-term gas supplies and transportation even for your long-term transactions, it probably won't want to agree to this language. Your goal as a gas contractor is not to break the deal; it is to negotiate reasonable contract terms. You do have choices in this situation—the first is to leave the language as it is and hope that nothing goes wrong. The second choice is to stonewall. When you are marking your suggested changes on the provided contract form, you could just cross it out, and note that your company never agrees to this type of language.

The final choice may involve different contract language altogether that would satisfy the consumer while protecting your organization. The additional language may already be found in other parts of the contract, and by massaging the terms, the buyer may no longer feel bound to the proposed language. Perhaps your organization would agree to provide the buyer with an officer's certificate (delivered at regular intervals) attesting to the fact that sufficient gas supplies and transportation are available for the time covered by the certificate. You might build more communication obligations into the contract so that the buyer would always know in advance that the transportation and gas supplies are there for stated periods of time. Your organization might agree to a number of other alternatives as well, but contract negotiations rarely are completed by surrender on the part of one party, but by retrenchment on both sides. The secret is to know the acceptable gray areas. Here are some guidelines to follow when faced with unsatisfactory contract language:

Do I understand the issues? It isn't uncommon for untrained contract personnel to be given responsibility for heavy-duty contract negotiation, but if you are to provide alternate contract language, you must have some knowledge of what is possible, both legally and operationally, prior to making any suggestions. The entire premise for this guide is that you must understand the issues in order to do your job. Contract negotiations, as you will learn in the final chapter, don't always involve fair trades. Some trades aren't fair—but they are necessary to make the deal work, and you must know which trade-offs work and which won't. A bibliography, reading list, and current Internet addresses for organizations providing information about the gas industry is included at the end of chapter 10. Many classes are available as well, so avail yourself of the tools necessary to understand the gas industry. Otherwise, you will have little chance to know what could go wrong in even a routine gas transaction.

Do I understand the proposed language? How many times have you read and re-read a proposed contract clause, only to realize that no matter how many times you read it, you won't understand it. That's your first good clue that the proposed language won't be acceptable to your organization. If you don't understand the language, you must ask your negotiation counterparty (1) what it means, and (2) why it's in the contract. As a rule of thumb, I've found that whenever I'm faced with this question, the most likely answer is that the language was added because of some previous contract dispute my negotiating counterparty had with another party. The new language was added to assure that whatever it was that happened will never happen again. Just ask.

Chapter 2 Gas Industry Players
(and why they play that way)

Do I understand my goals? What is your company's position on sales tax issues, on delivery obligations other than firm or interruptible, on which party should be the measuring party? Without knowing the position of your employer or client on contract issues, you are left in an untenable position of floating without direction. As has been said, if you don't know where you're going, it doesn't matter which road you take. It does matter in contract issues. You must be prepared and know before you ever engage in contract review or drafting what your ultimate goals are. Ask for guidance from your supervisor or another person in a position of authority, so you know the difference between what can be given up in contract negotiations and what must always stay (or be added) in the contract. You have absolutely no contracting leverage if you haven't first read the map to determine which road is best.

Do I have sufficient contract writing ability to avoid making a big mistake? In the chapter 10, you'll learn some of the tricks to successful contract writing. Before embarking on a contract review and re-write, you must understand the implications of words you are using. Experienced attorneys find themselves using an inappropriate word, or making an imprudent comma placement to the later dismay of their clients who found out too late just what that little word actually meant. As with any other matter in life, preparation is the key to success. Know the issues, know how to propose alternatives, and know how to correctly draft alternate contract language.

COMMON INTEREST OF ALL SELLERS
COMMON INTEREST OF ALL BUYERS

Let's break from differences and switch to the areas of common interest among all the players, depending on which side of the sale that player may find itself. Regardless of the identity of the two parties to a gas sale agreement (*i.e.*, a producer, marketer, etc.), additional differences in contract language occur, depending entirely on whether in a particular transaction that entity is the seller or the buyer.

All gas sellers have interests unique to *sales*, and all gas buyers have interests unique to *purchases*. A producer selling its gas has some of the same priorities as a marketer selling third-party gas. The industrial gas buyer has some interests in common with the LDC as well. Let's discuss some of the common bonds between all sellers and the common bonds between all buyers as expressed in their contract language.

Seller

The most fundamental interest of all sellers is, of course, receiving timely payment from the buyer.

Payment. From the seller's perspective, a buyer's ability to pay is first on the list of common interests. If payment for delivered gas is not made, the buyer may be in default of its contractual obligations, and the seller left trying to collect payment from an insolvent organization. Sellers have sole responsibility for determining whether their potential customers have the ability to pay for gas delivered. The first step in ensuring creditworthiness is typically a credit check of the buyer's organization to assure its necessary financial stability. Since billing and payment in the gas sales industry is still based on monthly business, the credit level of most buyers is determined on a 60-day basis.[25] Contractually, most gas sale contracts will require that the buyer continue to meet the seller's credit requirements through the life of the contract. Buyers will usually be required to provide routine proof of continuing creditworthiness and sellers will have the ability to demand adequate assurances of continued performance if the buyer's credit status drops significantly.

Many seller organizations are slow to place specific credit requirement language in their contracts. The usual reason for this is there are no industry standards for determining creditworthiness and setting credit limits. Every organization has its own method for evaluating the other's creditworthiness and establishing credit limits and most systems are propriety and confidential, so the buyer typically won't know the exact process used by its seller in setting limits. This is totally within the prerogative of sellers. If rigid credit standards were made contractual obligations, sellers might lose the flexibility they now have to use their own private credit evaluation.

Limiting liability. A seller also wants to assure that once the gas is no longer in its possession and control, it will no longer be liable for any damage that may be caused by the gas. Contract language that removes any responsibility for harm caused after the gas has left the seller's possession and control is usually in a combination of three contract clauses, *warranty disclaimer, responsibility,* and *indemnification.*

LIMITED WARRANTIES

As discussed earlier in this chapter, the seller will limit warranties and specifically disclaim any implied warranties. This is intended to avoid any potential liability for warranties as to the quality of the gas delivered.

Responsibility

These clauses identify the point in time and physical location where responsibility for the gas, including liability for harm caused by the gas, is with the seller and the buyer respectively. Generally, the seller is responsible for all liabilities, including physical injury and property damage, taxes, royalty payment, etc. upstream of the delivery point. The buyer is responsible for all such liabilities downstream of the delivery point.[26] As between the parties to the sale contract, the responsibility clause should state (define for all points of sale and custody transfer) which party bears the risk of loss and liability. Responsibility clauses for the seller and buyer are usually identical or nearly so, but sellers may add something more to assure that acceptance of the gas by the buyer or the buyer's transporter is the trigger for a complete and irreversible transfer of responsibility.

Indemnification

Indemnification is included in all gas contracts; and the best lay description of indemnification is that each party, in essence, tells the other that it has liability insurance. If the buyer:

- incurs claims
- is made party to a legal action
- suffers any property damage
- or any of it's employees suffers bodily injury related to performance of the contract

then the seller will agree to bear the costs of the damage, absent some misconduct of the indemnified party. The seller agrees to the same thing vis-a-vis the buyer, so an indemnification clause reflects a common understanding that each party will assume full responsibility for any damage that performance or the contract causes the other party.

Indemnification clauses vary with regard to specificity of the verbage used, but the general tone of all such contract language is parity of treatment. The differences in indemnification language exist in contracts drafted by organizations that own physical assets necessary for delivery of the gas. Industrial consumers may be very specific in the extent of indemnification given to protect their machinery and equipment from damage. For the same reasons gas suppliers frequently limit the scope of indemnification they give. The potential liability for injury to a buyer's employees, damage to equipment and facilities, and customer costs are enormous, so it wouldn't be uncommon to see language in a gas supplier's contract skewed to limit its liability.

Taxes

The seller also wants to know that any *taxes* assessed after the time title has been transferred will be the responsibility of the buyer (or someone else—but not the seller in any event). We'll review relevant taxes in chapter 8, but for the purpose of this discussion, it's important to distinguish between the various taxes from production to consumption. The first tax is assessed at the time the gas is separated from the ground. Most major production states have some form of *severance* tax, rather like a privilege tax for extracting the gas from the ground. The severance may be a flat tax per unit of production, a percentage-of-proceeds tax, or another variation. Most states that impose a severance tax require the producer to report and pay the tax, but others require payment from the first purchaser. Severance tax isn't necessarily the only tax imposed at the wellhead. Nearly every gas-producing state has some additional taxes associated with gas production typically paid by the producer.

After the wellhead, the most common tax is one on sale transactions. The three most frequently imposed taxes are:

- sales tax
- use tax
- gross receipts tax

A sales tax is imposed on the transaction itself, the sale. When multiple sales of the same product are made, only the final sale for consumption is usually taxed.

The seller of a product is responsible for collecting and remitting sales taxes to the taxing authority, as well as filing necessary forms. For every sale that is tax exempt, the seller should require adequate documentation of the exemption—usually through a valid exemption certificate provided by the buyer. That which makes a sale exempt is usually:

- the purpose for the sale
- the legal status of the buyer
- the intended use of the gas by the buyer

Since most states exempt "sales for resale," or wholesale transactions, and many states exempt sales of fuel used in the manufacturing process, the majority of gas marketing transactions are tax-exempt.

However, and this is an important caution, every state has its own system for determining the validity of a tax-exempt sale, and some require special filings and/or special exemption certificates. Even though there is a uniform sales tax exemption certificate, not all states accept it. It is important to maintain current information about the exemption certificate status in all states where gas is being sold and bought. In the discussion of taxes in chapter 8, we'll get into more detail on the peculiarities of gas sale taxation.

BUYERS

Gas buyers also share common concerns with all other gas buyers. The primary interest shared by buyers is receiving the gas. Any given buyer's interest in this regard may vary depending on the nature of the delivery obligation. For example, if the deal is *interruptible*, most savvy buyers understand that a seller will almost certainly stop deliveries if prices move against the seller. If a transaction is truly interruptible, either party can stop delivering or receiving the gas for any reason at all, or for no reason at all. Where there is no obligation to deliver, there should be no expectation of receiving gas. If the deal is *firm*, though, the buyer expects to receive the contract quantity, and the seller's deliverability is a major issue.

Deliverability

As noted earlier in this chapter, gas buyers use a variety of methods to ensure deliverability. Requiring the seller to *warrant* that it has sufficient gas to deliver through the term of the contract is fairly common. Another method to assure delivery when gas is being purchased from the marketing subsidiary of a production company is to ask for a *performance guaranty* from the parent. If the marketing subsidiary fails to perform its delivery obligations, the buyer can get delivery of the gas directly from the parent.

It is not uncommon for a buyer to require that the seller provide information on a regular basis about its reserves and/or gas supply contracts so that the buyer can determine for itself whether or not the seller can meet delivery obligations. Buyers taking this approach would also add a "what if" clause to penalize the seller for failure to meet the buyer's deliverability criteria.

Typical situations that might trigger contractually mandated action adverse to the seller are:

- seller's deliverability falling below a specified MMBtu level
- the ratio of seller's supplies to its own markets dropping to a certain number, by a given percentage, or pursuant to a formula
- the seller failing to maintain adequate long-term gas supply contracts or seller's reserves falling below some established point

It would be fairly uncommon to create a termination or default situation for failure to maintain adequate gas supplies, but the buyer might build into its contract a right to reduce its contractual commitment to take gas. An example would be a reduction of the buyer's obligation to take gas in the future in the same proportion as the drop in the seller's sources of supplies or reduction of reserves from the original numbers—a 20% reduction in

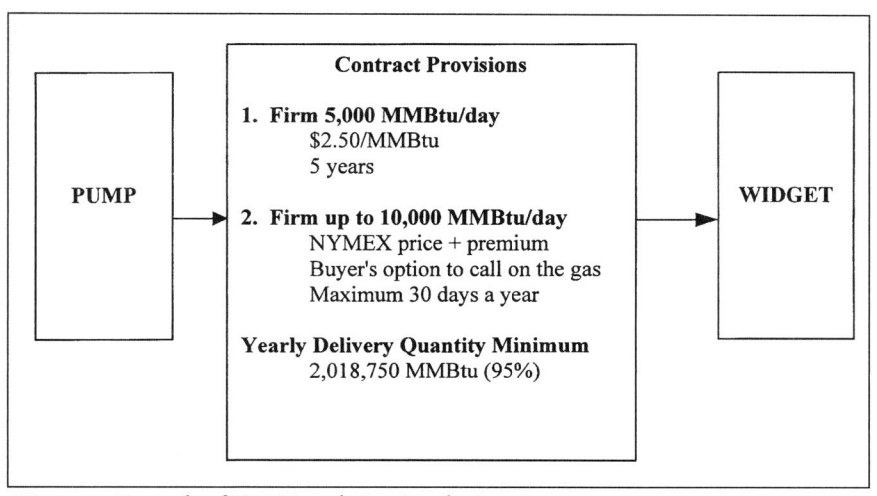

```
                    Contract Provisions

            1. Firm 5,000 MMBtu/day
               $2.50/MMBtu
               5 years
  PUMP                                              WIDGET
            2. Firm up to 10,000 MMBtu/day
               NYMEX price + premium
               Buyer's option to call on the gas
               Maximum 30 days a year

            Yearly Delivery Quantity Minimum
               2,018,750 MMBtu (95%)
```

Fig. 2-7 *Example of Two-Tiered Gas Supply Contract*

reserves equals a 20% reduction in the buyer's obligation to take delivery of gas under the contract, at buyer's option, of course.

More commonly, it is only after the seller has failed to deliver the contract quantity consistently for a stated period of time (*e.g.*, month, quarter, year), that the buyer may invoke any of a number of mini-remedies under the contract. Commonly, the buyer will have an ability to reduce its take commitments to the deliverability of seller. Let's say that PUMP is selling WIDGET gas for five years with a two-tiered delivery obligation, peaking and firm base load[27] (Fig. 2-7). If PUMP delivers at least 95% of the total yearly contract quantity, the contract will remain unchanged.

Title

Once a buyer has assured through contract language that the seller has sufficient incentive to deliver the gas, there are additional concerns unique to the buyer. The first is that the *title* be good and merchantable. Although I would like to say, in the big picture of gas sale transactions, this happens infrequently, it isn't unusual for a gas buyer to receive notice from some third party that the seller has not paid royalties owed to interest owners or that the seller has not paid taxes. Either of those events could put a *cloud*, or an encumbrance, on the gas title. Under certain circumstances the rightful owner can reclaim gas. But a potential result is that the buyer may have to pay twice for the gas—once to the seller and then again to the unpaid royalty owner—and then seek reimbursement from the producer for the royalty payment it paid on the producer's behalf.

Credit of seller

Buyers also have an interest in the creditworthiness of the seller, particularly if the buyer has arranged to purchase gas on a firm basis for a long term from one supplier. It isn't just the credit of the buyer that is important—both parties must maintain the other's creditworthiness standards. While the seller also has concerns regarding a buyer's potential failure to take gas, there is generally more harm to a buyer if the seller fails to deliver the gas. Typically, there will be provisions in a contract for calculation of the damages to either party if the other fails in its delivery/receipt obligation. If a seller doesn't deliver gas it has contracted to sell, the buyer could incur much more damage downstream than the seller would routinely incur upstream if the buyer fails to take delivery of the gas from a seller. (Remember, problems flow downstream.) A seller may have other markets for gas not taken by the buyer. Buyers would be less likely to have other options for reasonably priced supplies of gas if the seller doesn't show up with the product.

These veins of common interest are dispersed throughout the other different contract clauses from a producer, marketer, LDC, or industrial consumer. Now that you know who the players are and how each might approach gas contracting a bit differently because of its operations, it's time to turn to some of the legal issues. In chapter 4, you'll begin to explore the sales laws that govern each and every gas sale transaction in the U.S., excluding those involving the federal government, and in limited circumstances, the State of Louisiana. But first, transportation laws and regulations should be discussed.

---

1. The Gas Industry Standards Board has adopted a "Short-Term Base Contract for the Purchase and Sale of Gas." That contract in its entirety, and sometimes individual

Chapter 2 Gas Industry Players
(and why they play that way)

clauses, has been adapted for use by organizations representing all four sectors. The GISB Contract was developed for transactions with a delivery period of one month or less. This contract is used primarily for low-risk transactions where individual needs of the contracting parties are not so overwhelming that a special contract must be developed for the transaction. To the extent parties have elected to use the GISB Contract, they have indeed agreed to abide by certain contract standards. The GISB Contract will be discussed intermittently throughout this guide.

2. Force majeure, a French term originally meaning "act of God" includes any unanticipated event or occurrence that prevents either party's ability to deliver or take delivery of gas under the contract. A complete discussion of this topic can be found in chapter 8.

3. Natural gas produced with crude oil. If oil cannot be produced beyond certain amounts, the gas will be shut-in as well.

4. The process whereby owners of adjoining properties pool reserves from a single unit operated by one of the owners; production is divided among the owners according to the unitization agreement

5. See, for example, New York State Consolidated Laws, Environmental Conservation, Title 5, Well spacing in oil and natural gas pools and fields, § 23-0501, *et seq.*

6. The definition is established by transporters and may be on any basis. The GISB standard definition is 9:00 a.m. Central clock time until 9:00 a.m. Central clock time on the following day.

7. A term used to describe gas sales made from the wellhead, the first sale of gas in commerce.

8. This force majeure clause is based on, but not the same as Article 11 of the GISB short-term contract for the sale and purchase of gas.

9. States generally own the royalty on all gas produced up to three miles offshore.

10. The U.S. government owns all natural gas produced beyond the state-owned limit, essentially to international water boundaries. The area owned and controlled by the federal government is known as the Outer Continental Shelf (OCS) and was established by the OCS Lands Act of 1953 (43 U.S.C. 1331 - 1356, P.L. 212, Ch. 345, August 7, 1953, 67 Stat. 462) as amended by P.L. 93-627, January 3, 1975, 88 Stat. 2130; P.L. 95-372, September 18, 1978, 92 Stat. 629; and P.L. 98-498, October 19, 1984, 98 Stat. 2296. The 1953 statute defines the OCS as all submerged lands lying seaward of State coastal waters (3 miles offshore) which are under U.S. jurisdiction.

11. The Minerals Management Service (MMS) is a bureau of the U.S. Department of the Interior. The mission of the MMS is to manage the mineral resources of the OCS in an environmentally sound and safe manner and to timely collect, verify, and distribute mineral revenues from Federal and Indian lands.

12. This is distinguished from a gas *broker* that does not hold title to the gas. Brokers usually act as agents for either the seller or the buyer and typically receive a fee based upon the total quantity of gas brokered. Marketers may act in the capacity of brokers whenever the marketer is acting as agent for a third party. The important distinction between marketers and brokers is that brokers do not take title, while marketers do.

13. Natural gas sales made directly off an LDC system, rather than off a transmission system.

14. See chapter 1, fn. 19.

15. Tax exemption certificates are required by every state in order to remove certain types of sales from the requirements of paying sales tax, gross receipts tax, and others. This topic will be discussed fully in chapter 8.

16. A promise made regarding some aspect of the gas being delivered. This topic is extensively covered in chapter 4.

17. An indemnification, simply stated, is a promise to pay for any property damage, personal injury or other claim made by the other party or any of its employees, agents or assigns if such damage or injury was caused while performing obligations under the contract. See generally, the discussion of indemnification in chapter 4.

18. Warranties given by the seller may be disclaimed contractually. See a complete discussion of this in chapter 4.

19. This is a term used to identify a contract for both sale and purchase of gas, in which either party may in a given transaction, be either the buyer or the seller. As buyer or seller, both parties have identical obligations and rights.

20. Nearly all pipeline systems have policies to encourage shippers to make accurate and timely nominations. Failure to comply with these policies may result in payments to the pipeline. The most commonly understood imbalance remedy is known as the "cashout" procedure used by all interstate pipeline companies.

21. City Gate refers to the point at which physical possession of the gas is delivered into the LDC distribution system. Other terms for city gate are "town border" and "tap station."

22. The term "average" is being used as an indicator of what an index price represents. Index prices may be published in a variety of manners, some prices will be published on a daily basis for multiple delivery points. Others may be published for a week, and others reflect a monthly average. Each publication uses its own method for collecting data and for determining the price. Any given price may reflect an average, a range, a median or other variables, but each publication provides data collection and calculation methods used. Some major gas price publications include *Gas Daily*, a McGraw-Hill publication and *Natural Gas Intelligence*, an Intelligence press publication. Most of these index publications are available by mail, email, fax, or by accessing the publisher's home page.

23. A plant that produces two types of energy fueled by one type of energy. Natural gas is typically used to generate electricity that is sold to electric utilities and the steam produced as a byproduct of electricity generation is sold to steam customers.

24. See generally the discussion of futures contracts in chapter 1.

25. Credit limit determination varies widely in the industry. Some organizations are conservative in their approach to credit requirements of their customers and may use a sophisticated credit system to analyze exposure for the short term as well as overall exposure for long-term transactions and hedging activities. Others still use the 60-day cycle, which reflects typical spot business.

26. For the discussion of liability *at* the delivery point, see the discussion in chapter 8.

27. The term *base load* means many different things depending on its usage. In this context, the term is being used to designate a uniform daily delivery commitment. See generally the discussion of this term in chapter 6.

Chapter 3: Natural Gas Transportation Laws and Regulations

With the exception of sales made by LDCs to their customers, natural gas sales are not regulated. Gas transportation is. The gas sale business is private commerce and the government cannot interfere with private contracts. More than 95% of all natural gas sales involve transporting gas through a regulated pipeline, so the gas sale business is strongly influenced by governmental regulation.

The transportation service used may be gathering, transmission, or distribution, and the types of agreements used for each service vary. Gathering activities are exempt from FERC oversight, thus subject to whatever level of regulation the state may impose. Interstate transportation service agreements (TSAs) are approved by the FERC and cannot be changed by the pipeline company after such approval without subsequent review and approval by the FERC. Distribution, like gathering, is regulated by the state in which the distribution system is located. But, distribution is different from gathering, since gathering systems are not public utilities, and LDCs are. Varying levels of state oversight, from light regulation on a utility-by-utility basis to FERC-like regulation means that no national rules apply to gas distribution; every state is different, so every utility is subject to the regulation unique to it.

The majority of gas sales in the U.S., and those between U.S. and Canadian parties, are made on the interstate/interprovincial pipeline system, so this chapter will center on the laws and regulations affecting interstate and interprovincial transportation.

SIGNIFICANT U.S. FEDERAL GAS LAW

In the early years, when pipelines were made chiefly of cast iron and the back-breaking process of laying pipeline realistically and economically impeded laying pipe over long distances, pipeline construction occurred at a relatively slow pace. In chapter 1, we covered the general history of interstate transportation regulation. In this chapter, we'll go into the specific provisions of the major laws and FERC orders during the Twentieth Century.

Natural Gas Act of 1938 (NGA)[1]

Congress passed this act for the reasons indicated above during a decade of vast federal energy law activity. Under the NGA, the Federal Power Commission (FPC), predecessor of the FERC, was created as the regulatory overseer of interstate energy activities, including transportation of natural gas in interstate commerce, sales in interstate commerce for resale, and natural gas companies engaged in sale or transportation.

The NGA never regulated direct sales, local distribution of gas and necessary distribution facilities, gas production commerce, and related facilities, gas gathering and related facilities, *Hinshaw pipeline commerce*,[2] or intrastate pipelines.

NGA regulation included:

- certification and abandonment proceedings for bundled sales to begin and end
- the requirement that regulated pipeline companies collect only just and reasonable rates
- a prohibition against discriminating among customers
- significant reporting requirements

The law was burdensome because of its invasiveness into nearly every aspect of the natural gas industry. It isn't uncommon for legislatures contemplating totally new laws like the NGA to write their own fears into the bill through onerous requirements upon those regulated. In the process, nightmarish compliance strings may be attached.

This was true with the NGA. The interstate pipelines were in a new world of federal oversight and the effect on newly regulated interstate pipeline companies was traumatic, but they lived with it and coped. For example, for every bundled sale to an LDC, each pipeline was required to pursue a "certificate of public convenience and necessity" proceeding at the FERC to prove that the service was needed and would benefit the public good. This type of proceeding is common to public utilities and necessary for protection of the public but is cumbersome, time consuming, and costly.

Most of the original bundled transactions were for long periods of time, 20 years not being uncommon, and the certificate proceedings had to be completed and approved before the service could begin. When the long term contract was ready to expire, the pipeline company then went back to the FERC to get permission to stop the service through abandonment proceedings. Individual certificate proceedings are not required for open-access services, but most of the original act is still in effect. The NGA has not been repealed, only amended.

The NGA addressed the industry as it was in 1938 and contained no incentives for drilling and exploration. During the WWII years, new oil pipelines were built to assist with the war effort, but natural gas pipelines with few active markets generally didn't benefit from that effort immediately. After the war the pace of pipeline construction quickened, as noted in chapter 1 and industry observers quickly realized that the NGA no longer met the needs of the country in general and the gas industry in particular. It wasn't until an entire decade of energy crises that further Congressional action was taken. The next major congressional action was in 1977 with the National Energy Act,[3] (NEA) a mammoth piece of legislation enacted to reduce U.S. reliance on foreign energy—particularly on oil. It included creation of the FERC and abolishment of the FPC. The part of the Energy Act relating to natural gas was the Natural Gas Policy Act of 1978.[4]

Natural Gas Policy Act of 1978 (NGPA)

This law was passed, in part, to remove some of the commercial barriers that had been established with the NGA. To spur exploration and new production, the NGPA established eight categories of natural gas and maximum lawful prices for first sales in each category, including sales of intrastate gas (Table 3-1).

Category #	Description
102	New natural gas and certain natural gas produced from the outer-continental shelf
103	Natural gas produced from new, onshore production wells
104	Natural gas dedicated to interstate commerce on November 8, 1978
105	Natural gas sold under existing intrastate contracts
106	Natural gas sold under rollover contracts (both interstate and intrastate contracts)
107	High-cost natural gas
108	Stripper well gas
109	Natural gas not covered by any other section

Table 3-1 NGPA Pricing Categories

As mentioned in chapter 1, these price controls worked beyond the wildest expectations of congress, and while short-term results were positive, the industry ultimately suffered from the aftershocks. Other changes in the NGPA included:

- partial deregulation of prices on most first sales of gas supplies
- removal of first sales of gas and most gas from certificate and abandonment requirements of the NGA
- authorization of Section 311 transportation

Creation of Section 311 transportation was, in many respects, the critical element in formation of the spot market for natural gas. This new type of transportation opened new markets outside the physical boundaries of individual pipelines systems. Under Section 311 transportation, an interstate pipeline could transport gas on behalf of any intrastate pipeline or LDC, an intrastate pipeline could transport gas on behalf of any interstate pipeline or LDC, and an intrastate pipeline could sell gas to any interstate pipeline or any LDC served by an interstate pipeline. To this day, Section 311 transportation is used extensively.

Natural Gas Wellhead Decontrol Act of 1989

Toward the end of the 1980s, when high supplies and low demand again skewed the economics for natural gas producers, transporters, and consumers; congress removed maximum lawful prices in the Natural Gas Wellhead Decontrol Act of 1989 (NGWDA).[5] With passage of the NGWDA, each of the eight categories of gas created by the NGPA was phased out over the next four years. Effective January 1, 1993, all price controls for natural gas had been decontrolled and NGA filing requirements for natural gas producers were abolished. This act took the federal government out of the price-setting business, and today there aren't any controls on natural gas prices. This history has been included to give you an idea of what the federal government has done and what it can do in the future.

SIGNIFICANT FERC ORDERS

Congress sets policy through its laws, and the FERC enforces that policy through its orders. The two significant orders described in chapter 1 were Order No. 436 and Order No. 636. Now that you have the big picture of federal intervention through laws and orders, here are the specific elements of each of those orders

Order No. 436/500

Blanket transportation certificates. Every open-access interstate pipeline was issued a transportation certificate under which it could provide services to all qualified third-party shippers. This ended the long-standing practice of certification for each and every transaction undertaken by the pipeline. When blanket transportation certificates were issued to all open-access pipelines, some heavy burdens in Washington, D.C. were lifted and replaced with realistic opportunities all along the pipeline system.

Nondiscriminatory access to transportation. Any organization meeting the threshold credit and contractual requirements of the pipeline company was to be allowed equal access to the pipeline system.

Unbundled rates. When interstate pipeline companies were providing a bundled sales and transportation service to their customers, the product component (gas) and the service component (transportation, storage, gathering, processing, etc.) were not separately identified in the pipeline company's tariff.[6] The customer paid one price for all goods and services without a separate identification of the component costs. Order No. 436 required the pipeline companies to separate their transportation costs from sales costs so the services could be separately priced.

Interruptible transportation. A pipeline operating as a merchant had no need to define different levels of service. There was no such thing as *levels of service* because pipeline companies worked with their own gas suppliers and their gas customers to deliver gas when it was needed. After open access became a reality, the pipeline companies had to establish a priority system for their third-party shippers. In addition to firm transportation that included a demand fee and payment for the gas that actually flowed under firm TSAs, the pipeline companies were required to provide interruptible transportation, a less costly service available when the pipeline wasn't fully subscribed for firm capacity or wasn't filled with firm entitlement gas.

Capacity allocation. Could the pipeline companies sell the same capacity in their pipeline more than once a year? This was the concern facing the FERC for pipelines that had traditionally heavy seasonal loads. The pipelines carrying gas to the New York or Chicago areas have high winter load profiles, for example. The pipeline companies could obstensibly sell the space not being utilized during relative low-usage months to a manufacturing consumer having a summer production season. Capacity allocation is a process by which the pipelines report to the FERC in May of each year their firm capacity subscriptions for the reported period and they cannot increase that figure during the period.

Curtailment policy: interruptible first, then firm on a prorata basis. Interruptible transportation costs less than firm transportation because the pipeline's only obligation is to provide interruptible service to a shipper when and if the capacity is available. Because of the obligations pipelines have to their firm shippers who have paid demand charges to reserve pipeline capacity, the FERC required that interruptible transportation be curtailed first if there is insufficient capacity in the pipeline for all gas nominated at a given time.

Stranded costs. Four elements of Order No. 436/500 deal with the costs already incurred by pipelines in reliance on the system remaining the same—at least until termination of all the pipelines' supply contracts and sales contracts. Since Order No. 436 had a "snapshot" effect on the industry, where everything changed as of a day certain, the pipeline companies were left with massive costs and no one to pass those costs on to. Conversely, the FERC had concerns that some of the pipeline customers and producers who had been the pipelines' gas suppliers would be held captive to the old sales system if they were not given direct access to transportation. Here's how the FERC handled all of these problems.

Contract demand (CD) conversion. Based upon their customers' current sales entitlements (the CD service), the FERC required every open access pipeline company to annually offer the LDCs to which they sold bundled services the ability to convert a percentage of their current bundled CD service to transportation service. This was how the FERC assured sales customers that they would have a realistic opportunity to purchase at least a portion of their gas supplies from an organization other than the pipeline, and LDCs would have the ability to transport gas under firm transportation service agreements.

Take-or-pay credits. In the first chapter, I described the circumstances leading up to the take-or-pay clauses in many gas supply contracts pipelines had with their producers. The value of these contracts to the producers frequently added up to millions or billions of dollars. When Order No. 436

was issued, the pipeline companies and producers began negotiations to terminate the contracts containing the take-or-pay clauses and settlement discussions usually led to litigation. Without these long-term relationships at the well head, producers were concerned that they might not have access to the pipeline as shippers to sell their gas to other downstream customers. The FERC concluded that the fairest way to allow producers access to the pipeline as shippers while at the same time reducing the pipelines' take-or-pay obligations, was to require the producers to give take-or-pay credits. A producer who had a supply contract with an interstate pipeline was required to give one take-or-pay credit under its gas sale contract for every unit of gas it shipped on that pipeline. Assume that PUMP and PIPE had a gas supply contract with a take-or-pay clause and PUMP wanted to transport gas as a shipper on the pipeline's system after Order No. 436. PUMP was then required, as a prerequisite to providing this transportation, to credit PIPE's take-or-pay obligation for a like amount under the gas supply contract. If PUMP were to transport 420,000 units of gas on PIPE's system, PIPE would be given a take-or-pay credit of 420,000 units under its gas supply contract.

Gas inventory charges (GICs). Even after the pipeline companies were relieved of their take-or-pay burdens, they still had the obligation as gas merchants to provide the same bundled sales service to sales customers still wanting that service. This required new gas supply contracts so the pipelines would have sufficient gas to meet their sales customers' gas needs. A pipeline's customers shared in these new gas acquisition costs through payment of surcharges called GICs, a small charge on each unit of gas.

Passthrough of take-or-pay settlement costs. Because the take-or-pay commitments had been so large, the costs of settling the contracts were large as well. A pipeline was allowed to share with its customers a portion of the settlement costs and these were recouped through a surcharge added to a customer's total cost of service.

The transportation rules first established in Order No. 436 remain the foundation for open access transportation on interstate pipelines today. The surcharges added to cover the pipeline companies' stranded costs lingered well into (and in some cases, nearly to the end of) the 1990's.

Order No. 636

Because this order effectively changed the way an entire industry worked, implementation time may go well into the first decade of the year 2000. Pipeline companies are usually behemoth operations, consequently they may move with behemoth-like deliberation and care. The differences in configuration, ownership, operations, and system from company to company means that there will be different levels of ease in making the following major changes ordered in Order No. 636.

Unbundling sales services. In Order No. 436, the pipeline companies were required to unbundle their sales services from their transportation services. Order No. 636 required them to further unbundle their sales services to identify and enumerate all the components in their bundled sales to LDC customers. In essence, this unbundling meant that the interstate pipeline companies would no longer be gas merchants at the city gate but would only be transportation service providers. The terms *transportation service provider* and *TSP*, by the way, are the two most commonly used references for these organizations nowadays.

Comparability of access to shippers. From the beginning of open access transportation, the transportation-only customers of interstate pipelines have closely guarded their rights to receive equal treatment with other categories of customers receiving transportation services—in particularly the LDCs. In Order No. 436, the pipeline companies were required to provide equal access to transportation services, but once access had been achieved, many third-party shippers (primarily marketers) complained that they still weren't receiving a service equal to the embedded transportation in the LDC's bundled sales service. To remedy this, the Commission required comparability of access to transportation services, meaning that all shippers of the same class would receive essentially the same service.

Market centers. For anyone unfamiliar with market centers (also called hubs), an analogy to the airline industry might be useful. O'Hare International Airport in Chicago is a United Airlines hub, Bush International Airport in Houston is a Continental hub, Lambert Field in St.

Louis is a TWA hub, and so on. Each of these airports serving as hubs for one airline receive fees for the many services they provide to other airlines utilizing airport services.

A natural gas hub is comparable. As noted earlier, the most important U.S. natural gas hub is the Henry Hub in Louisiana, where Gulf Coast gas, Texas gas, and Louisiana gas enter and exit the facilities operated by the Sabine Pipeline Company in a very competitive environment. Like airlines, natural gas hubs operated by interstate pipeline companies provide a variety of services available to shippers moving gas through the hub. The services provided and fees assessed for hub services are all subject to FERC approval and relevant tariff proceedings.

In Order No. 636, the Commission prohibited any pipeline tariff provision that would inhibit development of market centers, thus announcing in a formal order that the FERC wants a vibrant and active hub system in the U.S.

Open access storage. When interstate pipelines were providing bundled city gate services, many of the pipeline companies acquired storage facilities themselves or contracted with third-party storage facilities to assure that sufficient gas would be available to the LDCs on the coldest day of the year. The very heart of Order No. 636 was unbundling the services being provided by the interstate pipelines. Unbundling essentially took the interstate pipelines out of the gas merchant business. Without a merchant function, the pipelines no longer had any need for most storage capacity. Much of the resulting dilemma involved the pipelines' potential stranded costs related to storage. The solution was to add a new level of service for the pipeline companies. Now natural gas storage is offered as a separate service by all interstate pipelines that previously held storage capacity for their own use. The types and levels of storage vary from pipeline to pipeline, dependent on the storage facilities available and customer usage.

Upstream capacity. While some of the major pipelines traverse the U.S. from production area to market area, others are located primarily in the field zone or in the market zone[7]. The downstream interstate pipelines—having limited access to gas for their customers—traditionally contracted with upstream pipeline companies for firm transportation of the gas necessary to serve the downstream pipeline's customers. After Order No. 636, this upstream capacity was largely unneeded by the downstream pipelines, so the FERC required those pipelines whose LDC customers were the primary beneficiaries of the upstream capacity to make any unused upstream capacity available directly to the LDCs. The pipeline companies benefited because they were able to pass on what would otherwise have been stranded costs of unused firm transportation to the LDCs and the LDC benefit was having access to what could otherwise have been unavailable firm transportation capacity on field zone pipelines.

Capacity reallocation. This part of the order is generally called *capacity release*. For many years, shippers and pipeline companies had been caught in a struggle over the issue of unutilized firm capacity. When a shipper holding firm transportation agreements with a pipeline company didn't use all or a part of its firm capacity, the shipper was still obliged to pay the demand charge. That unutilized firm capacity would then become available for other lower-priority uses such as interruptible transportation. Holders of firm capacity argued that some mechanism should be put into place to allow transfer or assignment of unutilized or unwanted firm capacity. So this section of Order No. 636 accomplished that goal, and the so-called *secondary transportation* market has become dynamic and important to the gas transportation industry. Here's how capacity release works.

Assume that WIDGET wants to release its firm capacity on PIPE during its low production times. WIDGET may wish to release all or a portion of its total maximum daily quantity (MDQ). WIDGET has several options that it may pursue. First, WIDGET may have knowledge that FLEX wants to acquire firm capacity utilizing the same or similar receipt and delivery points as those subscribed to WIDGET. It may take little more than a phone call to effectively negotiate a "prearranged release." Once an agreement has taken place, WIDGET, known as the *releasing shipper,* successfully releases the agreed capacity to FLEX, the *acquiring shipper.* If FLEX offers to take the firm capacity at the same receipt and delivery points and at the maximum tariff rate, WIDGET's only obligation is to inform PIPE. PIPE will then conduct a credit check if necessary, sign an agreement with FLEX,

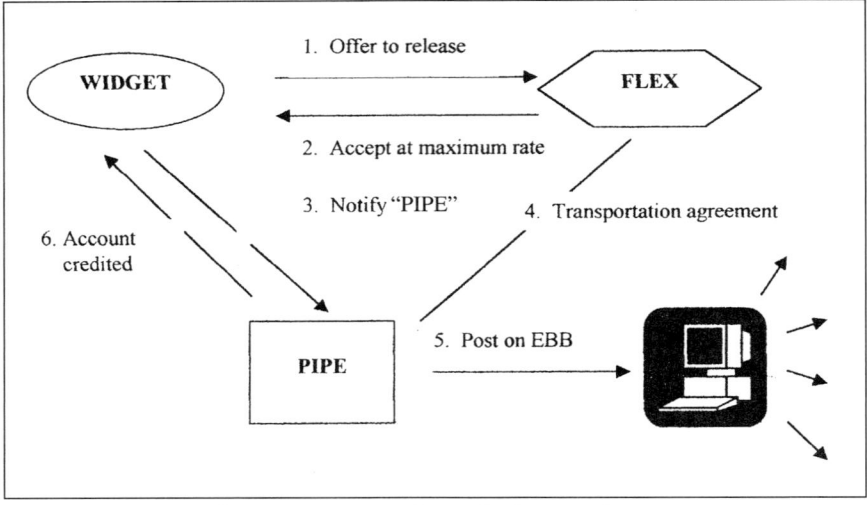

Fig. 3-1 Capacity Release Transaction—Privately Negotiated

and electronically post the transaction for informational purposes only. It is a relatively simple procedure (Fig. 3-1).

If FLEX is not willing to pay the maximum tariff rate, PIPE will electronically post the offering and other shippers may bid for the capacity.

The second option for WIDGET is to ask PIPE to post the information for all bidders. Any potential shipper viewing the posting may make its own bid for the service, and all tariffs specify that the pipeline company (usually in consultation with the releasing shipper) will accept the "best" bid (Fig. 3-2). The best bid determination varies from transporter to transporter, and will be described in the tariff provisions for capacity release. In either case, once PIPE has accepted the best bid, the practical result to WDGET is that whenever FLEX makes its demand charge payment to PIPE, WIDGET's account will be credited for the amount paid. Because WIDGET retains its contract with PIPE, if FLEX fails to make demand payments, PIPE may still collect its demand charges from WIDGET.

An active market now exists for capacity release, sometimes referred to as the *secondary* or *gray market* of transportation. Many marketers are now large capacity holders of firm transportation procured through capacity release.

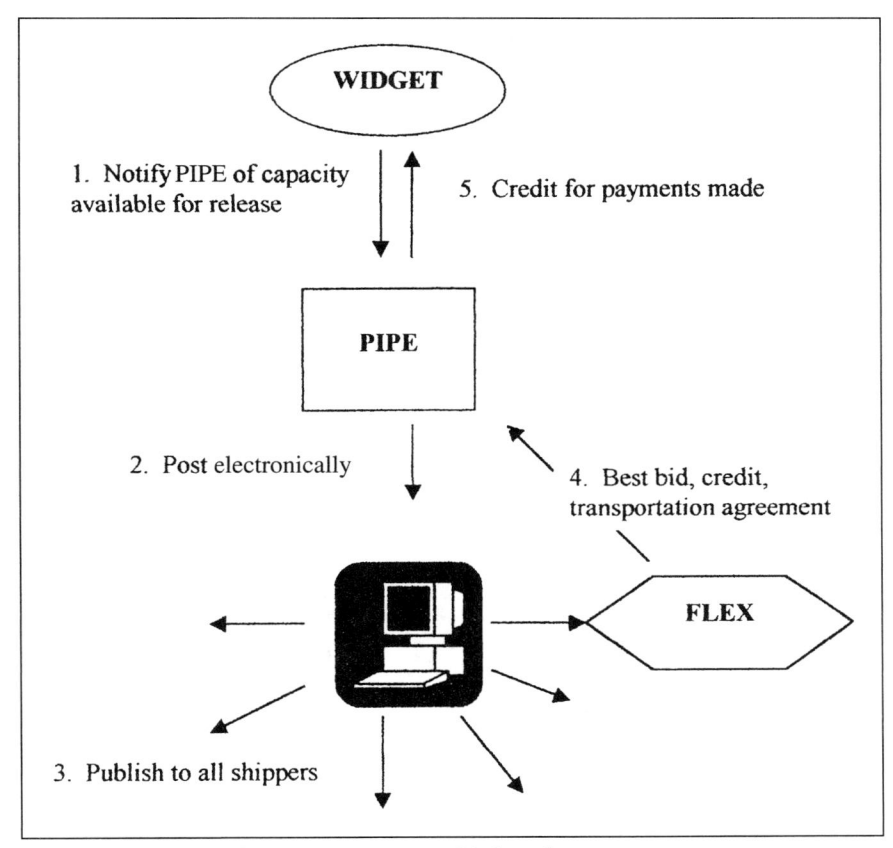

Chapter 3 Natural Gas Transportation Laws and Regulations

Fig. 3-2 Capacity Release Transaction—Publicly Bid

Straight fixed-variable (SFV) rate methodology. Interstate pipeline service rates have always been established based upon costs incurred by the pipeline plus a fair return on investment. The costs to own and operate a pipeline system are generally separated into two components, fixed costs and variable costs. As one would imagine, fixed costs are those that will remain unchanged regardless of the amount of gas actually flowing through the pipeline. Variable costs, on the other hand, fluctuate with pipeline usage and other variables. In earlier descriptions of pipeline systems, it was noted that every pipeline is different from every other pipeline. This is true in physical structure, load profile, capital requirements, ownership, and many other characteristics. The FERC, however, has attempted to homogenize these characteristics so all pipeline rate cases will be bound to follow the same rules.

In Order No. 636, the FERC selected the SFV rate setting methodology under which a pipeline would, in determining rates for firm services and interruptible services, allocate all fixed costs to the *demand rate* and all variable costs to the *commodity rate*. This cookie-cutter method has been challenged by a number of interstate pipelines since the order was issued, and time may bring some changes in this area as the FERC has struggled with individual cases and allowed some pipeline companies to vary from the rigid requirements of SFV rate-setting.

Blanket sales certificates. This order removed the merchant function from the pipelines' menu of services by only disallowing city gate bundled services. No longer can pipeline companies sell and transport gas through their own proprietary system and deliver to LDC customers at the city gate. The interstate pipelines were all given a blanket sales certificate to make sales on their own pipeline—just at the furthest practicable upstream point through its marketing entity. That upstream point needn't necessarily be at the pipeline's point of origination, but it must be far upstream in the field zone for trunkline pipelines. This part of the order resulted in new terminology to describe the interstate gas pipelines as providers of transportation services, hence the TSP description discussed earlier.

Receipt and delivery point flexibility. Under a firm transportation service agreement, shippers elect a primary receipt point and a primary delivery point.[8] The primary receipt point will, curtailments aside, always be available for that shipper to nominate gas up to the MDQ. Similarly, capacity at the primary delivery point is always available for nomination by the shipper. Some TSPs are more flexible in allowing shippers to nominate at places other than the designated point than others. The Commission has publicly stated that the TSPs should be more flexible in this regard, but with Order No. 636, commissioners restated this desire in the form of an order, charging all the interstate pipeline companies to work with their shippers and be more adaptable to shippers' needs.

Chapter 3 Natural Gas Transportation Laws and Regulations

With all the foregoing in mind, shippers that rely on the natural gas transportation industry must always be poised to change. Regulators change, executive administrations change, times change, and the industry must be capable of adapting to inevitable change. Many of those changes may effect gas sales, so the far-reaching power of this regulatory authority should not be overlooked by gas contracts, particularly when negotiating a long-term gas transaction.

TRANSPORTATION LAWS AND REGULATIONS IN CANADA

In chapter 1 you were given an overview of the provincial regulatory structure in Canada. The next step for anyone unfamiliar with gas transportation regulation in Canada is to examine some of the major policies instituted by the NEB to glean an appreciation for the nature of regulation in Canada. The NEB has different regulatory provisions and procedures for pipelines that are regulated by size. Group 1 pipelines are the 10 largest oil and gas pipelines in the country. The remaining companies under the board's jurisdiction have been designated as Group 2[9] companies.

To reduce the regulatory burden on smaller companies, the NEB regulates some of the Group 1 pipelines and all of the Group 2 companies on a complaint basis. Shippers are thus encouraged to work out any problems with the pipeline company. If this is unsuccessful, a complaint may be filed with the NEB.

Tariffs

A pipeline company's tariff contains the conditions under which transportation service is provided. The tariff includes conditions on accepting new shippers, on allocating capacity to shippers and on determining which position a prospective shipper will occupy on the waiting list for service. The NEB requires that pipeline companies operate according to the principle of open access. This means that all parties must have access to transportation on a non-discriminatory basis. In addition, tolls for services provided under similar circumstances and conditions with respect to all traffic of the same description, carried over the same route, must be the same for all customers. The NEB conducts compliance audits as part of its monitoring responsibility.

Tolls

Payment for transportation services rendered is in the form of tolls. Toll design is the process of deriving tolls from determining the cost of service or revenue requirement and throughput. Tolls should generate sufficient revenue to recover approved costs, and to fairly allocate charges to users in relation to the costs and benefits of different services. When establishing tolls for the Group 1 companies, the NEB traditionally examines their capital and operating costs to ensure companies shipping oil or natural gas are protected from unjustified high transportation costs. Tolls set by the NEB cover the cost of service plus a fair and reasonable return to investors.

Major toll applications normally warrant a public hearing. However, the requirement for lengthy and costly oral public hearings has been declining, in large part due to the advent of negotiated multi-year settlements. In 1995, the NEB republished its *Guidelines for Negotiated Settlements of Traffic, Tolls and Tariffs*. The guidelines are intended to facilitate a negotiated settlement process that will allow pipeline companies, producers, shippers, consumers, governments, and other interested parties to resolve toll and tariff matters through consensus and negotiation, without resorting to a lengthy hearing process. The NEB must still approve any negotiated settlements. Some of the largest pipeline companies have negotiated multi-year incentive toll settlements.

Federally regulated Canadian and U.S. gas transporters set their rates, or tolls as the case may be, by using essentially the same methodology—SFV, under which, as noted earlier, all of a pipeline's fixed costs will be designated as part of the demand component of the service charges and all commodity costs will be allocated to the variable component.

Natural gas exports

The NEB, under either long-term licenses or short-term orders, authorizes the export and import of natural gas. Following a public hearing, long-term licenses may be issued for up to 25 years subject to governor in council approval. Short-term orders for a maximum period of two years can be issued without a public hearing and do not require governor in council approval.

Natural gas exports occur at several major interconnection points along the Canadian/U.S. border. The volume exported depends upon market supply and demand as well as available pipeline capacity. Canada's imports of natural gas are relatively small, compared to its exports, and are used primarily to serve markets in southern Ontario and British Columbia. NEB approval is required for export of propane, butane and ethane by-products.

The NEB monitors the supply and demand of natural gas, including the performance under existing export authorizations. This ensures that the quantity of gas exported does not exceed the surplus remaining after Canadian requirements have been met. In fact, one of the guiding principles behind every export of natural gas is that any such authorized export must not impair the ability of Canadian citizens to receive adequate natural gas supplies.

Transportation Regulation in Mexico

After months of study, Mexico's state-run petroleum company began separating out transportation as a distinct service with its own price. On June 7, 1999 the CRE issued a permit under which Pemex was formally granted a 30-year renewable permit to transport gas through its vast network of pipelines. Under the new system, Pemex is obligated to transport gas for third-parties and may not discriminate in allowing access to its system.

The CRE established a ceiling on revenues that Pemex may not exceed under open access. Until the year 2005, that ceiling is $0.94 per gigacalorie. Shippers may negotiate lower rates with Pemex based upon the volume of gas shipped. The open access system will operate on a zone-rate basis through the 19 zones on the 8,700 kilometer system. Because open access is so new to Mexico, the CRE will undoubtedly further refine the parameters for open access transportation.

1. 15 U.S.C. §717-717w.

2. Interstate pipelines may deliver gas into intrastate pipelines if the gas is consumed within the state of delivery and the intrastate pipeline is regulated by the state. Interstate pipelines gain this type of delivery permission by designation as a Hinshaw pipeline.

3. National Energy Board Act. R.S., c. N-6, s. 1.

4. 15 U.S.C. § 3301-3432.

5. Natural Gas Wellhead Decontrol Act of 1989, Pub. L. 101-60, July 26, 1989, 103 Stat. 157.

6. Prior to 1984, each certificated service provided by a pipeline company resulted in a tariff just for that transaction. After 1984, pipeline companies had both sales and transportation tariffs.

7. The terms "field zone" and "market zone" are being used generically. Gas is introduced into the pipeline primarily in the field zone and removed primarily in the market zone.

8. The receipt and delivery point designation isn't always for one point only. Sometimes a shipper may be allowed to designate more than one primary point. Other times, any receipt point in a particular region may be designated as a primary point. Every transporter is different and the tariff for each provider will either allow multiple points or it will not.

9. See generally the NEB Internet homepage http://www.neb.gc.ca for information on Group 1 and Group 2 pipelines.

Chapter 4 U.S. Gas Sale Laws

Contracts are written to protect the interests of each party to a deal. Good contracts do, sloppy ones don't. The agreements made by buyers and sellers are molded into a form largely dictated by rigid legal rules of drafting, interpretation, and construction. Of course, anyone with knowledge of a deal can draft contract language capturing that bargain—in fact, I believe the person who makes the deal should assist in writing contract language. But without an understanding of the legal implications of the words used, the contract language may be self-contradictory, incomplete and outside legal boundaries, ultimately resulting in an unenforceable document.

Contract law is filled with paradox, and it's no wonder that contract drafters and negotiators who haven't had intensive legal contract training are sometimes at a disadvantage during contract review and negotiations. Non-attorneys needn't be at a great disadvantage though, and the two best ways to achieve contract drafting and negotiating competency are:

- to learn the contract law basics
- to speak the same contract *language* as your negotiation counterparty

Throughout this guide, I'll discuss both contract law basics and the importance of choosing your words carefully.

What is a Contract?

Contracting is as much a part of our daily life as having lunch. Nearly every time you spend money, you are doing so because of a contractual relationship with some other entity. When you buy goods, services, or property the general rules of contract law govern those relationships. Whenever one party gives *performance* or *something of value* in exchange for *performance* or *something of value* from another party, a contract has been made. A legally binding contract can be formed only if the elements of *offer*, *acceptance*, and *consideration* are present. When two parties attempt to enter into a contract, the significant determining factor of whether a binding contract has been formed is their intention behind the words they use. In other words, they must have *mutual assent* to the agreement. If one party coerces the other into a bargain, or if the parties are mistaken about a material element of the agreement, no contract will be formed.

To determine whether a contract has been formed, we look at the communications between the parties. Whether a bargain has been formed by mutual assent of the parties is determined by examining what a reasonable person in the position of each party would be led to believe by the words or conduct of the other party.

Offer

When one makes an *offer*, he or she must indicate a present willingness to enter into a bargain, done in such a way that a reasonable person in the shoes of the person to whom it is addressed would believe that the bargain could be concluded by giving assent or accepting the offer. A clear intent to agree, coupled with clarity of both subject and terms of the bargain, is critical. Either the seller or the buyer may make the offer.

Acceptance

Acceptance of an offer may take the form of words or actions—by a return promise or by conducting full performance. As will be discussed later in this chapter, acceptance by performance occurs frequently in gas sale transactions, and a routine practice of letting the paperwork catch up with reality is a source of many gas contract dilemmas. Acceptance is normally effective when the one making the offer receives it.

Consideration

A contract or agreement is not enforceable unless it is supported by *consideration*, usually thought of as a bargain, or an exchange of promises, or a promise for performance where each party gives something of value for what is received. Just about any form of consideration will do, but it cannot be nominal, based upon a preexisting legal obligation, or a promise to refrain from carrying out a legal claim (such as giving up a right to sue under another contract). The bargain must be exchanged for something of real value to each party. This is true when one neighbor sells a car to the other and it is true of agreements to sell and purchase gas.

While the general rules indicated above are the baseline for determining questions of contract law, some specialized types of contracting, such as contracts to sell goods, are subject to statutory law that alter the basic contract elements of offer, acceptance and consideration under the common law[1] and other statutory laws. Commercial laws coexist with, and in some cases, supplement other laws. Keep in mind in the following discussion that commercial laws are the primary, but not necessarily the only laws that may be applicable to a given sales transaction.

INTRODUCTION TO COMMERCIAL LAWS

Assume your organization is buying gas from another company and a dispute arises during the performance of the contract. You cannot find anything in the contract itself that addresses the issue in dispute. Where do you go for an answer? In the U.S., if one of the disputing parties is the federal government, federal law applies. In all other cases between U.S. parties, if your contract is subject to the laws of any state but Louisiana,[2] you will go to the Uniform Commercial Code (UCC or Code as adopted in the state

A Practical Guide to Gas Contracting

having jurisdiction over the contract.) The UCC came into being to address differences in commercial laws between the states, thus to encourage commercial dealings between parties in different states. The nine UCC articles currently in effect are the following listed in Table 4-1:

Article 1	General Provisions
Article 2	Sales
Article 2A	Leases
Article 3	Negotiable Instruments
Article 4	Bank Deposits and Collections
Article 4A	Funds Transfers
Article 5	Letters of Credit
Article 6	Bulk Sales
Article 7	Warehouse Receipts, Bills of Lading and other Documents of Title
Article 8	Investment Securities
Article 9	Secured Transactions: Sales of Accounts and Chattel Paper

Table 4-1 Current UCC Articles

In the sale contract dispute, then, you would look to Article 2 "Sales" as adopted in the state you have elected as the governing law in your contract.[3] If provincial law in Canada controls, you will turn to the "Sale of Goods Act"[4] in the relevant province for your answer. Québec, like Louisiana, adheres to a form of Napoleonic law[5] and Québec's sales laws vary from the standard acts of the other provinces and territories. If the two disputing parties are from different countries, the United Nations Convention on Contracts for the International Sale of Goods (CISG) may apply.[6] Canadian sales law and the CISG will be discussed in chapter 5.

You can turn to numerous legal sources for resolution of disputed issues. Even when you find an answer, the law might not be clear or precise thus leaving the question open to court interpretation. More commonly you will find that the law provides guidance but few absolute answers to your questions.

The law isn't always clear

Why are the answers to your legal questions so often neither black nor white? Any buyer or seller searching for a legal answer to a contract question has the same three hurdles to overcome:

- broadly worded or vague language in statutes
- lack of uniformity in the UCC from state to state, and
- inconsistent court opinions

First, statutes are usually drafted to address a multitude of different circumstances, so it would be unusual to find a legal answer specifically written to address your unique situation. Furthermore, the current UCC isn't always a model of clarity, as is true of many laws. Courts and legal scholars alike have wrestled for years with unresolved issues, sometimes haggling over the meaning of a word in the Code or a comma placement.

Second, the UCC isn't always uniform from state to state, even though the same version of the UCC was given to each state legislature for consideration. Once a bill, even a uniform code, is introduced into a legislative forum, the bill may be amended according to the desires of a majority of the legislators. So while the UCC is uniform in the whole, after legislators have tinkered with the parts, the lack of uniformity from state to state may leave a contract drafter uncertain of the final result of given contract language.

The process of drafting a uniform code like the UCC is deliberative and usually takes a number of years. A select committee of U.S. scholars and legal practitioners specializing in the law join together under the auspices of the National Conference of Commissioners on Uniform State Laws (NCCUSL or Uniform Law Commissioners). The organization is comprised of more than 300 lawyers, judges, and law professors, appointed by the states as well as the District of Columbia, Puerto Rico, and the U.S. Virgin Islands, to draft proposals for uniform and model laws and work toward their enactment in legislatures. Since its inception in 1892, the group has promulgated more than 200 acts, among them the UCC, the Uniform Probate Code, and the Uniform Partnership Act.

After the NCCUSL select committee has intensely studied the topic, drafted the code and fully discussed and debated its contents, the final version is given to the American Bar Association and to the American Practicing Law Institute for approval. Afterwards, the code is presented to the state legislatures (usually through the state bar association) for consideration and potential changes.

Third, in any lawsuit, the role of courts is to consider evidence and apply the law to that evidence. Questions of law and the facts of that particular case will be considered. The court must reach a decision based upon the facts, the evidence, and the law. Every set of facts is different and the law must oblige those facts, often resulting in conflicting decisions on the same code section by different courts, and sometimes by the same court.

The widget rules

Where the natural gas industry is concerned, we can identify yet a *fourth* reason for that uncomfortable, insecure feeling so many of us get when turning to Code sections for an answer. Natural gas isn't a widget. Sales laws were created primarily for sales of widgets—manufactured *things* that you can see, feel, hold in your hand and that you can identify as being in a specific place at a stated time. A major difficulty in applying many Code sections to a natural gas sale dispute is that we as humans can't see, feel, hold in our hand or identify natural gas in a specific place at any given time. Accurate identification of the goods (quantity or lot number) at any given location (at the factory, on the dock) at a particular time (October 1 at 1:00 p.m.) underscores the UCC as it was drafted.

Since gas molecules aren't widgets, the result obtained after applying some UCC provisions literally might be ridiculous, if not impossible. For example, UCC Section 2-308 specifies that if a specific place for delivery is not stated in the contract "the place for delivery of goods is the seller's place of business or if he has none, his residence." Try that with 100,000 MMBtus of gas a day.

Like shoes, paper, tractors, or other manufactured goods, natural gas sale and delivery is comprised of several components—production, transportation, and distribution. Unlike other industries, however, there is usually not a specific location or point in time that the goods sold can be *identified*. This fact may lead to both operational and contractual complications.

Widget rules and natural gas operations

The pipeline systems through which gas is transported are extremely sensitive to contaminants that could be in the gas stream. As stated earlier, too much water in the gas stream can foul equipment; measurement devices are sensitive to sediment, contaminants and water; and the operational results may range anywhere from inaccurate measurement to structural pipeline damage. If any measurement inaccuracies or other incidents occur, how do we know which part of the commercial gas system caused the problem? Getting to the source of these problems can be difficult with a non-widget commodity like natural gas.

Commerce rules and natural gas contract issues

As for contractual complications, it might be helpful to first take a look at a typical sale of goods by a manufacturer of shoes.

FACTS:

- WIDGET manufactures shoes and sells them to its distributor, ACME
- WIDGET has contracted to sell 1,000 pairs of men's brown loafers, #2345, to ACME and delivery is due on October 1
- After production of the shoes, CAB, a trucking company, has contracted to deliver the shoes to ACME, and will load the shoes from WIDGET's dock for delivery to ACME

SITUATION #1:

- WIDGET does not manufacture the shoes in time to meet the October 1 deadline

RESULTS:

1. As to ACME: The contract between WIDGET and ACME may provide a remedy to ACME because of the delivery failure
2. As to CAB: The failure to meet the contractual deadline may not cause WIDGET to have problems with CAB particularly if transportation was not contracted until the shoes were manufactured

SITUATION #2:

- CAB cannot deliver the shoes by October 1

RESULTS:

1. As to ACME: If CAB's failure to deliver was unexcused and ACME lost a sale or was otherwise financially harmed by CAB's failures, ACME may be able to collect damages from CAB.
2. As to WIDGET: If CAB caused the delayed delivery, it may owe penalties or late delivery fees to WIDGET. The truck can still be identified somewhere at some point in time with 1,000 pairs of size 10D loafers—delivery has just been delayed. The sale process moves in a predictable way from manufacturing through distribution—first the shoes are manufactured, then they are transported, then distributed. The natural gas industry doesn't work that way

Natural gas is unique

The major reason for uniqueness in this industry is that natural gas is generally considered to be fungible[7]. The shoes used in the illustration above are not fungible with all other shoes. All of the of size 10D brown loafers of a particular lot number may be fungible with all other size 10D brown loafers of the same lot number, but these shoes are not fungible with all the other types of shoes being manufactured, transported and distributed. In any case, those shoes of like kind may easily be identified upon visual inspection.

Unlike the shoe example, all the methane molecules and other elements making up the natural gas stream in a given pipeline at any given time are considered to be easily interchangeable unit for unit. Natural gas doesn't usually come out of the ground in a fungible state, but after it has been conditioned,[8] and processed, the natural gas stream is, for gas sale purposes, considered to be a homogenous unit.

Before the operator of a downstream pipeline interconnection[9] will allow gas into its system, the gas must meet the quality specifications of that pipeline. This usually means that non-combustible gases, water, sediment and other contaminants have been removed to specified levels and that the

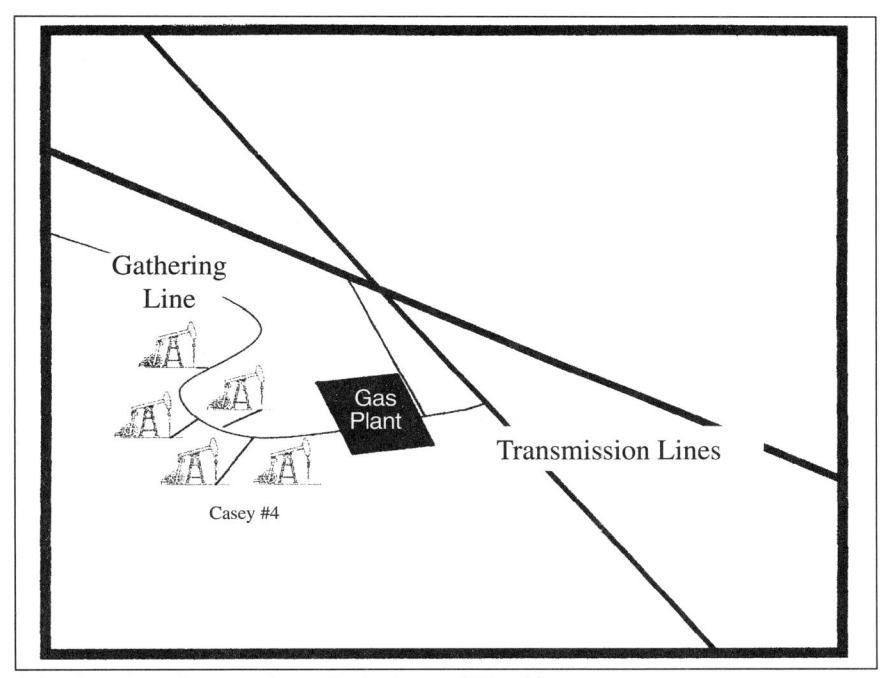

Fig. 4-1 Basic Layout of Gas Gathering and Trunkline System

heat content of the gas is within permissible limits. In some circumstances the pipeline will waive quality standards for a particular batch of gas if the commingled gas stream meets quality specifications, but as a general rule all gas must meet the quality standards before being admitted into the downstream system. Gas being introduced into the system at further downstream receipt points must meet the same specifications.

The critical difference between manufacturing and transporting shoes or other types of widgets and producing and transporting natural gas is based upon the characteristics of fungibility. Natural gas moves through a transmission pipeline at about 20 miles per hour by a combination of *operating pressure* (compression) and *displacement* of downstream molecules, so gas delivered into a pipeline will eventually reach a downstream delivery point. The unique features of natural gas in the pipeline (fungibility, compressibility, and displacement) permit simultaneous receipt by a pipeline system in the production or field zone[10] and delivery out of the system to a customer hundreds or thousands of miles downstream in the market zone (Fig. 4-1).

How gas producer sales may be affected

Let's look at some of the same issues raised in the shoe example above when the product being sold is natural gas. A producer might contract to sell its gas months or years in advance of production, or it might contract to sell gas on the spot market. Unlike the shoe factory that may have from one to three production shifts, gas is produced around the clock. There is usually no point in time at which a particular molecule of methane can be identified as having entered the commerce stream. Unlike the shoe manufacturer who orders truck trailers for loading at its docks after the shoes have been manufactured, the producer gives up gas production for sale on a continuing basis. The wells are producing, the gas is flowing and sales are being made.

FACTS:
- PUMP owns 5 wells in a production field of 60 wells (Fig. 4-1)
- PUMP has contracted with FLEX to sell all of its wellhead production for a period of two years
- PUMP has contracted with gathering company SHORT to collect the gas from its wells and deliver it into several long distance gas transmission systems
- PUMP has contracted with CLEAN to condition the gas

Contracts

Each of the services necessary to produce, gather, process, and sell the gas involves a contract. In addition to the *gas sale contract*, in this example there will be:

- a production contract, the most common type of which is called a *joint operating agreement* or JOA, necessary when multiple owners are involved

- a *gathering agreement* between PUMP and SHORT
- a *processing agreement* between PUMP and CLEAN

Before gas is delivered into a transmission pipeline system, four (perhaps more) contracts are already in place.

SITUATION #1:

A fitting on one of PUMP's wells (Casey #4, producing 500 MMBtu/day) fails and the well is shut-in. PUMP must make nominations into and out of the gathering system and then into the transmission pipeline. PUMP must also advise FLEX of the changes so FLEX can correctly nominate to its transporter(s). When PUMP makes its 2,500/day nominations into the gathering system and into the transmission pipeline system without knowing that Casey #4 has been shut-in and the nomination for all wells should be 2,000/day, PUMP may have (or cause) contractual problems under the gas sale contract, the gathering agreement, and the downstream transportation agreement. Failure to deliver the contract quantity will be the gas contract issue, and as to the transportation agreements, balancing[11] is the major issue. Both the gatherer and the downstream pipeline company may have contractual rights to impose fees and penalties upon their customers for failure to maintain balance requirements.

RESULTS:

1. As to FLEX:
   - If PUMP tells FLEX about the lower production before nomination deadlines, FLEX may still have a claim against PUMP for failure to deliver under the gas sale contract
   - If PUMP tells FLEX after the nomination deadline, then in addition to a potential failure to deliver, PUMP may cause FLEX to incur imbalances (and resulting charges) under FLEX's transportation service agreement

2. As to SHORT and CLEAN:
   - If PUMP over-nominates on the gathering agreement, the resulting

imbalance may mean penalties to PUMP, both for failure to nominate correctly and for failure to flow the nominated quantity
- The processing agreement may not be affected by this failure unless the agreement requires PUMP to deliver a minimum quantity of gas and imposes penalties for that failure

How midstream marketing may be affected

For marketers who aggregate supplies and markets, the same issues indicated in the producer example have a different type of effect. If a production problem occurs on the purchase side and the marketer has no prior knowledge of a gas flow change, the marketer will give inaccurate nomination information to its customer and to its own transporter. The aggregation marketer is truly in the middle, neither a producer nor a consumer. Most marketers don't contractually match specific packages of gas supplies to specific customers. Instead, the marketer will purchase a variety of gas supplies within its designated supply area and sell to a variety of customers within its market area. The supplies and markets are matched, sometimes on a long-term basis, but also on a daily basis to provide the greatest flexibility and opportunity for the marketer as well as its suppliers and customers.

So the marketer has exposure from both directions, and it must rely upon downstream and upstream nomination information. When that information is inaccurate, heart rates in the marketing operation jump precipitously. If multiple transactions occur from wellhead to burnertip, every contract along the purchase and sale chain could be adversely effected. These problems occur frequently. Most often the two affected parties (particularly if the parties are both marketing organizations) will reach a business solution and the situation won't disintegrate into a contract dispute. When the stakes are high enough, both parties go to the contract to see exactly what liabilities and remedies may apply.

How consumers and LDCs may be affected

As transactions get closer to the burner tip, it is more likely that the parties will not be able to negotiate a satisfactory business settlement when a problem occurs. The consumer or LDC is at the end of the line, so the language in these parties' contracts is understandably self-protective. There is typically nowhere left to pass supply problems.

During peak need times, a consumer receiving inaccurate upstream nominations may be in a perilous situation. If its gas supplier nominates 100,000 MMBtu and only delivers 80,000 MMBtu, the industrial consumer still needs the extra 20,000 MMBtu to fuel its manufacturing process. When that consumer keeps its valves open until it has received the full 100,000 MMBtu, it may be required to pay high penalties to the transporter or LDC for taking that extra 20,000 MMBtu of gas that was not nominated. The gas it took belonged to someone else, the transporter, the LDC, or another customer.

When an LDC selling gas to its customers relies on nominations that are greater than actual delivered quantities during peak periods, the LDC may also incur penalties from the transporting pipeline that delivered gas into its system. In return for the exclusive right to provide gas in the LDC service area, these utilities have an obligation to serve their customers, even on the coldest day of the year. Like the consumer mentioned above, an LDC will take the gas it needs or switch to another fuel. LDCs have additional concerns because their regulators review their performance, including additional costs the LDC may incur in penalties paid to their transporters.

This is not to say that the downstream operators cannot cause problems upstream. If a consumer or LDC takes more or less gas than it has nominated to its transporter(s), problems can just as easily go upstream. As an example, let's say that a consumer using natural gas to fuel its manufacturing operation has a boiler explosion. The consumer must stop taking gas sometimes without providing timely nomination change information to its supplier or transporter. When that happens the upstream supplier may incur an imbalance on its TSA. When the temperature falls 30° overnight, the LDC is going to provide necessary fuel to its customers whether or not timely notification can be made to transporters and gas suppliers. Problems can flow in every direction.

With just these few illustrations, you can see how unique the gas industry is and how interdependent gas organizations are on information supplied by others. From the point gas comes out of the ground to the point at which it is consumed, natural gas as a commodity demands some creative thinking for us in the gas industry whenever we need guidance to solve a contractual dispute. The laws are broad and not always clearly written, the UCC isn't uniform from state to state, courts sometime reach conflicting decisions, and natural gas isn't a widget. Given all that, when we are faced with contract issues, the Code does supply more answers than questions, so let's begin exploring the finer legal points of the UCC.

THE UNIFORM COMMERCIAL CODE

The UCC and similar commercial laws in other countries provide most, but not all answers to legal issues that could transpire incidental to a sale of goods. The common law[12] and other statutory law supplements matters not specifically addressed in the Code, such as capacity to contract, principle and agent, estoppel, fraud, misrepresentation, duress, coercion, mistake, bankruptcy, and others.[13] Keeping in mind that the UCC is not necessarily the only word in sales law, a totally prepared gas contractor will be familiar with UCC Article 1, Article 2, Article 5, and Article 9.

If you don't have a copy of the UCC or ready access to the version adopted in your state, there are several ways to access the Code, and some of those are free. First, for a small cost, the American Bar Association Section of Business Law has a series of paperback books for each section of the UCC, as well as "The Portable UCC" that contains all current UCC articles. Those can be ordered through the ABA Website at http://www.abanet.org, and are a particularly convenient source of information if an organization employs numerous contractors. Many states have posted the UCC as adopted locally on the Internet for viewing and downloading, and one of the best sites for these connections is http://www.cornell.edu.

Article 1 "General Provisions"

This article contains provisions applicable to all other Code articles, so the terms of Article 1 apply to sales under Article 2 as well as the other commercial transactions governed by Articles 3 through 9. The primary purpose for the UCC, as stated in Article 1, is threefold:

- to simplify, clarify and modernize the law governing commercial transactions
- to permit the continued expansion of commercial practices through custom, usage and agreement of the parties, and
- to make uniform the law among the various jurisdictions[14]

This lays the foundation for all commercial transactions, and courts deciding cases governed by the UCC will factor these concepts into their decisions.

Article 1 contains other essential baseline rules as well. Every commercial transaction covered by the UCC must be performed and enforced in good faith by both parties.[15] This means that all performance and enforcement activities must be undertaken with honesty in fact.[16] Definitions common to all articles are located here,[17] but remember that additional article-specific definitions are found in other code articles. UCC law generally follows the rule that the entire agreement of the parties signing the contract has been merged[18] into the written document. In later chapters, more attention will be devoted to the merger issue itself, but in Article 1 you will find the statutory terms that allow courts to look beyond the written evidence of a contract when the parties' intentions are not clear or complete in the record itself.[19]

Article 5 "Letters of Credit"

Because the stakes are typically quite high in a commercial natural gas transaction, all sellers should evaluate their buyers' credit to determine a trading limit. If a buyer's credit does not meet the anticipated level of business, or if the buyer's credit becomes inadequate during performance, it is common practice for the seller to require the buyer to provide credit enhancements satisfactory to the seller. One of the most frequently used forms of credit enhancement is an irrevocable standby letter of credit issued by a bank.

Article 5 sets forth the parameters for creation of the credit, presentation for payment by the beneficiary, and payment by the bank. A bank may delay payment for up to three days without dishonoring the credit.[20] Partial payments may be made[21] and the right to draw on a credit may not be assigned or transferred unless specifically allowed on the credit itself.[22] These issues are typically handled through an organization's credit department, but in some organizations all aspects of contract negotiations including credit enhancement issues are the responsibility of contractors. A general familiarity with Article 5 is useful in any event.

Article 9 "Secured Transactions"

Let's say that your organization has sold gas to a company not meeting your credit requirements for that particular sale. To make the sale happen, the buyer has agreed that it will provide cash collateral for the last 10 days of the delivery month to satisfy your organization's credit requirements. An appropriate security agreement has been signed, as well. If the buyer fails to pay on time, Article 9 of the UCC addresses the aspects of the sale transaction involving your security interest created in the cash held as collateral. Both Article 2 and Article 9 will have relevance to this situation, but to determine your rights as a secured seller to convert collateral to your own use, you may turn to Article 9.

One of the most pertinent parts of Article 9 *vis á vis* sales is Section 9-113, which states that a security interest arising under Article 2 is subject to Article 9 provisions except and to the extent that the debtor does not have possession of the goods. In that case, no security agreement is necessary to make the security interest enforceable, and no filing is required to perfect the security interest. Article 2 will control in that case. What the exception means to a party attempting to enforce its security interest is simply that the process will differ depending on whether the debtor has possession of the goods.

Article 2 "Sales"

This article covers only *sales of goods*, and not other types of sales or provision of services. *Goods* include anything moveable at the time of the sale except money and things in action.[23] Goods also include the unborn young of animals, growing crops[24] and minerals after severance from the ground.[25] Because of the movability aspect, courts have struggled with the issue of whether electricity is goods. The majority of courts still hold that electricity is not *goods*, thus sales of electricity are not typically subject to Article 2 provisions.

Every sale of goods has two components from the seller's perspective. First, the sale itself is defined as the *transfer of title* to the goods in return for the price,[26] but the additional component of the sale usually means that the seller will give up *possession* of the goods to the buyer. It's important for gas contractors to intellectually separate those two components since many fine points of dispute resolution will involve either the transfer of title or the change in possession, but not necessarily both. Article 2 deals with both of these concepts.

The article has been divided into seven practical sections that follow the natural progression of the agreement and contract from negotiation through remedies for breach.

Part 1. Short Title, General Construction and Subject Matter

The bulk of this part is concerned with definitions. The article was crafted to cover a variety of sales, everything from the new stove you purchase from an appliance store to the car you sell to your neighbor to sales of natural gas. Consumer protection laws, laws relating to farmers, and other categories of buyers may be addressed in other statutes, and Article 2 will not override those laws. For the most part though, Article 2 is the primary statutory source for sales law.

The concept of "Merchants." There is a presumption that the readers of this guide will typically be employed or retained by organizations in any one of the five major gas industry sectors—production, transportation, marketing, distribution, or consumption. For those to whom the presumption applies, it is extremely important to understand that gas buyers and sellers who are defined as "Merchants" by the UCC are subject to different standards, in some instances, in their dealings with other Merchants and with non-Merchants. An individual buying a new stove or selling a used car to his neighbor probably doesn't have any greater knowledge about stoves or cars than the other party to the transaction. Organizations that buy and sell gas as their business do have a high level of industry knowledge and the Code will not allow those organizations as Merchants to use that knowledge to exploit those with whom they do business.

"'Merchant' means a person who deals in goods of the kind or otherwise by his occupation holds himself out as having knowledge or skill peculiar to the practices or goods involved in the transaction or to whom such knowledge or skill may

be attributed by his employment or an agent or broker or other intermediary who by his occupation holds himself out as having such knowledge or skill."[27]

The term "person" refers either to an individual or to an organization,[28] so any company involved in buying and selling gas as its business is considered to be a Merchant by the UCC, at least as far as buying and selling gas is concerned. An organization may be a Merchant of one product and not of another, so a natural gas marketing company that deals in natural gas sales is a Merchant of natural gas, but when that same organization purchases 50 new office files, it is not a Merchant for that purchase. The furniture supplier probably is a Merchant of office files. Just how are Merchants treated differently under the Code? Some types of communications "between Merchants"[29] have unique connotations relating to firm offers under Part 1 and confirmations under Part 2. Confirmations will be discussed in chapter 6.

Firm offers. The Code recognizes that dealings between commercial entities having particular knowledge of the type of goods being sold should be handled differently than dealings between consumers. When a Merchant signs a written offer and delivers it to another Merchant, care must be taken in specifying the deadline for response, because if none is indicated in the offer, the offer is irrevocable and cannot be withdrawn for a reasonable period of time not to exceed three months.[30] Think about the letters of intent and proposals that are a routine part of the gas sale industry. If the author of either type of document is unaware of the Code provisions, the offer to buy or sell in the letter or proposal could be deemed to remain open, awaiting acceptance by the other party for *up to 92 days.* When considering the volatility of gas prices, even on a daily—and sometimes hourly—basis, one can easily see how dangerous it is to treat letters of intent or proposals with any less care than one would give to contract language.

Definitions. As noted earlier, Article 1 definitions are supplemented throughout the Code. A characteristic of article-specific definitions is that they may actually change the common definitions found in Article 1, so when relying on Code definitions, be sure to check both Article 1 and the other relevant article to be sure you are relying on the most correct definition. As an example, "good faith" for a Merchant under Article 2 goes beyond the "honesty in fact" definition of Article 1 to include the concept of *reasonable commercial standards of fair dealing in the trade*.[31] This enhanced language requires gas Merchants to abide by gas industry standards in their dealings with others. In later chapters, I'll discuss the ambiguity of gas "industry standards," but at the very least, the Code has given some guidance in this area. The type of behavior that constitutes reasonable commercial standards is usually decided on a case-by-case basis.

Contract or agreement. Ask yourself a question. When you use the word *contract* to what are you referring? Is it the relationship between parties, a specific transaction or deal, or could it be the paper memorializing a transaction or deal? Most commonly, the conversational reference is being made to the paper document, but in commercial law, the word contract means *"the total legal obligation that results from the parties' agreement as affected by this Act and any other applicable rules of law."*[32] The UCC definition of contract relates to a relationship rather than to a particular bargain or to the writing that evidences obligations. For anyone involved in contract negotiations, care should be taken to use UCC defined words in the correct context.

The term *agreement* on the other hand, is defined in the Code as: *"the bargain of the parties in fact as found in their language or by implication from other circumstances including course of dealing or usage of trade or course of performance as provided in this Act (Sections 1-205 and 2-208). . . . (Section 1-103)."*[33] To the gas industry, this means that two organizations through their traders reach a bargain with each other during the deal negotiation. The obligations of contract don't attach until a mutually beneficial agreement has been reached. So you can see that the two terms (*contract* and *agreement*) are not technically synonymous, although they are used interchangeably by many contract drafters.

A related issue is whether the document memorializing a transaction or set of transactions should be called a *contract* or an *agreement*. As a practical matter, anyone with significant contract experience is accustomed to a variety of titles for gas sale documents including both agreement and contract. Could adverse legal consequences result from using one term or the other? Maybe so, but in the daily world of gas contracting, those fine distinctions aren't recognized, and the typical gas sale document may be called either a contract or an agreement. However, I think it's worthwhile to note again that in any high-risk contract, extra care should be taken with every word used.

While the UCC itself distinguishes between the two words, the effect for Code readers is a shorthand method for understanding Code terms. To illustrate how the UCC drafters have utilized the two words, Section 2-204 addressing general contract formation states the following: *"(1) A contract for sale of goods may be made in any manner sufficient to show agreement, including conduct by both parties which recognizes the existence of such a contract."* The distinction is slight, but obvious in this language. Agreement is the bargain itself. Contract is the result of the bargain. (For the remainder of this book the term "contract" will mean the document memorializing gas sales.) This very broad view of contracting actually reflects the way gas is usually bought and sold, many times without formal contract negotiations.

Termination vs. cancellation. When one party is in breach of the sales agreement, the other can *cancel* the contract. When the contract ends earlier than the stated termination date, but not for breach, the contract may be *terminated.* These two words have importance because the concept of cancellation for breach allows the canceling party to retain any remedy for breach of the whole contract or any unperformed balance.[34] If the contract is terminated, executory[35] obligations are discharged but any rights based upon prior breach or performance survive.[36]

To illustrate the difference in contract language, take a look at how the two terms may be used to mean different things in the same contract.

1. Cancellation: "If either party is in breach of this contract pursuant to the provisions of Sections 9 or 13, then, in addition to any other remedies available to the non-breaching party at law or in equity, the non-breaching party may, with no less than two (2) days notice to the breaching party, immediately cancel this Contract."

Notice in this clause that the non-breaching party has the right to invoke cancellation of the *entire contract* (with UCC remedies). The contract may include a number of different transactions all subject to the same general terms and conditions. With the language provided here, the entire contract ends, including all individual transactions under the contract.

2. Termination: "*If any Tax that was not in effect as of the Effective Date of this Contract is imposed on the seller, the buyer, or any transaction hereunder at any time subsequent to the Effective Date, then the party responsible at law for reporting and paying such new Tax to the taxing authority (the "Affected Party") may demand re–negotiation of the Contract Price. If (i) the other party refuses to re-negotiate or (ii) the parties cannot agree to a new Contract Price pursuant to negotiations, then the Affected Party may, upon no less than two (2) days notice to the other party, immediately terminate this contract.*"

In this termination clause, a no-fault incident (the new tax) has occurred depriving the affected party of the benefit of its original bargain, so it has been given the opportunity to stop the contract—but no damages are payable for the unexpired term. There may be a number of reasons why parties would not wish to contractually allow termination for any reason affecting a loss of value to one party, but you can see how the two words should be carefully used, depending on what you are trying to achieve in your contract.

Referring back to the difference between the meaning of contract and agreement in UCC § 1-201, what if the tax change affects only some transactions and not others? Under the UCC, termination relates to the contract, so in drafting this type of clause, be very clear in your description of what is being terminated. Do you want to terminate only affected transactions or all transactions? Without applying a reference to "all affected transactions" or something similar in termination language, your organization could be weighing the benefits and disadvantages of terminating *all* transactions, with no option to terminate just the transactions affected by the no-fault event.

Part 2. Form, Formation, and Readjustment of Contract

The typical process of buying or selling gas is not always conducive to the maintenance of perfect records, and this is true from the wellhead to the burnertip. Let's say that a producer has 1,000 MMBtu/day of unexpected production from one of its wells and no current customer willing to buy the gas at the producer's price. A manufacturing concern can slow or stop the production line when production exceeds demand, but gas producers encounter even more difficulties in weighing whether or not to shut-in a well. Certainly, if prices are good, the producer won't want to consider shutting-in production. The producer's sales representative may learn of a market willing to pay a good price for 1,000 MMBtu/day. If the producer and potential buyer don't have a signed gas sale contract in place, will they forego the deal? Most likely not. The deal will be struck to sell the gas, and the paperwork (that is, the written contract) will be negotiated and signed (if at all) after gas has begun to flow.

From a marketer's vantage, multiple sales and purchases of gas are made daily, that being the very essence of gas marketing. Day trading activities are founded in abundant opportunities to buy or sell gas. It is quite common for deals to be done by sales representatives before or at the same time the underlying contract is being negotiated by someone other than the sales representatives. For a marketing organization, this happens on the supply side and on the market side. Routine deals don't wait for contracts. For relatively short-term, low risk deals, the paperwork usually catches up to gas flow in the gas marketing business.

A consumer or LDC needing additional gas supplies for peaking needs may find itself in a situation similar to that of the producer mentioned above. With no practical options available, the consumer or LDC may have no choice other than taking unauthorized gas from the transporter's system (some might call this stealing gas). But that is costly and upsets the folks at the pipeline company. Like every other segment of the industry, the industrial consumer or LDC must do what it needs to operate most efficiently and fulfill its obligations. If the choice is between buying gas at market price from a supplier with no signed contract in place or taking unauthorized penalty gas from the transporter's system, if time is not of the essence, pure economics will always drive the decision to buy gas from the supplier.

In all of the examples given above, as long as the deals run smoothly, there will never be any questions about when the contract was made or what the precise terms of the agreement were. Part 2 of this article is useful, though, when problems do arise and the parties must determine either (1) whether or when a contract was made; or (2) the relevant terms of a contract.

Making an agreement and negotiating contract terms. We know from the discussion of Article 1 that the Code distinguishes between the agreement and the contract. We also know that the two terms are used interchangeably in gas contracting, usually with no adverse results. Some *contract formation* issues in Part 2 can be separated into agreement and contract components, and understanding that separation will help in understanding the UCC more clearly.

Here is a typical scenario for a gas supplier, gas marketer, or gas consumer. The sales and purchase representatives from two companies converse by telephone and agree to a transaction by jointly designating the quantity, price, delivery period, and delivery point. For long-term, high-risk transactions, the representatives may be actively involved in all phases of negotiation, but in *routine* gas sales and purchases, few elements beyond quantity, price, delivery point and delivery period are negotiated.

Once the deal has been made, each representative will convey the relevant information to someone else in the company. This is usually someone away from the trading floor—the gas contractor, who may be a contract administrator, the contract analyst, or an attorney. The counterparty identification, quantity, price, delivery period, and delivery point information may be the only information given to the contractor. When the information is received, the contractors for *each* organization may send the company's house gas sale or purchase contract to their negotiating counterparty for review.

If only one organization has sent its house contract, the contractors will begin negotiations on every issue *except* quantity, price, delivery period and delivery point. If each contractor has sent a house contract to the other, agreement must first be reached on which organization's contract will be used. There are no rules for resolution if a conflict arises. Economics of the deal usually prevail in these matters. The organization needing the deal more will usually agree to use the other's form contract. In chapter 6, the difficulties of determining which contract controls in a situation like this will be discussed.

The information transfer scenario given in the last paragraph points to one extremely important fact. In many major gas organizations, gas contractors are responsible for most negotiated issues in gas contracts after the gas sale and purchase representatives have reached an agreement for a particular quantity of gas, at a specified price, to be delivered at a designated delivery point for an agreed period of time. The representatives reach an agreement and the contractors negotiate the terms of the resulting contract. The reason this issue is so important is that many contract disputes involve matters negotiated by gas contractors rather than those negotiated by sales and purchase representatives. Gas contractors and marketing representatives should consult with each other on all material matters in dispute during negotiations so that the contract itself truly reflects the intentions of both organizations.

The formation issues addressed by this part are so critical to gas contracting that chapter 6 is dedicated to the types of uncertainties that can abound when:

- gas flows without any written evidence of the deal
- each party provides its own form contract or confirmation to the other and the terms vary
- the only evidence of agreement is contained in tape recordings of trader telephone conversations

Chapter 6 will also reveal how courts will look to the parties' relationship and performance in determining the intention of the parties in a contract dispute.

Part 3. General Obligation and Construction of Contract

This part of Article 2 neatly packages the contract between two parties, and provides a number of rules that will apply to all transactions for the sale of goods. Both the seller and the buyer have *dual* obligations under Article 2. The seller must transfer and deliver, and the buyer must accept and pay.[37] These obligations will apply whether or not the written contract specifically states as much. Contracts are filled with tradeoffs. The seller trades its product and title to the product in return for the purchase price. The buyer trades its money for title to and possession of the gas. The very essence of contracting is give-and-take and courts will look very critically at any contract if the burdens and rights of the parties are so asymmetrical that the court is shocked. This concept is called "unconscionability of contract" and in the commercial setting, a court may find that, as a matter of law (without regard to factual evidence) a clause or contract was unconscionable when it was made.[38] The result of this finding may be that the court will not allow enforcement of part or all of the contract. This is an important issue for Merchants who must always use care when dealing with non-Merchants.

A finding of unconscionability would be rare in a commercial gas transaction unless one of the parties had a great bargaining advantage over the other. Courts have found *procedural unconscionability* exists when one party used fine print to hide contract terms, exercised disparate or unequal bargaining power, or abused an obvious lack of understanding on the part of the other party.[39] *Substantive unconscionability* goes to the question of the bargain's general fairness. A common example of this would be an excessive sales price—rare, but not impossible in the context of natural gas sales. Assuming a "gotcha" position doesn't go over well in commercial sales litigation.

This is also the part of Article 2 that gives courts broad judicial oversight powers. Even if the contract fails for indefiniteness in a number of areas, a court reviewing a contract with unclear or even absent terms may fill in the blanks. The power of the courts is collectively referred to as *gap-filling* powers, a concept unique to sales of goods as expressed in the UCC. This Part 3, coupled with the basic purpose for the Code elucidated in Section 1-102, is why I frequently comment during my contract lectures and seminars that a court being asked to decide commercial contract issues will actively attempt to find a contract that exists between the parties.

Unless it can be shown to a court that the parties clearly did not intend to agree and contract, the court may utilize one of its extraordinary gap-filling powers under the UCC. Courts may fill in the gaps if parties fail to settle on a price,[40] delivery point,[41] or delivery time.[42] The only issue that a court cannot decide if the parties were not clear in their intentions is quantity. Even then, if it can be determined that the parties agreed to a "full output" or "full requirements" sale, or had an "exclusive dealing" relationship, the courts may fill that gap. Wholesale purchases and sales of gas are usually precise as to quantity, price, delivery point, and delivery period. Both producers and consumers finding themselves in court because of a contract quantity dispute may have to live with the court's interpretation of quantity if it appears that one of the relationships indicated above apply.

Many of the terms of Part 3 are widget-relevant, but organizations buying or selling gas should note Code terms regarding (1) warranties given by the seller and (2) failure by a buyer to provide a letter of credit prior to gas flow. Let's address the letter of credit issue first because it is simple and straightforward.

Failure to provide required letter of credit. If the buyer fails to provide a required letter of credit "seasonably," the buyer is in breach of the contract.[43] The term "seasonably" is not further defined in the Code, but if the seller has required the letter of credit prior to gas flow and the buyer fails in that obligation, the seller might be well advised to suspend its performance until the letter of credit is in place. This is an extraordinarily important issue from the seller's perspective because any breach resulting from this failure is a breach of the *entire* contract, and the onus is on the buyer to comply with the seller's requirements. Remembering also that both the seller and the buyer have an obligation of good faith and commercial reasonableness in performance and enforcement of contract rights, each should carefully weigh the consequences of acting hastily and in an unreasonable manner.

Warranties given by the seller. What is a warranty and what happens if a warranty is breached? A warranty is "an assurance by one party to an agreement of the existence of a fact upon which the other party may rely, intended to elicit reliance by the purchaser or to relieve the other party of any duty to ascertain the facts for himself and amounting to a promise, if the fact warranted proves untrue, to indemnify the other party for any loss or to answer in damages for any injury proximately caused."[44] That was the legal version. More simply, a warranty is akin to a guaranty or a promise made by the seller relating generally to the quality or nature of goods being sold. The chief reason warranty is included in the UCC is to hold sellers responsible for inferior products being sold to unsuspecting buyers.

Warranties may be given by words *or by conduct*, and not all warranties are the same. When you purchase a used car and the salesman tells you that the car has 50,000 miles and has never been in an accident, you will probably rely on that information if you choose to purchase the car. Later, when you take the car to your mechanic for repairs and discover that the odometer has been set back or that the chassis is bent, probably from a severe collision; you may have a cause of action against the dealership to collect your economic loss for breach of the oral warranty.

What damages may be collected if a warranty is breached? If a buyer can prove that:

- the goods carried a warranty
- the goods did not conform to the warranty, and
- damage was caused by the nonconformity

the buyer may:

- refuse to accept the goods and recover the price already paid, or
- accept the goods and collect the difference between "the value of the goods accepted and the value they would have had if they had been as warranted," plus incidental and consequential damages in the proper case [45]

Title Warranty

Warranties flow with the gas. Even if there is no title warranty language in the gas contract, the seller automatically warrants that "the title is good, that transfer of the gas is rightful, and that the gas will be delivered free from any security interest or other lien or encumbrance of which the buyer at the time of contracting has no knowledge."[46] Furthermore, a Merchant who regularly sells gas warrants that the goods will be delivered free of any rightful claim by a third party.[47] The most common instance of a rightful claim by a third party occurs at the wellhead. An operator selling gas for many gas interest owners is required to pay those owners their share of gas proceeds from gas sales. This may be done through a variety of arrangements with the interest owners. If royalty payments are not being made or not being made correctly, the title is no longer clear. There is a cloud over the title that will follow the gas downstream, and title warranties given by every downstream seller of that gas will be in jeopardy. In the discussion of title warranties in chapter 8, you'll learn how downstream buyers may be directly affected by such failure to pay at the wellhead. The seller may modify or disclaim any of these warranties[48] in the contract, but it is almost inconceivable that a gas buyer would agree to such a term.

Merchantability/Fitness for a particular purpose

Many gas contracts contain a clause similar to the following:

> "Seller hereby disclaims any warranty, express or implied, not specifically given in this Contract. This includes warranty of merchantability and warranty of fitness for a particular purpose."

This simple clause tells the reader several things. First, warranty of *merchantability* and warranty of *fitness for a particular purpose* mean something in sales law. Second, these warranties must be important because the seller is specifically trying to avoid them.

This clause is frequently found in gas sale contracts because of Article 2, Part 3 warranties. A seller may create so-called *express* warranties through its actions and words during the bargaining process.[49] An express warranty might be made by a gas producer simply claiming to sell high MMBtu content gas at the wellhead. A prospective buyer holding a downstream processing agreement might actually be purchasing this particular gas because of the added value it may recoup through processing. If the gas later proves to be relatively lean, the buyer might have a successful breach of warranty claim against the seller if the buyer suffers economic loss because of reliance on the express warranty. In downstream sales, a warranty as to gas characteristics is probably much less important simply because, once in the pipeline, gas is considered to be fungible. Keep in mind though, that there is no such thing as "never" in the gas industry, so even downstream sales could be subject to similar express warranties.

Sellers may also give *implied* warranties to their buyers. These may be given without ever uttering a word of guaranty or warranty. For example, gas merchants give an implied warranty of merchantability with every gas sale.[50] Merchantability means that the gas:

- passes without objection under the contract description
- is of fair, average quality within the description

- is fit for the ordinary purpose for which it is sold
- is of even kind quality and quantity within each unit, and
- is adequately contained

The second implied warranty is fitness for a particular purpose, and it comes into play when a:

- seller knows the purpose for which the product sold will be used, and
- the buyer relies on the seller's judgment in providing goods that will meet the buyer's needs

In the gas industry we see this issue most frequently when gas is being sold for industrial consumption. Natural gas high in sediment, water, contaminants, or liquids can damage highly sensitive equipment.

Those of you familiar with the industry might be thinking, "aren't these really all matters between a shipper and transporter?" The answer is, yes and no. Transportation service providers have nearly total control over the gas in a pipeline system. The important question is, " Who is the shipper?" If the seller's transporter delivers gas directly to a buyer's plant, and that gas directly causes property damage and perhaps personal injury, the buyer has no direct line of redress against the seller's transporter. Because the buyer in this example is not a shipper, it has no contractual relationship with the seller's transporter.

The harmed buyer in this instance has only one real option—to try to collect damages from the organization that sold it the gas. The gas seller may receive a proportionate share of damages from its transporter. The transporter will, in turn, investigate whether one of its shippers was responsible for putting non-spec gas into its system and many ultimately collect from that shipper. As to the buyer and seller under any gas contract, the gas sale contract is completely separate from transportation service agreements. We can't just force our way into a contract we have not signed. This is so because of the rule of *privity of contract.*

Parties that have contracted with each other are said to be *in privity* and those who have not contracted are *not in privity.*[51] As a general rule, only those in privity may enforce rights under the contract because only those in privity *have* rights under the contract. The industrial consumer in the previous example is not in privity with the seller's transporter, so it can only seek compensation for its loss from the buyer with which it has privity under the gas contract.

The gas industry isn't perfect, and the gas itself is rarely pure methane. Sometimes things happen that result in damage. Many careful contractors will routinely insert a warranty disclaimer in their organization's gas sale contract to protect against potential liability caused by any of a number of things that we hope never will happen. Contractors for organizations that are buying gas will probably be just as certain that important warranties are not disclaimed by the seller.

Warranty disclaimers

The Code specifically allows sellers to disclaim warranties, both express and implied.[52] If a seller breaches a warranty, that doesn't necessarily mean that the seller acted with evil intent. Innocent sellers—*i.e.*, sellers who don't intend to mislead and may have no knowledge of true facts—who give warranties then breach those warranties, will be subject to the remedies allowed buyers by the UCC.

All warranty disclaimers must be conspicuous. Many products we routinely purchase will have very conspicuous warranty disclaimers placed on the packaging or in the user's manual. In the gas industry, the conspicuousness requirement is seen in contract language that is bold, in caps, or in a different color than the print in the rest of the contract. Warranty disclaimer language that is not easily set apart from other contract language does not meet the Code's conspicuousness requirements of Section 2-316(2).

Warranties to third parties

You may be familiar with one of the boilerplate clauses in most standard gas contracts, the so-called *third-party beneficiary* clause. A typical third-party beneficiary clause might simply state: *"There are no third-party beneficiaries to this contract."* The clause may also be more detailed, but the heart of the matter is stated as above. The reason for this clause is found in Code Section 2-318.

Think back to the discussion of privity. A more complete examination of that topic reveals that two types of privity exist—*vertical* and *horizontal*. A *vertical non-privity party* is a buyer who did not buy directly from the defendant seller, but rather from another downstream seller—your customer's customer, for example. A *horizontal non-privity party* is not necessarily in the sales chain, but is one who uses, consumes, or is affected by the goods.[53] Under the UCC, in some circumstances, a party not in privity with the defendant seller may nonetheless be able to collect damages from the seller under a third-party beneficiary breach of warranty claim.

Assume that your organization is selling gas to a marketer that is re-selling the gas to a hospital, at the same delivery point. Your sales representative learned during contract negotiations that the gas is being delivered directly to the hospital. It is January, gas supplies are tight, and the transporter curtails all non-primary point firm transportation. Your transportation is firm but utilizing a secondary point, so it is curtailed. The alternate source for fuel is heating oil, but a car careening out of control at the oil storage facility has caused immediate danger and loss of fuel. Are you beginning to get the picture? Your organization has become the only source of fuel for this hospital. When the pipes burst, delicate machinery is damaged, children get sicker, and the hospital must find other locations for patients. The personal injury and property damage costs could skyrocket. If the hospital can prove that it was a third-party beneficiary of your gas sale contract to the marketer, be ready to spend lots of money.

Sellers obviously want to avoid those downstream problems, so gas sellers, particularly gas merchants, will almost always include a third-party beneficiary clause in their contracts, in addition to warranty disclaimer language.

The original Article 2 drafting committee took an unusual approach to Section 2-318. They presented three forms of Section 2-318 to states considering enactment of Article 2, and each state has made its own choice of Alternative A, Alternative B, or Alternative C. Gas sellers need to know which option has been elected by the governing law state. The three alternatives are listed below, so you can see how consumer-oriented and widget-oriented all three are. Notice that the seller cannot disclaim the third-party beneficiary warranty in any state. The only way sellers can attempt to protect themselves from third-party claims is to state in the contract that there are no such beneficiaries. If there are no third-party beneficiaries, there is no one to whom this warranty can possibly go.

Alternative A (to family and guests)

> "A seller's warranty whether express or implied extends to any natural person who is in the family or household of his buyer or who is a guest in his home if it is reasonable to expect that such person may use, consume or be affected by the goods and who is injured in person by breach of the warranty. A seller may not exclude or limit the operation of this section."

Alternative B (to any *natural* person)

> "A seller's warranty whether express or implied extends to any natural person who may reasonably be expected to use, consume or be affected by the goods and who is injured in person by breach of the warranty. A seller may not exclude or limit the operation of this section."

Alternative C (to any person)

> "A seller's warranty whether express or implied extends to any person who may reasonably be expected to use, consume or be affected by the goods and who is injured in person by breach of the warranty. A seller may not exclude or limit the operation of this section with respect to injury to the person of an individual to whom the warranty extends."

Alternatives B and C are obviously of greatest concern in wholesale gas operations. As the gas industry continues in its quest for unbundling behind the city gate, marketers may be confronted with this new potential liability in states that have chosen Alternative A, as well.

Part 4. Title, Creditors, and Good Faith Purchasers

Title to goods cannot pass until the goods have been *identified*.[54] The only relevance this term usually has for gas sales is that a producer cannot transfer title to gas *as goods* until the gas is severed from the ground. Title transfers concurrently with possession and once the seller has tendered the gas and transferred title to buyer, the only right or reservation that may be retained by the seller is a security interest in the gas sold. Depending upon the possession issue discussed earlier, Article 9 will interplay with this Part 4 when the seller has retained a security interest in the gas sold until payment is made.

Except in rare circumstances, the rights of a seller's unsecured creditors are generally inferior to the rights of the buyer.[55] The buyer acquires the same title that the seller had to transfer.[56] In the previous discussion of title warranty, I noted that any cloud on the title would travel with the title. A seller can only give that which the seller has to give, and this example illustrates how every downstream party will be affected.

FACTS:

There are three organizations involved in selling the same gas—PUMP, FLEX and REFLEX. PUMP sells gas to FLEX in Texas, FLEX sells the "same" gas to REFLEX at a delivery point in South Carolina, and REFLEX ships the gas from South Carolina to its customer in New York. The legal counsel for REFLEX receives a call from an individual in Texas claiming that he has not received royalty payments on that gas from PUMP.

RESULTS:

1. As to REFLEX:
   - Setting aside any other issues that may be relevant, REFLEX has been put on actual notice that the title to gas it purchased from FLEX was not clear
   - REFLEX cannot give a clear title to its customer because it did not receive clear title from FLEX
   - REFLEX may have a title infringement action against FLEX if REFLEX incurs damage

2. As to FLEX:
   - FLEX has most likely breached its title warranty to REFLEX
   - Upon receiving notice from REFLEX, FLEX must now investigate the allegation

3. As to PUMP:
- The unpaid royalty owner may have an action for price and warranty against PUMP
- FLEX may have a title infringement case against PUMP
- Ultimately PUMP will have to pay any royalties due.

In chapter 8, you'll learn about contract language that can protect unsuspecting downstream buyers as well as some of the practical results of transferring title that is not "good."

Part 5. Performance

Insurable interest and risk of loss. Do you know whether your organization has insurance to cover any damage by or loss to natural gas that it owns? Amazingly, my experience has shown that many gas contractors asked to make decisions about liability of their client or employer don't know much about the insurance aspect of that organization's risk management program. If you don't know the answers, find out. It will help.

Part 5 is essentially widget-based in its approach to sales as well, so the insurance issues are most easily understood in that context. When does the seller have an *insurable interest* in the gas it is selling? Is it while the seller holds title to the gas, or while the gas is in the possession, or deemed possession, of the seller? When does the seller lose that interest and the buyer gain it? The first question to be answered clearly is "what is an insurable interest?" Holding an insurable interest is a prerequisite to insuring the life, property or object being insured. I can insure my life or my husband's life, but can I insure my neighbor's life? Unless my neighbor is also my child or other close family member, probably not. Having an insurable interest means if that which is insured is lost or damaged, I would suffer direct loss as a result.

The seller has an insurable interest in natural gas as long as it retains title or any security interest in the goods.[57] On the other hand, the buyer has an insurable interest in gas already produced when the contract is made, and for the sale of future gas when the gas is shipped or transported.[58] Typical gas contract language states that the seller maintains all responsibility for natural gas while in the seller's possession (or its transporter's possession) and the buyer is responsible for the gas while in its possession, so this part of the Code has most relevance in the gas industry in atypical or unusual transactions where change of title and possession of the gas don't occur simultaneously.

The Code specifically states that the *risk of loss* passes from the seller to the buyer when the carrier tenders the goods for delivery.[59] In the natural gas context, this means that the risk of loss passes from the seller to the buyer when the transporting pipeline has the immediate ability to deliver gas to the buyer. Since natural gas is simultaneously received upstream and delivered downstream, the risk-of-loss issue is a difficult one for sellers and buyers.

Part 6. Breach, Repudiation, and Excuse

In the imperfect world of buying and selling natural gas, the Code has its own rules to determine the rights a party may exercise if the other party fails in its contractual obligations. This Part 6 lays the groundwork for the remedies included in Part 7. The first section speaks to the actions a buyer might take "if the goods or the tender of delivery fail *in any respect* to conform with the contract, . . ."[60] Think about the implications of this section. It's rather easy to imagine how this term applies if a truck trailer supposedly loaded with men's size 10D shoes is actually filled with men's size 9E shoes. Obviously the goods are not in conformance with the contract specifications, and the buyer may:

- reject the whole
- accept the whole
- or accept any commercial unit or units and reject the rest

Now think about this same situation where, instead of shoes loaded onto a truck, the product is natural gas in a multiple receipt and delivery point pipeline filled with fungible natural gas. It's a little more complicated then because of the buyer's inability to inspect the gas before it is accepted. Gas buyers that might have an occasion to utilize this right include industrial consumers and LDCs, both of which sample gas and will probably know very quickly whether the gas meets contract specifications. Read UCC Section 2-601. It may be a very powerful tool for gas buyers in the proper situation. Buyers and sellers should be familiar with Part 6 because the issues raised here usually aren't found in the gas contract.

Right to adequate assurance of performance. Most gas contracts contain a clause similar to the following*:*

> *"When reasonable grounds for insecurity of payment or title to the gas arise, either party may demand adequate assurance of performance. Adequate assurance shall mean sufficient security in the form and for the term reasonably specified by the party demanding assurance, including, but not limited to, a standby irrevocable letter of credit, a prepayment, a security interest in an asset acceptable to the demanding party or a performance bond or guarantee by a creditworthy entity."*[61]

If a clause similar to this one is not included in the contract, each party still has the right to adequate assurance of performance by the other party.[62] One might naturally think that the seller has more to worry about if the buyer doesn't pay for gas delivered, but one of the major reasons for the UCC is to expand commerce, so neither party is given preferential treatment in this regard. The seller has the right to know that its buyer can take the gas and pay for it, and the buyer has a right to know that the seller has the ability to deliver gas. Either party invoking its rights under this section must honestly believe that the other party's performance is in jeopardy. The most frequent use of this demand is by the seller if a buyer's credit limits no longer sustain the level of business between the two.

From the buyer's perspective, this type of clause should be included in the gas contract. If you carefully read Section 2-609, you will find that no mention of risk as to *title* of the goods is made, but in the clause above, title is included so that the buyer can demand that the seller provide it with evidence that the title is good. For certain types of transactions, counterparties may wish to expand the grounds for demanding adequate assurance beyond payment and title as the only two reasons the demand may be made in the clause above. The Code allows the demand for assurance by either party if there is a question about continuing *performance* of the other party, a much broader usage.

Anticipatory repudiation. Assume that your organization is selling gas under a 5-year contract. Two years after performance has begun, your customer starts making noises about having cash flow problems and mutters about bankruptcy. Now, on the other hand, assume that your organization is buying gas under a 5-year contract and the seller starts begging for, but not demanding, a higher price for the gas. Neither of these is clearly a contract breach, but may be contract *repudiation* by the muttering buyer and the begging seller even though neither the mutterer nor the beggar actually used the word, repudiation.

Repudiation of a contract usually occurs after a demand for adequate assurance of continued performance has been made with an unsatisfactory response from the other party. Repudiation generally must be inferred from action or inaction of the allegedly repudiating party. The most recognizable form of repudiation is when the buyer or seller does something to make its own performance impossible. Assume that a producer has dedicated its production to one buyer. If the producer subsequently contracts to sell the same gas for a higher price to another party, it has clearly repudiated the contract with the first buyer. A consumer purchasing its full fuel requirements from one supplier that subsequently contracts to buy cheap gas from another supplier has clearly repudiated the first supplier's contract. Most issues in gas sales aren't quite as clear as these two examples, however.

If either party believes to the best of its knowledge that the other party has repudiated the contract, it may undertake any of the following:

- Await the other's performance for a commercially reasonable time
- Resort to remedies for breach of the contract, or
- In either case, suspend its own performance[63]

The danger awaiting the non-repudiating party is that if it invokes any of its rights listed here *and it is wrong*, it will then be in breach of its contractual obligations.

Substituted performance and excuse by failure of presupposed condition. Sections 2-614, 2-615 and 2-616 collectively address issues related to failure by a transporter to deliver the goods and force majeure. These topics will be discussed in detail in the force majeure discussion in chapter 8.

Part 7. Remedies

As with force majeure and several contract formation issues referenced earlier, the damages and remedies issues in gas contracts are used so frequently that a separate "Remedies and Damages" section will include extensive discussion of these issues in chapter 6.

---

1. The law that has been made through court decisions, based upon traditional legal precepts and interpretation of statues.

2. Louisiana's civil law system is based upon the French civil code. While other sections of the UCC have been adopted for use in Louisiana, Article 2 "Sales of Goods" has not. If Louisiana law governs your contract, you will go to L.S.A. - C.C. Book III, Title VII Sales § 2438-2659.

3. UCC citations for all states may be found at the following web site: http:/www.cornell.law.edu/uniform/ucc.html.

4. Alberta: Sale of Goods Act, R.S.A., 1980, c. S-2; British Columbia: Sale of Goods Act, R.S.B.C. 1979, c. 370; Manitoba: The Sale of Goods Act, R.S.M. 1987, c. S10; New Brunswick: Sale of Goods Act, R.S.N.B. 1973, c. S-1; Newfoundland: Sale of Goods Act, R.S.N. 1990, c. S-6; Nova Scotia: Sale of Goods Act, R.S.N.S. 1989, c. 408; Ontario: Sale of Goods Act, R.S.O. 1990, c.S.1; Prince Edward Island (PEI): Sale of Goods Act, R.S.P.E.I. 1988, c. S-1; Saskatchewan: The Sale of Goods Act, R.S.S. 1978, c. S-1; Northwest Territories: Sale of Goods Act, R.S.N.W.T. 1988, c. S-2; Yukon: Sale of Goods Act, R.S.Y. 1986, c. 154.

5. Civil code of Québec, Book five-Obligations: Title two-Nominate contracts: Chapter 1-Sale: Section I-Sale in general.

6. treaty.html at www.cisg.law.

7. of such a kind or nature that one specimen or part may be used in place of another specimen or equal part in the satisfaction of an obligation, Webster's New Collegiate Dictionary. "Fungible" with respect to goods or securities means goods or securities of which any unit is, by nature or usage of trade, the equivalent of any other like unit. Goods which are not fungible shall be deemed fungible for the purposes of this Act to

the extent that under a particular agreement or document unlike units are treated as equivalents." UCC § 1-201(17).

8. The process of removing unwanted water, sediment and non-combustible gases from the gas stream.

9. Any point where two pipeline facilities join together to allow for transfer of gas from one pipeline system to another. Interconnections may be into gathering systems, between gathering systems and transmissions systems, between two transmission systems, etc.

10. The point of interconnection into a pipeline system where natural gas is produced.

11. When nominations equal measured amounts, the pipeline system and transportation service agreements are in balance. When nominations and actual measurements vary, an imbalance has occurred.

12. Like statutory law, common law may vary from state to state, but on any given common law issue, a recognized majority position among courts can usually be discerned. In the U.S. and Canada, English law principles are followed. In the State of Louisiana and the Province of Quebec, French civil law principles are followed.

13. UCC § 1-103.

14. UCC § 1-102(1).

15. UCC § 1-20.

16. UCC § 1-102(19).

17. UCC § 1-201.

18. The concept giving rise to a court's usual reluctance to look beyond the signed document for evidence of the parties' intent. This concept may also be referred to as *integration* or *entire agreement of the parties*.

19. UCC § 1-205.

20. UCC § 5-112(1).

21. UCC § 5-110(1).

22. UCC § 5-116(1).

23. UCC § 2-105(1).

24. Id.

25. UCC § 2-107(1).

26. UCC § 2-106(1).

27. UCC § 2-104(1).

28. UCC § 1-201(30).

29. UCC § 2-104(3).

30. UCC § 2-205.

31. UCC § 2-103(1)(b).

32. UCC § 1-201(11).

33. UCC § 1-201(3), emphasis added.

34. UCC § 2-106(4).

35. Obligations that have not yet been performed. Future obligations.

36. UCC § 2-106(3).

37. UCC § 2-301.

38. UCC § 2-302.

39. *The ABCs of the UCC, Article 2: Sales,* Henry D. Gabriel and Linda J. Rusch, Section of Business Law, American Bar Association, U.S.A., 1997, p.46.

40. UCC § 2-305.

41. UCC § 2-308.

42. UCC § 2-309.

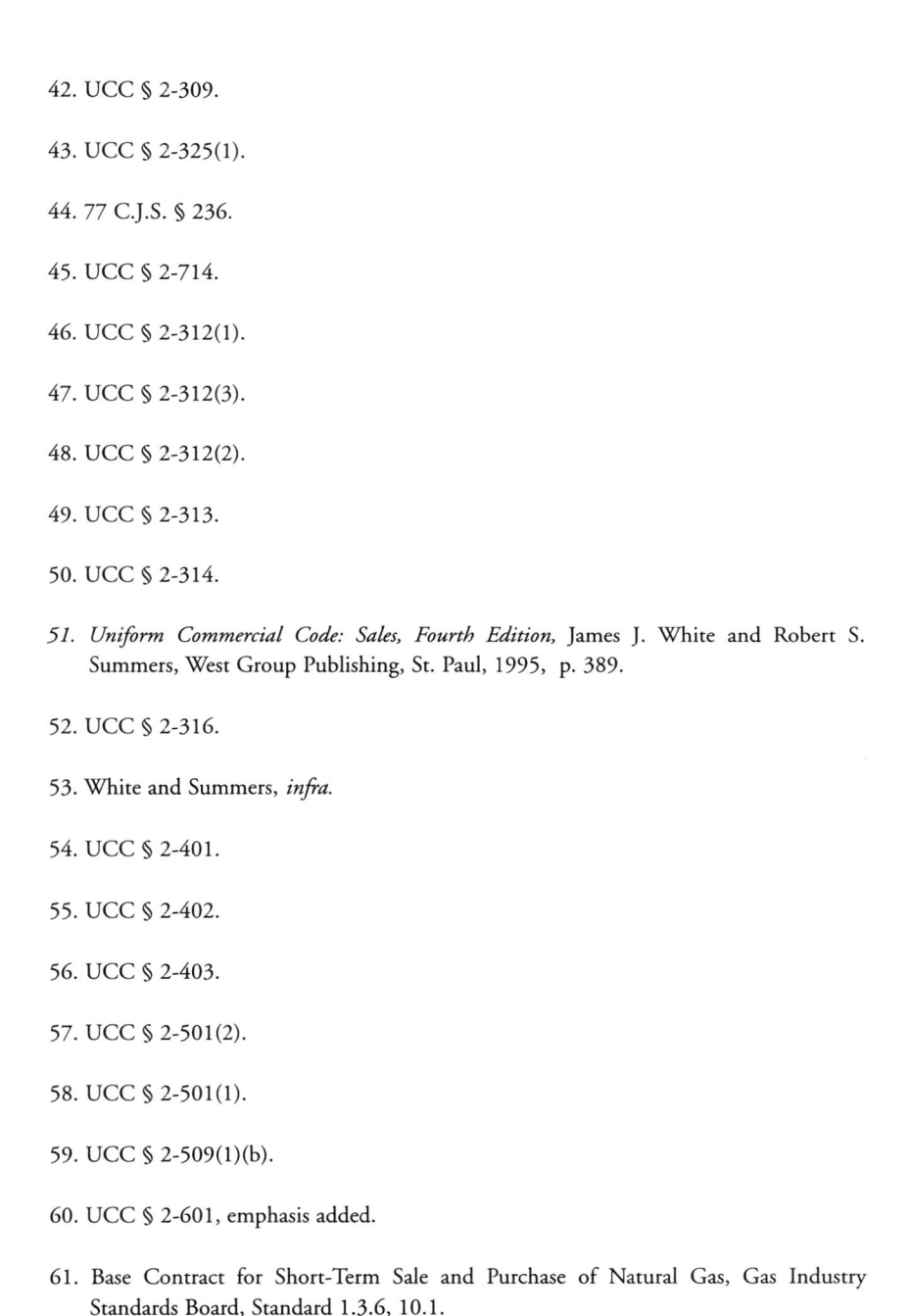

43. UCC § 2-325(1).

44. 77 C.J.S. § 236.

45. UCC § 2-714.

46. UCC § 2-312(1).

47. UCC § 2-312(3).

48. UCC § 2-312(2).

49. UCC § 2-313.

50. UCC § 2-314.

51. *Uniform Commercial Code: Sales, Fourth Edition,* James J. White and Robert S. Summers, West Group Publishing, St. Paul, 1995, p. 389.

52. UCC § 2-316.

53. White and Summers, *infra.*

54. UCC § 2-401.

55. UCC § 2-402.

56. UCC § 2-403.

57. UCC § 2-501(2).

58. UCC § 2-501(1).

59. UCC § 2-509(1)(b).

60. UCC § 2-601, emphasis added.

61. Base Contract for Short-Term Sale and Purchase of Natural Gas, Gas Industry Standards Board, Standard 1.3.6, 10.1.

62. UCC § 2-609.

63. UCC § 2-610.

Chapter 5: North American Sale of Goods

When U.S. and Canadian parties contract for sales, either the law of a state or the law of a province may control. Contract negotiations between U.S. and Canadian parties typically involve resolution of which law should govern the contract and its interpretation. Knowledge of the laws of another country is a fairly new concept for the wholesale gas industry, and many gas contractors simply agree to the laws of the other country since they are frequently told, "your laws and ours are just about the same."

U.S. laws, and particularly the UCC, were covered in the preceding chapter. The first part of this chapter is included to acquaint gas contractors in the U.S. with the sale laws of Canada. The second part will provide information on the United Nations CISG,[1] which may be the controlling law for gas sale transactions between residents of the U.S., Canada, and Mexico. As you will learn, state law, provincial law, or international treaty could be applicable in transactions between parties located in different countries.

Because of the plethora of transactions between U.S. and Canadian organizations, gas contractors in both the U.S. and Canada must appreciate the consequences of accepting the laws of the other country to govern contract interpretation and performance. The Uniform Commercial Code is addressed throughout this guide, and this chapter alone is intended to familiarize gas contractors with the basic sales law prevalent in Canadian provinces. As you might assume, since my experience as an attorney is predominately with U.S. laws, the approach taken in this chapter will understandably be from the U.S. perspective. The information relating to provincial Sales of Goods Acts in this chapter is not intended to be a learned treatise—just an introduction and overview to provincial laws governing sales of goods in Canada.

While similar, the Sales of Goods Acts in the provinces have not been made uniform to the extent the UCC has made commercial law somewhat uniform among the 50 states, the District of Columbia, Puerto Rico and the Virgin Islands. The Uniform Law Conference of Canada, a similar group to the Uniform Law Commissioners in the U.S., is devoted to harmonizing the Canadian statute law in certain respects. It does so by preparing uniform statutes that it recommends for enactment by the provinces and territories and sometimes by the federal government as well. It also does so by adopting uniform statutes dealing with choice of law.[2] On occasion it promotes particular provisions for statutes or publishes guides to uniform legal procedure. In criminal matters it usually adopts proposals for changes or additions to the Criminal Code and other federal legislation.

The conference has also recommended that all provinces adopt a uniform sales act, similar to the UCC in the U.S. The purpose of the proposed package of commercial statutes is to reduce the legal risks inherent in doing business across the country in a twofold way: first, by eliminating uncertainty about which law applies; and second, by making the law suit the needs of modern commerce. In addition, as Canadian borders are opened to international trade, the conference has indicated its collective belief that consistent legal rules within Canada will increase Canadian business competitiveness.

The process of making commercial laws uniform in Canada has not been completed, but even as the different laws now stand, more similarities than differences are seen in sales acts from province to province. In chapter 4, you were given legal citations for the Sale of Goods Acts in the Canadian provinces, but the majority of natural gas activity in Canada remains in the vast production regions of the west, and Alberta is the hub of that activity. Because of that, this review will be of the Alberta Sales of Goods Act.[3] In cross-border transactions between U.S. and Canadian parties, the law of Alberta is frequently selected as governing law for the contract. In content, the Alberta Sales of Goods Act is similar to the UCC, although the structure is different.

The Alberta Sale of Goods Act (the "Act")

The Act is divided into six parts and 59 sections. Like the UCC structure, each part encompasses only one broad subject matter. The six parts are:

- Formation of Contract
- Effects of the Contract
- Performance of Contract
- Rights of Unpaid Seller Against the Goods
- Actions for Breach of the Contract
- Supplementary

The drafters of this Act have provided some interpretative notes, three of which may be quite helpful as background when reviewing any part of the Act. *Good faith* for example includes anything done "when it is in fact done honestly whether it is done negligently or not."[4] Under the Act, a person is considered to be *insolvent* when "he either has ceased to pay his debts in the ordinary course of business or cannot pay his debts as they become due whether he has committed an act of bankruptcy or not."[5] Finally, the definition of *warranty* could raise numerous issues with regard to damages. "*Warranty* means an agreement with reference to goods that are the subject of a contract of sale but collateral to the main purpose of the contract, the breach of which gives rise to a claim for damages but not to a right to reject the goods and treat the contract as repudiated."[6] During this chapter, we'll revisit all three of these terms—good faith, insolvent, and warranty.

Part 1. Formation of Contract

Statutes like the UCC and the Sale of Goods Acts typically address contract formation issues to protect the parties and to specify the process that must take place before any sale contract will be enforceable in a legal venue. The issue of contract formation in general, and under the UCC, was discussed in chapter 4.

Sale. The first sections of this part distinguishes between a *sale* and an *agreement to sell,* for example, such that when the goods are transferred, a sale has occurred, but when transfer of the property is to take place in the future, an agreement to sell has been formed. The agreement to sell becomes a sale, in essence, when the goods are transferred.[7]

Statute of frauds. Contracts having a value of greater than fifty dollars ($50.00), to be enforceable, must be evidenced in "some note or memorandum in writing . . . made and signed by the party to be charged or his agent in that behalf."[8]

Acceptance of the goods. Goods will be deemed to have been accepted by the buyer when the buyer "does any act, in relation to the goods, that recognizes a pre-existing contract of sale whether there is an acceptance in performance of the contract or not."[9] In the natural gas industry, contract acceptance is evidenced by performance on a routine basis, and even nominating gas for flow when no contract has been signed is a significant act.

Price. The price paid for goods may either be specifically stated in the contract or left to be fixed at a later time, or may be agreed according to the course of dealing between the parties. The course of dealing reflects the two parties' dealings with each other in past transactions. If company A and company B have been doing business together for some time and gas bought and sold has been priced in a consistent manner, the prior dealings may have relevance to determining a price under the current contract. If price is to be determined by a third party valuation, such as a published index price, and the third party can not or does not make the valuation, the agreement is avoided.[10] In any instance, if the buyer has taken delivery of any portion of the goods, it has an obligation to pay for the goods received, even if the contract is avoided.

Conditions and warranties. The word "month" means prima facie[11] a calendar month. This section might come into play when no specific time periods are given for notice in a contract or if a contract clause calls for "*one month's* notice." The careful contractor will harmonize the contract language with the Act for any notice provisions in the contract, including renewal, termination or default.

When contract is subject to conditions. The buyer may waive any conditions to be fulfilled by the seller.[12] The Act distinguishes between contractual *conditions* and *warranties* and states that the facts of every case will determine what the nature of stipulations made by the contracting parties will be.[13] If the stipulation is a warranty, the non-breaching party may sue for damages, but will not have the right to reject goods and consider the contract to have been repudiated.[14] On the other hand, if a stipulation is a *condition* (even if that stipulation may have been called a *warranty* in the contract), the non-breaching party's remedy is solely contract repudiation.[15]

Implied conditions and warranties. The Act distinguishes again; this time between implied conditions and implied warranties. There exists an implied condition on the seller of goods that it has the right to sell the goods when delivered.[16] The implied warranties given by every seller of goods are:

- the buyer will have the right to "enjoy quite possession of the goods[17] and
- no third party claims to title, of which the buyer is not aware, exist as to the goods[18]

The buyer's knowledge at the time the contract is made determines whether the buyer may rely upon implied warranties.

Sale by description. Whenever goods are sold by description, an implied condition exists that the goods will correspond with the description. For gas sales, whenever the term "gas" or "natural gas" is defined, a careful seller's contractor will ensure through appropriate contract language that acceptance of the gas by the buyer's transporter will be deemed to have been acceptance of "gas" for all purposes under the contract.

Implied conditions regarding quality. There are no implied warranties of *merchantability* or *fitness for a particular purpose*[19] under the Act unless specific situations occur.

If the buyer makes known the particular purpose for which goods are being purchased, there is an implied condition that the goods are reasonably fit for that purpose.[20] Gas consumers that require very specific quality and heat content in purchased gas should be careful in both negotiations and drafting to include any unique or limited needs so the seller will be aware of the necessity for providing gas meeting whatever those unique needs may be. Of course, gas sellers want to know as little as possible about any specific use for the gas being sold.

Whenever an organization that deals in goods of a certain type sells those goods by description, there is an implied condition that the goods are of a merchantable quality.[21] If the contract language establishing a warranty or condition is clearly inconsistent with the conditions and warranties provided by the Act,[22] the contract will prevail.

Part 2. Effects of the Contract

Transfer of property. A transfer can only be made once the goods are "ascertained," so any contract to sell production in advance of actual deliveries does not give rise to the concept of transfer from seller to buyer. This means that when a producer sells gas that has not yet been produced or *severed* from the ground, the goods have not been ascertained. Only when the gas is severed may the transfer for sale of goods occur.[23]

Time of transfer. Transfer will occur when the parties so intend. To determine the parties' intent, courts may look at:

- the contract terms
- the conduct of the parties, and
- circumstances of the case[24]

The Act provides some rules for ascertaining intentions. For example, when an unconditional contract for the sale of goods in a deliverable state is made, the property transfer occurs when the contract is made, even when delivery and payment may be postponed.[25] If the contract is for goods in a deliverable state, but the seller is bound to measure, test, or do some other act for the purpose of ascertaining the price, the property doesn't pass until the act is done and the buyer has notice that the measurement or test is completed.[26]

Because the Act covers both current and future sales, much of the language in this section is included to identify the various differences in approach to *sales* and *agreements to sell*. In that regard, if an agreement to sell has been made, then when the seller delivers goods to the buyer or to a carrier, such as a transporting pipeline, the seller is deemed to have unconditionally appropriated the goods to the contract.[27]

Risk transferred with property. Sellers retain all risk until the property is transferred, and once the transfer has occurred, the seller no longer has the risk, even if the goods have not been delivered.[28] As a practical matter, the transfer of risk will almost always accompany delivery in the gas sale industry, but there may be situations when that will not be the case. Gas contractors for both sellers and buyers should understand that the concept of risk transfer, which is becoming an increasingly important issue for the gas industry, particularly where pipeline pools are concerned.

Transfer of title. When a sale is being made by a seller that does not hold title, regardless of the reasons (and might include some level of misconduct), the buyer only receives the title that was actually held by the seller *unless* the real title holder, by its own conduct, is precluded from denying the seller's authority to sell.[29] Gas brokers have a significant position in the gas industry, either acting as agents on behalf of producers, for example, or as finders, who bring sellers and buyer together. In either event, the actions of the third party gas owner may be significant in determining the type of title received by an innocent gas buyer. If a gas seller holds a "voidable" title, but the title has not been avoided at the time of sale, a good faith buyer without notice of the title defect will take good title to the gas.[30]

Part 3. Performance of Contract

Duties of seller and buyer. Sellers have a dual obligation to deliver the goods and the title, and buyers have a dual obligation to take the goods and pay for them.[31] Unless the contract specifies otherwise, the buyer must be ready and willing to make payment when the goods are delivered.[32] The contract should specify when and how transfer of the goods from seller's transporter or facilities will take place and when the buyer will take receipt of the goods. When the goods are in the possession of a third party, such as a transporting pipeline, no delivery occurs until the third party acknowledges to the buyer that it holds goods on the buyer's behalf.[33] The seller has an obligation to pay any expenses required to put the goods into a deliverable state.[34] In the gas industry this probably has most relevance to processing so that gas will meet the quality specifications of the buyer's transporter.

Rules regarding delivery. If a seller delivers less than the contract quantity, the buyer may reject the goods, but must pay for any portion actually taken.[35] If the quantity delivered is in excess of the contract quantity, the buyer has two options:

- accept the contract quantity and reject the rest, or
- reject the whole[36]

Installment deliveries. Many gas sale agreements are agreements to sell in installments. An example of this would be a one-year transaction, with daily delivery requirements and monthly billing and payment. When that is the case, the Act allows the parties to structure the sale contract so that the parties can agree whether a breach of one installment by either the seller or the buyer will be treated as a breach of the entire contract or a severable breach, giving rise to compensation payments instead.[37]

Delivery to carrier. When a seller delivers goods to the carrier for transport, the delivery is prima facie deemed to have been delivery of the goods to the buyer.[38] Interprovincial pipelines are designated as *carriers* pursuant to regulations of the NEB.[39]

Part 4. Rights of Unpaid Seller Against the Goods

The seller that has not been paid has several options to explore under the

Act. A seller is deemed to be an "unpaid seller" when the entire contract price has not been paid.[40] When a seller is classified as an unpaid seller it has a lien on the goods while the goods remain in the seller's possession;[41] upon buyer's insolvency, it has a right to stop goods in transit after the seller has given up possession,[42] and it has a limited right of resale.[43] If the property in goods hasn't yet transferred to the buyer, the seller also has a right to withhold delivery of the goods.[44]

When the unpaid seller has made partial delivery of the goods, it may exercise its lien rights or rights of retention on the remaining goods unless the partial delivery has been made in such a way to show the seller's intention to waive the lien or right of retention.[45]

The unpaid seller loses its lien or right to retain goods (i) when it delivers goods for transportation without reserving the right to dispose of the goods,[46] (ii) when the buyer obtains possession of the goods,[47] or (iii) by waiver.[48]

Stoppage of goods in transit. When a buyer of goods becomes insolvent during performance of a contract, the unpaid seller has a right to stop the goods in transit, take possession of the goods and hold them until payment or tender of the price is made.[49] Goods are deemed to be *in transit* from the time they are delivered to the carrier until the buyer takes possession.[50] The unpaid seller may exercise its right to stop goods in transit by actually taking possession of the goods or by giving notice to the carrier of its claim.[51] The expenses incurred because of such stoppage and any redelivery will be borne by the seller.[52]

Resale by buyer or seller. The rights of an unpaid seller regarding lien, retention or stoppage in transit are not affected by any sale of the goods that the buyer may have made unless the seller assented to such sale by the buyer.[53] But if a document of title to the goods has been lawfully transferred to any buyer, and that buyer again transfers the goods to another buyer who takes the title document in good faith and for valuable consideration, then the unpaid seller's rights mentioned above are defeated.[54] If the seller does stop in transit and resell the goods, however, that buyer acquires a good title as against the original buyer.[55]

Part 5. Actions for Breach of the Contract

Remedies of the seller. When goods have been delivered to the buyer and the buyer has neglected or refused to pay, the seller may bring an action against the buyer for the price of the goods.[56] This is true when payment is due on a specific date even if property in the goods has not passed and the goods have not been appropriated to the contract.[57] Sellers may collect interest on late payments from the date payment was due.[58]

If the buyer wrongfully refuses or neglects to take delivery and pay for the goods, the seller may maintain an action for damages for non-acceptance.[59] The measure of damages is limited to direct damages that result from the buyer's contract breach.[60] If a market exists for the goods not taken, then the measure of damages for buyer's non-acceptance will be the difference between the contract price and the market price at the time seller tendered delivery of the goods.[61]

Remedies for buyer. If the seller wrongfully refuses or neglects to deliver the goods, a buyer may maintain an action against the seller for non-delivery.[62] The measure of damages will be limited to direct losses resulting from the seller's failure.[63] The damages will be calculated as the difference between the contract price and the market price at the time goods were to be delivered when there is an available market for the goods.[64]

In circumstances considered appropriate by a court deciding a dispute against a non-performing seller, specific performance may be allowed.[65] In such a case, damages may not be collected. The court may render an unconditional judgment or its judgment may be made on terms and conditions deemed appropriate by the court.[66]

If the seller breaches any warranty or if the buyer treats a breach of condition as a breach of warranty, the buyer, in addition to rejecting the goods, may:

- sue the seller under a diminution of price theory[67] or
- maintain an action against the seller for breach of warranty[68]

The measure of damages for breach of warranty is the estimated direct loss.[69] If the warranty breach is a breach of quality, the measure of damages is the difference between the value of the goods at the time of delivery and the value the goods would have had if they had not breached the warranty.[70] Both the seller and the buyer may collect special damages and interest in appropriate cases and may recover money paid.[71]

Part 6. Supplementary

The parties to a contract may disclaim warranties that would otherwise automatically apply under the Act. This may be accomplished by express contractual agreement, or by a pattern of dealing between the parties, or by usage.[72]

The common law as it applies to law merchant, principal and agent, fraud, misrepresentation, duress or coercion, avoidance, mistake or other invalidating cause, continue to apply to the extent they are inconsistent with specific provisions of the Act.[73]

United Nations Convention on Contracts for the International Sale of Goods ("CISG" or "Convention")

The CISG Preamble states that the three purposes for the Convention are to:

- reduce the search for a forum with the most favorable law
- reduce the necessity of resorting to rules of private international law and
- provide a modern law of sales appropriate for transactions of an international character

The CISG was crafted to bring some certainty to international sales of goods between countries ratifying the convention. These countries are referred to as "states" in the CISG, and the U.S., Canada, and Mexico are all "contracting states," meaning all three have ratified the treaty, thus bringing the citizens of all three countries under the ambit of the Convention. The CISG is separated into 101 articles, and like the UCC and Sale of Goods Acts, the 101 articles follow fairly predictable patterns regarding topics—formation, performance, breach, and remedies.

Whenever private parties in different ratifying states sell and purchase goods, the CISG will override other law and be applicable unless the parties have followed the appropriate rules and desire that the Convention not apply. The rules for disclaiming the Convention may vary from contracting state to contracting state. In the U.S., the Convention must be specifically disclaimed in the sale contract itself. In Canada, the individual provinces have taken different approaches to this matter, as will be discussed in Article 6 of the CISG.

Part I. Sphere of Application and General Provisions

Chapter I. Sphere of application. Article 1 provides the general rules for determining whether the Convention is applicable to a contract of sale of goods as well as to its formation. The general application is to parties having their places of business in different states, but even if the parties have their places of business in different states, the Convention applies only if:

- the states in which the parties have their places of business are contracting states or
- the rules of private international law lead to the application of the law of a contracting state

If two parties from different states have designated the law of a contracting state as the law of the contract, this Convention is applicable even though the parties have not specifically mentioned the Convention. This is probably the most far-reaching aspect of the Convention—that parties merely by inaction or inattention to detail may be explicitly bound by the terms of the CISG.

Whether the Convention is applicable to a contract of sale of goods is determined primarily by whether the relevant "places of business" of the parties are in different Contracting states. The relevant "place of business" of a party is determined not by the nationality of the parties, but rather by the location of the business itself. If a contracting party has places of business in more than one country, the most relevant place of business is used to determine whether the Convention applies. For example, when the buyer is from the U.S. and the seller is Canadian, the Convention applies. If the Canadian party also has a place of business in the U.S., the Convention may or may not apply, depending upon which place of the business is *relevant* to those dealings. If the actual business with the U.S. party is done through its Canadian office, the Convention may not apply. Every situation is different.

In some legal systems the law relating to contracts of sale of goods is different depending on whether the parties or the contract are characterized as *civil* or *commercial*. No such distinction is found in the CISG, except under rare circumstances.[74]

Article 2 excludes certain types of sales from the provisions of the Convention. This Convention does not apply to sales:

> *"(a) of goods bought for personal, family or household use, unless the seller, at any time before or at the conclusion of the contract, neither knew nor ought to have known that the goods were bought for any such use*
> *(b) by auction*
> *(c) on execution or otherwise by authority of law*
> *(d) of stocks, shares, investment securities, negotiable instruments, or money*
> *(e) of ships, vessels, hovercraft, or aircraft*
> *(f) of electricity"*[75]

Article 3 deals with two different situations in which the contract includes some act in addition to the supply of goods. The first situation involves sellers that also supply materials for manufacture or production. The second is when services are also being provided under the sale contract. The Convention takes the approach that if the "preponderant part" of the obligation of the seller consists in the supply of labor or other services, the contract is not subject to the provisions of this Convention.[76]

Article 4 states that "*This Convention governs only the formation of the contract of sale and the rights and obligations of the seller and the buyer arising from such a contract. In particular, except as otherwise expressly provided in this Convention, it is not concerned with:*

> *(a) the validity of the contract or of any of its provisions or of any usage*

(b) the effect which the contract may have on the property in the goods sold"

In common practice, this article may be difficult to apply. Since the Convention does not address issues of contract validity, we must ask ourselves which law will be used to determine any validity issues. The only article in which the possibility of such a conflict is apparent is Article 10, which provides that a contract of sale of goods need not be in writing and is not subject to any other requirements as to form. In some legal systems, including those in the U.S. and Canada, the requirement of a writing for certain contracts of sale of goods is considered to be a matter relating to the validity of the contract. A contracting state whose legislation requires a contract of sale to be concluded in or evidenced by writing may make a declaration that, Article 10 will not apply where any party has his place of business in a contracting state that has made such a declaration.[77] The U.S., Canada, and Mexico have not made such declarations.

The Convention does not apply to liability of the seller for death or personal injury caused by the goods to any person,[78] and the parties may exclude the application of this Convention or, *subject to Article 12*, derogate from or vary the effect of any of its provisions.[79] When the U.S. ratified the Convention, it chose to accept this article as written, the result of which is that U.S. sales contracting parties must specifically disclaim the Convention in a relevant sales contract, or it will apply. Canada's ratification provided that the parties could exclude the application of the Convention "in accordance with the terms of the Convention and, in particular, by providing in the contract that other law applies in respect of the contract."[80] When individual provinces in Canada ratified the Convention, some of them altered this requirement.

The Alberta, New Brunswick, and Ontario Acts require the contract to state "that the local domestic law of (the enacting jurisdiction) or other jurisdiction applies to it or that the Convention does not apply to it." The Manitoba Act indicates that the parties may exclude the Convention "by expressly providing in the contract" that the Convention does not apply to it. Newfoundland allows the parties to exclude the Convention by expressly providing in the contract that the law of the province or another jurisdiction applies to it or that the Convention does not apply.[81] A lingering issue with regard to the federal and provincial actions taken in Canada is whether or not a court outside the relevant province would enforce the governing law clause that does not specifically exclude or disclaim the CISG. Article 98 of the CISG states: "No *reservations* are permitted unless expressly authorized in this Convention."

Chapter II. General provisions. Article 7 states: "*In the interpretation of this Convention, regard is to be had to its international character and to the need to promote uniformity in its application and the observance of good faith in international trade.*"[82] National rules on the law of sales of goods are subject to sharp divergence in approach and concept. Thus, it is especially important to avoid incompatible constructions of the provisions of this Convention by national courts, each dependent upon the concepts used in the legal system of the country of the forum. Courts must have due regard for the international character of the Convention and for the need to promote uniformity.

The obligation of *good faith* under the CISG is broadly written and applies to all aspects of contract interpretation and application. The concept is specifically used throughout the document in the following areas:

- Article 16(2)(b) on the non-revocability of an offer where it was reasonable for the offeree to rely upon the offer being held open and the offeree acted in reliance on the offer
- Article 21(2) on the status of a late acceptance which was sent in such circumstances that if its transmission had been normal it would have reached the offeror in due time
- Article 29(2) in relation to the preclusion of a party from relying on a provision in a contract that modification or abrogation (termination) of the contract must be in writing
- Articles 37 and 38 on the rights of a seller to remedy non-conformities in the goods
- Article 40 which precludes the seller from relying on the fact that notice of non-conformity has not been given by the buyer in accordance with Articles 38 and 39 if the lack of conformity relates to facts of which the

seller knew or could not have been unaware and which he did not disclose to the buyer
- Articles 49(2), 64(2) and 82 on the loss of the right to declare the contract avoided and
- Articles 85 to 88, which impose on the partys obligations to take steps to preserve the goods[83]

Article 8 on interpretation furnishes the rules to be followed in interpreting the meaning of any statement or other conduct of a party that falls within the scope of application of this Convention. Interpretation of the statements or conduct of a party may be necessary to determine (i) whether a contract has been concluded, (ii) the meaning of the contract, or (iii) the significance of a notice given or other act of a party in the performance of the contract or in respect of its termination.

Article 8 provides the rules to be applied in terms of interpreting the unilateral acts of each party, *i.e.*, communications in respect of the proposed contract, the offer, the acceptance, notices, etc. Nevertheless this article is equally applicable to the interpretation of "the contract" when the contract is embodied in a single document. Analytically, this Convention treats such an integrated contract as the manifestation of an offer and an acceptance. Therefore, for the purpose of determining whether a contract has been concluded, as well as for the purpose of interpreting the contract, the contract is considered to be the product of two unilateral acts.[84] Article 8 cannot be applied if the party who made the statement or engaged in the conduct had no intention on the point in question, or if the other party did not know and had no reason to know what that intent was. In such a case Article 8(2) provides that the statements made by and the conduct of a party are to be interpreted according to the understanding that a reasonable person (of the same kind as the *other party*) would have had in the same circumstances.

In determining the intent of a party or the intent a reasonable person (of the same kind as the other party) would have had in the same circumstances, it is necessary to look first to the words actually used or the conduct engaged in. However, the investigation is not to be limited to those words or conduct even if they appear to give a clear answer to the question. It is common experience that a person may dissimulate or make an error and the process of interpretation set forth in this article is to be used to determine the true content of the communication. If, for example, a Canadian party offers to sell a quantity of gas for $2.50/MMBtu and it is obvious that the offeror intended Can. $2.50/MMBtu and not U.S. $2.50/MMBtu, and the offeree knew or could not have been unaware of it, the price term in the offer is to be interpreted as Can.$2.50/MMBtu.

In order to go beyond the apparent meaning of the words or the conduct by the parties, Article 8(3) states that *"due consideration is to be given to all relevant circumstances of the case."* It then goes on to enumerate some, but not necessarily all circumstances of the case that are to be taken into account. These include (i) the negotiations, (ii) any practices that the parties have established between themselves, (iii) usages and any subsequent conduct of the parties.

Article 9 states that *"(the) parties are bound by any usage to which they have agreed and by any practices which they have established between themselves."*[85] Unless otherwise agreed, the parties are considered to have impliedly made applicable to their contract or its formation a usage of which the parties knew or ought to have known and which in international trade is widely known to, and regularly observed by, parties to contracts of the type involved in the particular trade concerned.[86]

The determining factor whether a particular usage is to be considered as having been impliedly made applicable to a given contract will often be whether it was *widely known to, and regularly observed by, parties to contracts of the type involved in the particular trade concerned*. In such a case it may be held that the parties "ought to have known" of the usage. The important aspect of this provision, from the point of view of an organization bound by the terms of the CISG is that the usages are based upon international standards, which may differ from local standards.

Article 11 specifies that sales contracts need not be in writing and that any type of proof may be used to determine whether the contract exists, including the testimony of witnesses. As stated earlier in this discussion, some states consider the requirement that contracts for the international sale of goods be in writing to be a matter of important public policy. Accordingly, Article 12 provides a mechanism for contracting states to prevent the application of the rule in Article 11 to transactions where any party has a place of business in their state.

Part II. Formation of the Contract

Article 14 states the types of communication necessary to constitute an offer—*i.e.*, the form of the offer, and the Convention differentiates between offers that are made to one person and offers made to multiple persons. In order for the proposal for concluding a contract to constitute an offer it must indicate "*the intention of the offeror to be bound in case of acceptance.*" Since there are no particular words that must be used to indicate such an intention, it may sometimes require a careful examination of the "offer" in order to determine whether such an intention existed. This is particularly true if one party claims that a contract was concluded during negotiations that were carried on over an extended period of time, and no single communication was labeled by the parties as an "offer" or as an "acceptance." Whether there is the requisite intention to be bound in case of acceptance will be established in accordance with the rules in Article 8 discussed earlier.

An offer must indicate the type of goods, the quantity, and the price. Quantity needn't be a total unit representation if the contract is for all production or for all fuel needs, for example. To constitute an offer, a proposal must expressly or implicitly fix or make provision for the price. It is not necessary that the price could be calculated at the time of the conclusion of the contract. For example, the offer, and the resulting contract, might call for the index price on the date of delivery, which date might be months or even years in the future.[87]

An offer becomes effective when it reaches the offeree,[88] and even if it is irrevocable, may be withdrawn if the withdrawal reaches the offeree before or at the same time as the offer.[89]

Until a contract is concluded an offer may be revoked if the revocation reaches the offeree before he has dispatched an acceptance. However, an offer cannot be revoked:

- if it indicates, whether by stating a fixed time for acceptance or otherwise, that it is irrevocable or
- if it was reasonable for the offeree to rely on the act in reliance on the offer[90]

Once an offer has been made, the Convention indicates the form of rejection or acceptance that will be allowed. An offer, even if it is irrevocable, is terminated when a rejection reaches the offeror.[91]

A statement or other conduct of the offeree indicating assent to an offer is an acceptance. Silence or inactivity does not in itself amount to acceptance.[92] However, under Article 18(3) if, by virtue of the offer or as a result of practices that the parties have established between themselves or of usage, the offeree may indicate assent without giving notice to the offeror by performing an act, such as nominating gas for flow or payment of the price. In that instance, the acceptance is effective at the moment the act is performed. Since the acceptance is effective and the contract is concluded at the moment the act is performed, the right of the offeror to revoke his offer terminates at that same moment.

Article 17 provides that an offer is terminated when the rejection reaches the offeror. In a case where the offer is accepted by a written indication of assent (such as by signing and returning a transaction confirmation), Article 16(1) provides that acceptance is effective at the moment the offeree has *sent* the confirmation, and not at the moment the confirmation reaches the offeror.

The Convention follows other rules of revocation as well. Article 16(2)(a) provides that an offer cannot be revoked if it indicates that it is irrevocable. It should be noted that this provision does not require a promise on the part of the offeror not to revoke his offer nor does it require any promise, act, or forbearance on the part of the offeree for the offer to become irrevocable. It reflects the judgement that in commercial relations, and particularly in international commercial relations, the offeree should be able to rely on any statement by the offeror that indicates the offer will be open for a period of time.

The offer may indicate that it is irrevocable in different ways. The most obvious is that the offer may state it is irrevocable, that it will not be revoked for a particular period of time, or that it is irrevocable by stating a fixed time for acceptance.[93]

What happens under the Convention when an offer is accepted, but the offeree changes some of the terms? A reply to an offer that purports to be an acceptance but contains additions, limitations or other modifications that materially alter the terms of the contemplated contract is a rejection of the offer and constitutes a counter-offer.[94] However, a reply to an offer that purports to be an acceptance but contains additional or different terms that do not materially alter the terms of the offer constitutes an acceptance. This will be so unless the offeror, without undue delay, objects orally to the discrepancy or notifies the offeree to that effect. If the offeror does not so object, the terms of the contract are the terms of the offer with the modifications contained in the acceptance.[95]

Additional or different terms relating, among other things, to the price, payment, quality and quantity of the goods, place and time of delivery, extent of one party's liability to the other, or the settlement of disputes are considered to alter the terms of the offer materially and will result in a rejection and counteroffer.[96]

Part 3. Sale of Goods

Chapter 1. General provisions. *"A breach of contract committed by one of the parties is fundamental if it results in such detriment to the other party as substantially to deprive him of what he is entitled to expect under the contract, unless the party in breach did not foresee and a reasonable person of the same kind in the same circumstances would not have foreseen such a result."*[97] The definition of fundamental breach is important because various remedies of buyer and seller,[98] as well as some aspects of the passing of the risk,[99] rest upon it.

The basic criterion for a breach to be fundamental is that "it results in substantial detriment to the injured party." The determination whether the injury is substantial must be made in the light of the circumstances of each case, *e.g.*, the monetary value of the contract, the monetary harm caused by the breach, or the extent to which the breach interferes with other activities of the injured party.[100]

Once this basic criterion is met, a breach is fundamental unless the party in breach can prove that it "did not foresee and had no reason to foresee such a result," *i.e.*, the result that did occur. It should be noted that the party in breach does not escape liability merely by proving that it did not in fact foresee the result. He must also prove that he had no reason to foresee it.[101]

If, in accordance with the provisions of the Convention, one party is entitled to require performance of any obligation by the other party, a court is not bound to enter a judgement for specific performance unless the court would do so under its own law in respect of similar contracts of sale not governed by this Convention.[102]

Unless the contract specifically requires amendments or termination to be in writing, the Convention allows both modification and termination upon agreement.[103] In addition, Article 29(1) is applicable when the terms in a confirmation from or in an invoice sent by one party to the other modify the contract where those terms are additional or different from the terms of the contract as it was concluded. If it is found that the parties have agreed to the additional or different terms, Article 29(1) provides that they become part of the contract.[104]

Chapter II. Obligations of the seller. Article 30 provides that *"(the) seller must deliver the goods, hand over any documents relating to them and transfer the property in the goods, as required by the contract and this Convention."* One of the major distinctions between the CISG and other sale of goods laws is found in the anticipation of international delivery of goods. Under both the UCC and Sale of Goods Acts, the seller's obligation is to "sell and deliver" the goods. In this context, the concept of documents necessary for international trade is introduced. This discussion will focus only on the obligation to sell and deliver.

The seller will typically have a contractual obligation to deliver natural gas at a specified delivery point. Once the seller has given goods—in this case natural gas—to the first transporter, it has completed its delivery obligation.[105] The seller must give the buyer notice that it has given the goods to the transporter, which will be accomplished through the nomination process.[106]

The seller must deliver the goods on the date(s) specified in the contract[107] and the goods must be of the quantity, quality, and description required by the contract.[108]

The parties can contractually agree otherwise, but if they do not do so, the goods do not conform with the contract unless they:

- *"are fit for the purposes for which goods of the same description would ordinarily be used*

- *are fit for any particular purpose expressly or impliedly made known to the seller at the time of the conclusion of the contract, except where the circumstances show that the buyer did not rely, or that it was unreasonable for him to rely, on the seller's skill and judgement*
- *possess the qualities of goods which the seller has held out to the buyer as a sample or model*
- *are contained or packaged in the manner usual for such goods or, where there is no such manner, in a manner adequate to preserve and protect the goods*"[109]

The seller will not be liable for any nonconformity breach of the preceding paragraph if at the time of the conclusion of the contract the buyer knew or could not have been unaware of such lack of conformity.[110]

A seller will be liable in accordance with the contract and the Convention for any lack of conformity existing at the time the risk passes to the buyer, even though the lack of conformity becomes apparent only after that time. In addition, the seller will be liable for any lack of conformity occurring after the time risk of loss passes if due to a breach of any of the seller's obligations. Those may include a breach of any guarantee that for a period of time the goods will remain fit for their ordinary purpose or for some particular purpose or will retain specified qualities or characteristics.[111]

The buyer loses the right to rely on a lack of conformity of the goods if it does not give notice to the seller specifying the nature of the lack of conformity within a reasonable time after it has discovered it or ought to have discovered it.[112] In any event, the buyer loses that right if notice is not given to the seller within two years after the non-conforming goods were delivered.

Article 41 provides that the seller must deliver goods free from any third-party claim unless the buyer knew of the claim and took delivery of the goods anyway. The buyer loses any right to place the seller in breach of its Article 41 obligations if it fails to provide timely notice to the seller after it has learned (or ought to have known of) of the third-party claim.[113]

Seller's breach. If the seller fails to perform any of its obligations under the contract or this Convention, the buyer may exercise a number of remedial rights. Those may include specific performance,[114] requiring substituted performance if the goods are nonconforming,[115] agreeing to an additional time for the seller to perform its obligations,[116] terminating the contract if the seller's breach was fundamental,[117] or reducing the price for nonconforming goods if it has not utilized other available remedies.[118]

After a breach of an obligation by the seller, the buyer's principal concern is often that the seller perform the contract as originally promised. Legal actions for damages cost money and may take a considerable period of time. Moreover, if the buyer needs the goods in the quantities ordered, it may not be able to make substitute purchases in time.

Although the buyer has a right to the assistance of a court or arbitral tribunal to enforce the seller's obligation to perform the contract, Article 28 limits that right to a certain degree. If the court could (or would) not give a judgement for specific performance under its own law in respect of similar contracts of sale not governed by this Convention, it is not required to enter such a judgment in a case arising under this Convention. This is true even though the buyer had a right to require the seller's performance under Article 46.

One measure that may be available to a buyer to enforce the seller's contractual performance obligation would be a clause requiring the seller to pay the buyer a specific sum of money, *e.g.*, $50,000.00, if the seller fails in a delivery-related obligation. This type of clause sometimes referred to as a "liquidated damages clause" serves the function of estimating the damages that the buyer would suffer as a cause of the breach to ease the problems of proof. All legal systems appear to recognize the validity and social utility of a clause which estimates future damages, especially where proof of actual damage would be difficult.[119]

Since the parties may agree to alternative methods for remedying the harm caused by a seller's failure, the buyer is precluded during any agreed period for seller's cure, from exercising its rights enumerated in the Convention.[120] When the buyer has resorted to a remedy for seller's breach of the contract, a court or other legal body hearing the dispute may not grant any grace period not specified in the contract.[121]

If the seller delivers more than the contract quantity, the buyer may accept the excess delivery and payment will be at the contract rate.[122]

Chapter III. Obligations of the buyer. Article 53 specifies that the buyer must pay the price for the goods and take delivery as required by the contract and the Convention. Article 54 states that, as part of the buyer's obligation to pay the price, it must take the steps and comply with any formalities required by the contract and by any relevant laws and regulations to enable payment to be made. These steps may include (i) applying for a letter of credit or a bank guarantee of payment, (ii) registering the contract with a government office or with a bank, (iii) procuring the necessary foreign exchange or (iv) applying for official authorization to remit the currency abroad. Unless the contract specifically places one of these obligations on the seller, it is the buyer who must take the steps.[123]

The buyer's obligation under Article 54 is limited to taking action to comply with formalities. The buyer is not required to ensure that its efforts will result in the issuance of a letter of credit, the authorization to procure the necessary foreign exchange, or even that the price will finally be paid.[124] The major significance of Article 54 lies in the fact that taking such actions and complying with such formalities that may be required to enable payment to be made is considered to be a current obligation, breach of which may give rise to remedies indicated in Articles 61 to 64, discussed later.

Article 59 requires the buyer to pay the price on the date fixed by or determinable from the contract without the need for any request or compliance with any formality on the part of the seller. Strictly interpreted, this means that the seller is not required to submit an invoice prior to the obligation of buyer to make timely payment.

Article 62 states that if the buyer fails to perform any of its obligations, the seller may require the buyer to pay the price, take delivery, or perform its other obligations, unless the seller has resorted to a remedy that is inconsistent.[125] This differs from the law of many countries (including the U.S.) in which the seller's remedies in respect of the price are restricted. In the U.S., for example, even though the buyer may have a substantive obligation to pay under the contract, the general principle is that the seller must make a reasonable effort to resell the goods to a third party and recover as damages any difference between the contract price and the price it received in the substitute transaction. The seller may recover the price if resale to a third person is not reasonably possible. The concept of "cover" is not a part of the CISG.

Buyers breach. If the buyer breaches, the seller may also fix an additional period of time of reasonable length for performance by the buyer of its obligations,[126] or declare the contract avoided if the buyer's breach is fundamental or if any extended time for performance granted by the seller has expired without performance.[127]

If the buyer is contractually required to take measurements and fails to provide measurement information either on the date agreed upon or within a reasonable time after receipt of a request from the seller, the seller may, without prejudice to any other rights it may have, use its own measurements.[128]

Part IV. Passing of Risk

If the seller is bound to hand the goods over to a carrier (a transporter) at a particular place, the risk does not pass to the buyer until the goods are handed over to the carrier at that place. This may occur at the designated delivery point but may also occur at the seller's receipt point into the pipeline unless the contract specifically designates the point at which risk passes. Loss of or damage to the goods after the risk has passed to the buyer does not discharge the buyer from its obligation to pay the price, unless the loss or damage is due to an act or omission of the seller.[129] The risk in respect of goods sold in transit passes to the buyer from the time of the conclusion of the contract.[130] This article may raise some interesting questions when natural gas is bought and sold midstream, rather than at the wellhead or at a point of consumption. But, if the circumstances so indicate—usually through specific risk-transfer language in the contract, the risk is assumed by the buyer from the time the goods are handed over to the transporter.[131]

Provisions Common to the Obligations of Buyers and Sellers

If, after a contract has been made, it becomes apparent to either the seller or the buyer that the other will not perform its obligations, either through a failure of creditworthiness or by insufficient preparation to perform, that party may suspend its performance.[132] But, if the other party gives assurance of its ability to perform, the suspension must be lifted.[133]

In the case of installment contracts, let's take the example of a two-year gas sale commitment that is nominated, delivered, and paid for on a monthly basis. A fundamental breach of one installment may be either a breach of that installment or a breach of the entire contract, depending on the circumstances.[134]

Article 72(1) provides for the special case where prior to the date for performance, it is clear that one of the parties will commit a fundamental breach. In such a case the other party may declare the contract avoided immediately. The future fundamental breach may be clear in either of two circumstances:

- because of the words or actions of the party that constitute a repudiation of the contract or
- because of an objective fact, such as the destruction of the seller's plant by fire or the imposition of an embargo or monetary controls which will render future performance impossible[135]

A word of caution is wise at this point. Even though a party has the right to declare the contract avoided under the indicated circumstances under Article 72, this should be done with caution since an error in this regard may result in the declaring party breaching the contract itself.

Damages for breach of contract by one party consist of a sum equal to the loss, including loss of profit, suffered by the other party as a consequence of the breach. Any such damages may not exceed the loss that the party in breach foresaw or ought to have foreseen at the time of the conclusion of the contract. This is viewed in the light of the facts and matters of which breaching party then knew or ought to have known, as a possible consequence of the breach of contract.[136]

A party who relies on a breach of contract must take reasonable measures to mitigate the loss, including loss of profit, resulting from the breach. If the non-failing party does not mitigate its damages, the party in breach may claim a reduction in the damages in the amount by which the loss should have been mitigated.[137]

Articles 79 and 80 address force majeure. Under Article 79 a non-performing party is exempt from liability if it proves that:

- the failure to perform was due to an impediment beyond its control
- it could not reasonably be expected to have taken the impediment into account at the time of the conclusion of the contract
- it could not reasonably have been expected to have avoided the impediment or its consequences and

- it could not reasonably have been expected to have overcome the impediment or its consequences

The fourth element is the most difficult for the non-performing party to prove. As stated in the force majeure discussion in chapter 8, all potential impediments to the performance of a contract are foreseeable to one degree or another. Wars, storms, fires, government embargoes, and the closing of international waterways have all occurred in the past and can be expected to occur again in the future. As is common in the natural gas sale industry, the parties will have explicitly stated whether the occurrence of the impending event would exonerate the non-performing party from the consequences of the non-performance. In other cases it is clear from the context of the contract that one party has obligated itself to perform an act even though certain impediments might arise, as evidenced in the exclusion clauses common to force majeure. If such contractual stipulations are made, they will be enforceable under the Convention.

Gas contractors need to know the basic structure and the parameters of the Convention. While the provisions of the CISG are similar to both the UCC and the provincial Sales of Goods Acts, the underlying principle regarding the Convention is a little different. The CISG relates to international sales of goods and international (rather than local) usages, customs, and characteristics regarding sales of goods will implicitly be a part of the contracting parties' understandings.

---

1. United Nations Convention on Contracts for the International Sale of Goods. 52 Fed. Reg. 6262, 6264, 6280 (March 2, 1987); 15 U.S.C.A., Appendix (Supp. 1987). Also see Chapter 4, fn. 5.

2. See chapter 3 discussion and fn. 5.

3. An excellent overview of the work of the Uniform Law Conference of Canada can be found at http://www.law.ualberta.ca.

4. Sale of Goods Act, R.S.A., 1980, c. S-2.

5. Sale of Goods Act, § 2 (1).

6. Sale of Goods Act, § 2 (2).

7. Sale of Goods Act, § 1 (n).

8. Sale of Goods Act, § 3 (4) and (5).

9. Sale of Goods Act, § 7 (1)(b).

10. Sale of Goods Act, § 7 (3)

11. Sale of Goods Act, § (12) (1). When an agreement is avoided, it is cancelled.

12. At first sight; a fact presumed to be true unless disproved by some evidence to the contrary, *Black's Law Dictionary*, Revised Fourth Edition, 1978.

13. Sale of Goods Act, § 14 (1)

14. Sale of Goods Act, § 14 (3)

15. Repudiation means a rejection or disclaimer; the rejection or refusal of any offered right or privilege. *Black's Law Dictionary*, Revised Fourth Edition, 1978. See also the discussion of contractual anticipatory repudiation under the UCC in chapter 4.

16. Sale of Goods Act, § 14 (2).

17. Sale of Goods Act, § 15 (a).

18. Sale of Goods Act, § 15 (b). Quiet possession means that the possession will be unmolested and undisturbed.

19. Sale of Goods Act, § 15 (c).

20. See the discussion of implied warranties of fitness for a particular purpose and merchantability in chapter 4.

21. Sale of Goods Act, § 17 (2).

22. Sale of Goods Act, § 71 (4).

23. Sale of Goods Act, § 17 (7).

24. Sale of Goods Act, § 19.

25. Sale of Goods Act, § 20.

26. Sale of Goods Act, § 21 (2).

27. Sale of Goods Act, § 21 (4).

28. Sale of Goods Act, § 21 (7) (b).

29. Sale of Goods Act, § 23 (1).

30. Sale of Goods Act, § 24 (1).

31. Sale of Goods Act, § 25.

32. Sale of Goods Act, § 28.

33. Sale of Goods Act, § 29 (b).

34. Sale of Goods Act, § 30 (5).

35. Sale of Goods Act, § 30 (8).

36. Sale of Goods Act, § 31 (1).

37. Sale of Goods Act, § 31 (2).

38. Sale of Goods Act, § 32 (2).

39. Sale of Goods Act, § 33 (1).

40. See the discussion of pipelines in Canada in chapter 3.

41. Sale of Goods Act, § 39 (1)(a).

42. Sale of Goods Act, § 40 (1)(a).

43. Sale of Goods Act, § 40 (1)(b).

44. Sale of Goods Act, § 40 (1) (c).

45. Sale of Goods Act, § 40 (2).

46. Sale of Goods Act, § 42.

47. Sale of Goods Act, § 43 (1) (a).

48. Sale of Goods Act, § 43 (1) (b).

49. Sale of Goods Act, § 43 (1) (c).

50. Sale of Goods Act, § 44.

51. Sale of Goods Act, § 45 (1).

52. Sale of Goods Act, § 46 (1).

53. Sale of Goods Act, § 46 (4).

54. Sale of Goods Act, § 47 (1).

55. Sale of Goods Act, § 47 (2) (a).

56. Sale of Goods Act, § 48 (2).

57. Sale of Goods Act, § 49 (1).

58. Sale of Goods Act, § 49 (2).

59. Sale of Goods Act, § 49 (3).

60. Sale of Goods Act, § 50 (1).

61. Sale of Goods Act, § 50 (2).

62. Sale of Goods Act, § 50 (3).

63. Sale of Goods Act, § 51 (1).

64. Sale of Goods Act, § 51 (2).

65. Sale of Goods Act, § 51 (3).

66. Sale of Goods Act, § 52 (1).

67. Sale of Goods Act, § 52 (2).

68. Sale of Goods Act, § 53 (1) (a).

69. Sale of Goods Act, § 53 (1) (b).

70. Sale of Goods Act, § 53 (2).

71. Sale of Goods Act, § 53 (3).

72. Sale of Goods Act, § 54.

73. Sale of Goods Act, § 55.

74. Sale of Goods Act, § 59 (1).

75. CISG Article 1 (3).

76. CISG Article 2.

77. CISG Article 3 (2).

78. See Secretariat Commentary to the 1978 draft of the CISG, Article 4, n.3.

79. CISG Article 5.

80. CISG Article 6.

81. Bill C-81, 1992 session.

82. Jacob Ziegel, "Canada Prepares to Adopt the International Sales Convention", 18 *Canadian Bus. L.J.* (1991).

83. CISG Article 7 (1).

84. See Secretariat Commentary to the 1978 Draft of the CISG, Article 7, n. 3.

85. See Secretariat Commentary to the 1978 draft of the CISG, Article 8, n. 2.

86. CISG Article 9 (1).

87. CISG Article 9 (2).

88. See Secretariat Commentary to the 1978 draft of the CISG, Article 14, n. 14.

89. CISG Article 15 (1).

90. CISG Article 15 (2).

91. CISG Article 16.

92. CISG Article 17.

93. CISG Article 18 (1).

94. See Secretariat Commentary to the 1978 draft of the CISG, Article 16, n. 7.

95. CISG Article 19 (1).

96. CISG Article 19 (3).

97. CISG Article 19 (3).

98. CISG Article 25.

99. See CISG Articles 46(2), 48(1), 49(1)(a), 51(2), 64(1)(a), 72, 73(1) and 73(2).

100. See CISG Article 70.

101. See Secretariat Commentary to the 1978 draft of the CISG, Article 25, n. 3

102. See Secretariat Commentary to the 1978 draft of the CISG, Article 25, n. 4.

103. CISG Article 28.

104. CISG Article 29.

105. See Secretariat Commentary to the 1978 draft of the CISG, Article 29, n. 4.

106. CISG Article 31 (a).

107. CISG Article 32 (1).

108. CISG Article 33 (a).

109. CISG Article 35 (1).

110. CISG Article 35 (2).

111. CISG Article 35 (3).

112. CISG Article 36.

113. CISG Article 39.

114. CISG Article 43 (1).

115. CISG Article 46 (1).

116. CISG Article 46 (2).

117. CISG Article 47 (1)

118. CISG Article 49 (1) (a).

119. CISG Article 50.

120. See Secretariat Commentary to the 1978 draft of the CISG, Article 46, n.10.

121. CISG Article 48 (2).

122. CISG Article 45 (3).

123. CISG Article 52 (2).

124. See Secretariat Commentary to the 1978 draft of the CISG, Article 54, n.2.

125. See Secretariat Commentary to the 1978 draft of the CISG, Article 54, n. 3.

126. CISG Article 62.

A Practical Guide to Gas Contracting

127. CISG Article 63 (1).

128. CISG Article 64 (1).

129. CISG Article 65 (1).

130. CISG Article 66.

131. CISG Article 67.

132. CISG Article 68.

133. CISG Article 71 (1).

134. CISG Article 71 (3).

135. CISG Article 71 (1) and (2).

136. See Secretariat Commentary to the 1978 draft of the CISG, Article 72 (1), n. 1.

137. CISG Article 74.

138. CISG Article 77.

Chapter 6: Essentials of Contracting

When is a deal a deal? What terms bind the parties to perform once the deal is done and a contract made? The most common contract dilemmas facing gas sellers and buyers today relate to one of the "terrible two" questions—"Do we have a deal?" or "What are the contract terms?" Once a situation giving rise to either question occurs in your organization, the answers may not be to management's liking. The first part of this chapter gives you the tools to respond when either of the terrible two questions is asked. These tools include good business practices, good contract and pre-contract planning, and knowledge of the UCC and evidentiary rules.

The second part of this chapter focuses on general issues relating to contract performance. What are the obligations and rights of the parties during performance? What would a court scrutinize if any of the performance terms of your contract do not clearly indicate the parties' intentions? This section will also include some recommendations for sound business practices and good contract drafting.

Finally, in the last section of this chapter you'll learn the rules traditionally followed when contract performance unravels. What is a breach of contract and what happens when one occurs? What are your rights and remedies available under the UCC? How may you change the UCC approach through your own contract language? These and other questions will be answered and I'll give you some helpful hints for resolving many of the problems that may occur in contract formation, performance, and remedies. The UCC generally provides good guidance in this area, and with good contract drafting, nearly every type of contract failure may be anticipated and addressed accordingly.

Some problems are easy to solve, but others are not, and may leave the designated problem-solver wondering where to begin. For a complex contract issue, the solution may have a business (or operations) component and a contractual component, so try this. Divide the problem into three categories as follows:

- Identify the source of a problem
- Determine the nature of the resulting contractual problem
- Plan your path for resolving all issues

During the following discussions, we will return to this formula to illustrate its use for your own difficult or complex situations.

Six Routine Practices that Cause Contract Problems

Some of the most common practices in the gas sales industry also cause a majority of its contracting problems. The industry is prone to mistakes and adjustments, but to the extent possible, gas contractors can make the necessary adjustments less severe through solid contract language. By understanding and then following some basic contract rules, you may be able to steer your organization through even the most dangerous contract territory. Do any of the following practices sound familiar to you?

1. Day traders in your organization buy and sell gas on short-term bases, usually limited to less than a week. The only recorded evidence of the deals done is found in tape recordings of trader telephone conversa-

tions. While they have been given a list of items to include in all transactional telephone conversations, sometimes the traders forget. They may turn the recording device on and off, depending on the nature of the telephone conversation. The actual tape records are maintained for up to six months, then reused.

2. One of the traders in your organization, habitually "does deals" without putting anything in writing. This trader may be successful and innovative, highly motivated, and may generally feel that he or she isn't bound by the rigors of the established contract process because the most important thing is to do the deal.

3. Your *origination group*[1] typically spends several months negotiating potential multi-year sales to cogeneration facilities, industrial consumers, and LDCs where gas restructuring has not occurred. During the negotiation process, it is common to utilize *letters of intent*[2] that specify the nature of preliminary discussions and specifically state that no *contract* will be formed between the parties until final management approval of both organizations has been given and a formal contract document signed.

4. Your organization requires written confirmations of all transactions and also tapes telephone conversations between traders. The form contract you use states that in the event of a conflict between the written confirmation and a taped recording of the transaction conversation, the tape will prevail for evidentiary purposes.

5. Your organization has an electronic energy management system, facilitating multiple applications of transactional information entered into the system. Once entered, the information for a transaction is automatically made available for various applications, including accounts receivable or payable, volume administration, dispatching, contract administration, credit, etc.

6. The middle office and back office staff has been reduced but the number

of transactions done on a daily basis by your organization has risen dramatically during times of downsizing and layoffs. You use fax confirmations as the record of those transactions, but the sheer number of confirmations coming across your fax machine has become overwhelming. You do not have enough employees to assure timely confirmation checks.

The list of six could be much longer, but these practices are all common and seem to cause a majority of routine and frequent problems. Significant contract disputes could occur as a result of any of the six practices. In the following pages, you will learn how the law (particularly the UCC) could be used to resolve problems arising out of the six practices.

Two Most Common Contract Issues

For all contracting processes, your primary guiding principle, whether writing, negotiating, or revising a contract, is that your contract be *enforceable*. Two basic enforceability issues occur in gas contracting, and nearly every contract dispute is based on one of the two basic issues (or a variation of either). The first is *contract formation*, and relates to the legal requirements for a contract to have been formed. If a contract hasn't been legally formed, nothing exists to enforce. In that instance, disputes usually arise when one party thinks there was a deal and the other one doesn't.

The second problem occurs when a contract was formed, but the exact *terms are unclear*, either because contract clauses were poorly written, or more commonly, when the written communications between contracting parties don't agree. In the former case, the parties, and any court hearing a disputed matter, will look to the intent of the parties when the contract was made. In the latter case, the question becomes "Which terms are applicable?"

In chapter 4, you learned that the three essential elements of offer, acceptance, and consideration must be present for any contract to exist. You also learned that general contract formation rules under Article 2 of the UCC may be different for Merchants and for non-Merchants.[3] In addition to essential contract elements and the role of Merchants under the Code, a number of additional considerations should guide a gas contractor, both in pre-contract stages and when the contract is negotiated.

When I am advising my clients on gas contracts, my responsibility is to craft contract language that works—clients want the deal to work, and contract language can help pave the way to accomplish that goal. Admittedly, one of the principle motivators behind most of my suggestions for specific contract language is based upon the proposition that the contract may someday be litigated or arbitrated. For the rest of the negotiation team, a general mood of optimism generally reigns, anticipating that all will run smoothly, while I concentrate on what happens when something goes wrong. That involves taking into account not only contract laws, but also the rules of evidence that will be followed in any subsequent litigation. With the UCC and basic evidence rules in mind, here is how I approach contracting. Hopefully, you will benefit from my experience, gained in part through the lessons I've learned from my own mistakes through the years.

Contract Formation

Enforceability

Do contracting parties have legal capacity: In addition to the formal requirements of offer, acceptance, and consideration required for a contract, contracting parties must have the *legal capacity* to enter into contracts. Without having the required capacity, contracts signed by such parties will not be enforceable. In fact, there is no contract when one of the parties does not have (or neither of the parties has) the legal capacity to enter into that transaction. In the business world, legal capacity is a twofold concept—*power* and *authority*.

Power: The rules governing the legal *power* of natural persons relates to the age of that person, the mental capacity of that person, and other attributes directly related to an individual's ability to comprehend the transaction itself, and natural persons having legal capacity may freely contract, usually without restraint. Power to contract is not without bounds for any entity created solely pursuant to laws authorizing establishment of such entities. This group includes corporations, partnerships, and government subdivisions.

To have the required legal capacity, all such non-natural "persons" must have been legally created and continue to validly exist. Corporations, partnerships, governmental entities, and other artificially created bodies satisfy the capacity requirement as long as the legal formation and status requirements are fulfilled. Corporations, partnerships, and other entities that exist only because of organizational documents or statutory authority may be limited in the ways they can contract, both in the type of contracts that may legally be executed and in the level of commitment to which the corporation may be bound. The latter is especially true with governmental units.

Corporations. Corporations exist in different forms. There are not-for-profit corporations that exist to provide an educational, social, religious, or other community service and that are usually (but not always) tax exempt. For-profit corporations exist for the purpose of providing value to shareholders and the liability of such organizations is limited to corporate assets. Many states have passed limited liability corporation laws that allow corporations authorized under such laws to further limit their liability, as well.

Corporations become legal entities through a formal incorporation process in just one state, even though the company may have offices in many states. In every state where it maintains a sufficient business presence, a corporation is required to register, usually through the office of secretary of state and to maintain that registration annually. It is common for a corporation to be incorporated in the state of its home office, but the state of Delaware has attractive corporation laws, so many organizations incorporate in Delaware, with no other connection to that state.

Any corporation's total power is found in its *articles of incorporation*, *bylaws*, and any subsequent action taken by the board of directors such as *corporate resolutions*, and unanimously agreed actions taken outside the parameters of a regular or special meeting. For-profit corporations are typically given a blanket right to enter into contracts in the article of incorporation. Thus, the organizational documents that must be filed with the secretary of state in the state of incorporation reveal the extent of original corporate powers.

Don't expect to find anything too unusual in public corporate documents. Much of the incorporation process is subject to strict statutory requirements regarding content and form, and most such documents are extremely broad in terms of the corporation's powers. If a power isn't specifically given through formal documents, the corporation doesn't have that power.

Corporations typically exist on a perpetual basis, but must follow filing and fee-payment requirements for maintaining their corporate status, both in the state of incorporation, and in states where they are doing business. How do you determine whether a corporation has maintained its legal status? First, you may contact the secretary of state's office (sometimes this is handled through a corporate division of the secretary of state) in the state of incorporation and ask whether the corporation is currently registered and in good standing in that state.

If your dealings with the corporation are through an office in a state other than the state of incorporation, or if the home office is located in another state, contact the secretary of state in the relevant location to see if the corporation is currently registered and in good standing. This seems like a time-consuming matter, so would you do this every time your organization does business with a new company? Probably not, particularly if the new company is an affiliate or subsidiary of a company you recognize and for which your credit department may already have the information you need.

When you are negotiating a contract with a corporation with which you are not familiar, take the time to call—or check various services providing access to the same information through the Internet.[4] It may be well worth your time to spend those few minutes on the phone or Internet in the long run, because you can't enforce your contractual rights if you cannot identify or find the other party. If a corporation fails to maintain its status, you may be left with essentially nothing to sue.

Partnership. A partnership's total power is found in the *partnership agreement*. Partnership agreements are usually confidential documents and will be subject to the partnership laws in the state of organization. All of a partnership's powers are created in the partnership agreement, meaning that specific authority to enter into contracts must be given somewhere in the partnership agreement. A critical point of distinction between corporations and partnerships is this: while corporations exist on a perpetual basis, partnerships typically do not. Some state laws require a partnership agreement to have a stated expiration date. When doing business with a partnership, it is wise to know the life expectancy of that organization. Because of the confidential nature of many partnership agreements, the documents won't be publicly available. But your negotiation counterparty should be willing to provide a redacted[5] version of the agreement, showing the relevant sections. You may be required to sign a confidentiality agreement to take a look at the agreement.

Government entities. The federal government's authority to contract is found first in the U.S. Constitution, and authority for specific branches of government is found in the statutes creating and controlling those agencies, departments, administrations, etc.

Authority for any entity created under the laws of a specific state will be found in the state constitution and in individual statutes as well. For example, without a statute authorizing creation of a municipal gas agency or public power pool, such creatures would not exist. A municipality or other statutorily created governmental entity is limited in its powers to those specified in the statutes and by the regulations under which it operates.

State-created public entities may be limited in their contracting practices—consequently limiting to the same extent the parties that contract with them. Many states have statutes limiting the right to sue public entities, including strict time limitations for filing lawsuits as well as requirements that administrative procedures be exhausted prior to filing a lawsuit against the entity. And, many states have passed laws limiting the amount of damages that the other party may be awarded in specific types of legal proceedings. Even though it would be uncommon to limit the amount of damages that may be collected for contract damages, other types of damages may be available in conjunction with a contract-based claim, and these may be subject to limitation or prior approval of the legislature.

If your organization is one of these statutorily created entities, assist the other party through the negotiation process by providing the legal citations for the relevant statutes and regulations under which the entity was created and under which it operates. If your organization contracts with public entities, you should do at least cursory research of the authorizing statute(s) and implementing regulations. Don't assume that just because a public entity is buying gas from your company or selling gas to your company that no legal or regulatory hurdles exist to impede your enforcement rights.

Authority

The second part of legal capacity relates to individuals having the *authority* to bind the company to certain types of contracts. What person or persons in your organization have the authority to sign a contract? *Presidents* and v*ice-presidents* typically have the right to contract on behalf of a corporation. The *managing general partner* is given the power to bind a partnership contractually through a designated representative—usually an officer of the partner. Any statutorily created entity must *follow the statutory or regulatory rules* regarding authority to sign on behalf of that entity, and these rules will vary from state to state.

Sometimes the waters of authority for private businesses become muddied. This usually happens when someone without specific authority to do so signs a contract on behalf of an organization. Most corporations require strict procedural compliance in formal corporate documentation through records found in the official corporate records, sometimes just called the *minutes book*. Every organizational document, minutes of all meetings including notices and waivers of notice, resolutions passed by the board of directors, and all action taken outside a regularly scheduled meeting (usually through *unanimous consent* actions), are included in the minutes book.

Maintenance of corporate records in strict compliance with the articles of incorporation and bylaws is critically important to every corporation because of the need to comply with the law for shareholder information. Any action taken outside the boundaries of the articles of incorporation, bylaws, any board resolutions, or unanimous consent actions, is *ultra vires*.[6] Because the action is outside the authority granted in one of the official documents, it is void. One level of interest that every corporation should have is maintaining internal discipline through corporate records.

The corporate records issue has relevance to your gas contracts, also. If, as is usually the case, the articles of incorporation specify that the president and vice-presidents have authority to sign contracts, then signature authority extends only to the president and vice-presidents. It isn't unusual in the fast-paced gas sale industry for directors or managers to sign contracts, so how do we make the leap from authorization traditionally extended only to senior management to non-officers? The official corporate documents may allow delegation of certain authority, so employees below senior management level will have the authority to sign contracts if a formal *delegation of authority* has been developed, signed by the individual with delegating authority, and maintained in the official corporate books.

Otherwise, a director, manager, or other non-officer employee may be acting outside his or her authority by signing gas contracts. When the title below the signature line of your contract isn't president or vice-president, can you trust that the person signing actually has the power to bind the other company? For this, you get a typical attorney response. It depends—on the facts of each and every situation.

As a gas contractor looking for guidance, you probably want something more than this cautious but accurate response, so here are some ways to identify those situations where reliance may be aptly placed—even when the other organization's representative doesn't have the formal requisite authority to contractually bind the company. Your organization can usually rely on the *apparent authority* of that individual, simply because he or she was given a position to sell or buy gas within the organization whose business is buying or selling gas.

If any action is beyond an individual's authority in his or her relationship with the employer, the dispute will ultimately be between the company and its potentially errant employee acting outside the scope of his or her employment. If your organization, in its erroneous reliance upon the apparent authority of that individual, did so in a reasonable manner and without knowledge that the employee was acting outside his or her authority, you probably have an enforceable contract with the other entity.

Try not to be too blasé about this, though. These are not black and white issues, and every fact situation may be different. Even though your organization may ultimately prevail if any legal dispute arising from your reliance on the apparent authority of the other party's representative surfaces, you can avoid time and money spent proving reasonable reliance and good faith by doing some advance work. During contract negotiations, ask for *incumbency certificates* for all parties authorized to sign contracts on behalf of the other company who may be signing contracts with your organization. An incumbency certificate is issued by the corporate secretary or an assistant secretary and certifies:

- that the named person began employment with the corporation on a certain date and is currently an employee of the organization, and
- that the named person has the right to contractually bind the organization.
- A specimen signature of that person may be included

You may want to require the other party to provide this document when negotiating a long-term, high-risk transaction. For routine business using a base agreement, an alternative to asking for incumbency certificates is to require a *representation* similar to the following made by both parties to the contract:

"*Each party represents and warrants to the other the following: (i) that it has the full and complete power and authority to enter into and to perform this Contract, (ii) that the person signing this Contract has full and complete authority to do so, and (iii) that the person(s) entering into any agreement for this Contract, executing this Contract, or both, (has/have) full and complete authority to do so.*"

Remember that it is your obligation to satisfy your own curiosity about the other party's legal capacity and not an obligation of the other party to prove its legal capacity to you.

Must this contract be in writing?

Under the UCC, agreements can result in contracts in any manner sufficient to show the agreement, including conduct of the parties.[7] It isn't necessary that all of the terms be included, only that it can be shown that the parties intended to make a contract and there is some reasonable basis for applying a remedy if one of the parties fails in its contractual obligations.[8] Even if a specific point in time at which agreement was reached cannot be ascertained, if the necessary elements of intent are found, a contract will have been made.[9]

With that in mind, why are attorneys always so adamant that gas sale contracts should be recorded in some reproducible form? Part of the answer is found in UCC Section 2-201, which follows the centuries-old *statute of frauds*. During the middle ages in Europe, the monarch was the arbitrator of disputes between landowners on contractual matters. When the monarch heard these disputes, the requirement of a sealed document became necessary just to ascertain which party was telling the true facts. The purpose for a writing and a signature was then, and is now, to prevent fraudulent testimony regarding the matter in dispute.

Even though contracts are formed, sometimes by conduct alone, if you must resort to a court of law to resolve any dispute, you will need to provide evidence to substantiate your position. Section 2-201 states that any contract for a sale of goods valued at more than $500.00 must be in *writing* and *signed* by the party against whom enforcement is sought. The writing will not be insufficient if it omits or incorrectly states an agreed term, but courts will not enforce quantity obligations beyond those found in the writing.[10] This entire subject will be discussed at length in the next section of this chapter.

The UCC also addresses the situation where no writing exists, stating that a contract not in writing will still be enforceable if:

- the goods are specially manufactured under conditions indicating that a special order has been authorized,
- the party against whom enforcement is sought admits in court that a contract was made,
- performance has begun or payment made[11].

When performance begins, a contract has been made and will be enforceable. What happens when a dispute occurs during performance , and the written contract is not clear as to the parties' intentions? If litigation results, the terms of the contract will be determined by evidence establishing the intentions of the contracting parties introduced during the lawsuit.

What role should Rules of Evidence play during contract negotiations?

If your organization files a suit against the other party, the proceedings may be held in either a federal or state court. The rules of evidence that those courts and litigants are bound to follow provide the second basis for the attorney's insistence that contracts be written and signed. The rules for admission of evidence in federal and state courts may differ.

Federal courts, including federal district courts, are located in all major cities and in most state capitals, and federal appeals or circuit courts that hear appeals from the federal district courts are regional. A set of common rules, known as the Federal Rules of Evidence, governs admission of evidence into any federal court. The definition of *writings* in the Federal Rules of Evidence is *Writings and recordings—'Writings' and 'recordings' consist of letters, words, or numbers, or their equivalent, set down by handwriting, typewriting, printing, photostating, photographing, magnetic impulse, mechanical or electronic recording, or other form of data compilation.*[12]

This extensive list means that if a federal court is deciding a dispute, your organization could present evidence from telephone tape recordings, faxes, and e-mails to found your position. But under both the Federal Rules of Evidence and state rules of evidence as well, nothing will be admitted unless it would tend to prove the matter being asserted—all evidence must be *relevant*, consequently, non-relevant evidence will not be admissible.

THE RECORD

Automatic fax confirmations

The *signature* requirement of the Code has been satisfied in the federal court system through logos, letterheads, encrypted digital signatures, and, a handwritten signature.[13] After a deal has been done, the transaction confirmations used by many organizations are automatically faxed when the transaction has been entered into an electronic gas management or electronic contract management system. No hard copy of the fax may exist, so no one signs it and it will be received by the other party without a signature from someone having authority to bind the corporation contractually. However, organizations using this form of electronic faxing system are identified, either by corporate letterhead or corporate logo on the faxes sent. Gas contract language sometimes specifically adopts letterheads and logos as *signatures* for Section 2-201 purposes, both to avoid any misunderstandings during contract performance and to lessen the impact of any evidentiary objection that the other party may make during formal legal proceedings in a state court.

Tape recordings

The major contractual problems faced by organizations relying on tape recordings as evidence of the required *writing* arise from the fact that telephone taping is quite new to the gas industry and many traders are not sufficiently trained or managed to accurately complete contracts over the phone. The natural gas industry began taping conversations *en masse* when it learned how financial brokers operated. After the NYMEX started trading natural gas futures contracts in 1990, financial traders assumed many futures contracts positions through financial brokers that routinely tape recorded of all transaction conversations. It didn't take very long for most gas organizations to realize that those brokers' tapes would usually work to the disadvantage of the gas organization and to the benefit of the brokerage if only the broker was taping. The tape would be available to the marketer when its contents would prove the position being asserted by the broker, but not otherwise. So the gas industry began taping phone trading activities in the early 1990s, and that practice has continued to grow.

If a tape recording is the *written* evidence of a contract, and no written confirmation has been sent, then the tape *is* the contract and will be admissible as any other contract would be. But what if a tape is used only for backup purposes and a confirmation is the written proof of the contract? The tape may be nothing more than a *business record.* Whether or not a tape recording would be admissible under this circumstance would depend on a number of variables. The most significant issue is whether the written confirmation clearly evidenced the intentions of the parties. But putting that issue aside for a moment, let's assume that the confirmation was not clear and that it would be to your organization's advantage to have the tape recording admitted. If your organization is involved in litigation about the matter and the indeterminate confirmation is clarified by evidence on a tape recording made by your trader, you would want it to be heard by the court, while the other party would want it excluded. One of the arguments your opponent may use in its objection to your offer of evidence is that the tape as a business record in its objection must be excluded from evidence because the tape was not properly maintained and kept secure.

Business Records

Many organizations store tape recordings, either on their own premises or offsite, and in order to introduce business records evidence in the federal court system, the proponent must show that the records are maintained in accordance with routine business practices.[14] The reason for this is found in the "hearsay" rule, defined by the Federal Rules of Evidence as *a statement, other than one made by the declarant while testifying at the trial or hearing, offered in evidence to prove the truth of the matter asserted.*[15] The hearsay rule exists to prevent admission into evidence of any matter that cannot be cross-examined.

Business records cannot be cross-examined but may be admitted as an exception to the hearsay rule if "*kept in the course of a regularly conducted business activity and if it was the regular practice of that business activity to make the memorandum, report, record, or data compilation, all as shown by the testimony of the custodian or other qualified witness.*"[16] For organizations new to taping and relying upon trader conversations, caution is advised. These issues are largely untested in the courts so most companies are establishing policies for a tape maintenance structure to follow at all times.

Authentication. To be admitted, evidence must also be *authenticated.* This means that the proponent of any evidence must also provide proof that the evidence is indeed what it purports to be.[17] Both voice identification[18] and telephone conversations[19] may be used as forms of authentication, so at least in the federal court system, tape recordings may be authenticated as evidence of the *writing*. Your taped trader conversations may, if relevant, be admitted in a legal proceeding, providing that all other prerequisite evidentiary rules are followed. For organizations that use fax confirmations and maintain tape records of day trading activities, the fact that those tapes may be challenged under rules of evidence is crucial. In gas contracts that recognize use of both confirmations and tape recordings as business practice, care should be taken when deciding which form will be the "contract."

Where would my client like to litigate a contract dispute?

Given that the Federal Rules of Evidence are the same in all federal courts regardless of the state in which the court is located, you might assume that any litigation involving your company will be in the federal courts. There may be problems associated with that assumption.

All issues regarding federal law must be heard in a federal court. Access to the federal court is also possible when disputes arise between parties domiciled in different states. To gain access to the federal courts for non-government contract disputes, the *matter in dispute* must be at least $75,000.00.[20] Second, the parties must have complete *diversity of citizenship*.[21] This means that the principal place of business and state of organization of both companies must be in different states. If a Delaware corporation (with its principal place of business in Texas) is involved in litigation with an Oklahoma corporation (with its principal place of business in Kansas), complete diversity exists because the principal places of business are different and the incorporating states are different. If the Oklahoma corporation had its principal place of business in Texas, the federal court system would not be available to those litigants. In chapter 8, during the discussion of *governing law*, you'll find more information on the diversity issue.

A second problem with acceptance into the federal court system is that all disputed matters arising under federal law must be heard in the federal court system. The system is completely overworked, burdened by ever-increasing numbers of federal criminal cases, and even if you are successful in filing your lawsuit in a federal district court, the backlog of most federal courts means that you will probably face delays in having your case heard.

If your case is not appropriate for the federal court system, your forum will automatically be in a state court. In some circumstances, your organization could find itself in litigation in a locale that doesn't allow introduction of certain types of evidence. States have adopted the Federal Rules of Evidence in their own court system, a variation of the federal rules, or their own set of evidentiary rules that may be out-of-date and inconsistent with the fast-paced gas sales environment.

You should have some familiarity with the admissibility restrictions in evidentiary rules in the states that could be forums for any subsequent litigation before you agree to specific contract language regarding fax confirmations, tape recordings, and electronic communications.

Enforceability II

Once the legal prerequisites of offer, acceptance, and consideration have been met, the contracting parties have the required legal capacity, and a contract satisfies the statute of frauds, you are over the first hurdle of enforceability. The second issue of enforceability deals with the contents of the document itself. If your organization cannot, through contract language, force the other party to do something it doesn't want to do at the time, you will not have succeeded in crafting a solid contract. If a contract is enforceable, you can take it into a court of law or into some other accepted form of alternate dispute resolution system (such as arbitration) and the decision-maker will enforce the contract according to its terms. Your contract will be enforceable if it meets certain criteria.

Let's go back to the six practices introduced at the beginning of this chapter.

Applying Practice #1

These practices happen all the time. In fact, some of the most prevalent problems in the gas sales industry occur because of one of those six practices. The "terrible two" questions: "When is a deal a deal?" and "What are the terms of the contract?" are usually asked because of some variation of the six described practices.

Use what you have learned so far, including the problem-solving method described at the beginning of this chapter, to develop the issues and form some potential solutions when routine operational practices give way to contract disputes. As a refresher, the three building blocks identified as problem-solving mechanisms earlier in this chapter are:

- identify the source of a problem
- identify the nature of the resulting contractual problem
- identify the path for resolving all issues

Let's work through the described practices and determine both potential problems and how those problems might be solved.

Day traders in your organization buy and sell gas on short-term bases, usu-

ally limited in time to less than a week. The only recorded evidence of the deals done is found in tape recordings of trader telephone conversations. While they have been given a list of items to include in all transactional telephone conversations, sometimes the traders forget. They may turn the recording device on and off, depending on the nature of the telephone conversation. The actual tape records are maintained for up to six months, then reused.

Practical cause of a problem

- Tape recordings are used as the only evidence of day-trading transactions
- Tape recordings may be activated and deactivated by the traders
- Tape recordings are not regularly maintained as business records

In this situation, telephone tapes are the only recorded evidence of multiple transactions being undertaken by your day-traders, yet without any of the controls commonly used with fax confirmation. When written confirmations are used, traders are normally required to provide pieces of information, including name of the other organization, identity of the other trader, the gas quantity, price, delivery point, and delivery period. That information is then distributed internally, either through the electronic gas management system or by routine paper routing, to every department with a need to know and use the relevant facts.

If no adequate records are maintained, how can an organization invoice or pay for gas? How can nominations be made? How does the credit department know the current limit of trading partners?

Nature of any resulting contract problems

- Has a contract been created?
- What are the terms of that contract?
- What evidence of the contract will be admissible in court?

Any of the three potential contract issues could occur in a given situation. When no words of agreement are used during the phone conversation, how can one organization prove or disprove that a deal was made, absent performance? Even if the deal is made, in some cases it will be difficult to prove the elements since the traders don't always follow the established rules. Finally, because the traders can activate and deactivate the recording system, the potential for human error is so large as to be unacceptable. Trading floors are pits of activity, and in the hubbub of daily trading, who wouldn't, at one time or another, forget to turn the recorder on or off?

The path to problem resolution

- Traders must follow the rules and face consequences if they don't.
- The recording system should be automatically activated for every phone conversation. If traders are making non-transaction calls, other non-recording phones should be used.
- The tapes should be maintained according to a plan, with due regard for the time limits that other business records are maintained. For tax purposes, records are typically maintained for seven years. For contract purposes, records are typically maintained for the applicable *statute of limitations* period, typically four to six years. For GISB business practice purposes, pipelines maintain their records for three years.

Applying Practice #2

One of the traders in your organization, habitually "does deals" without putting anything in writing. This trader may be successful and innovative, highly motivated, and may generally feel that he or she isn't bound by the rigors of the established contract process because the goal is to do the deal.

Now, let's use the same analysis above, and work through the first practice, that of the trader who has found a way to elude established procedures.

Practical cause of a problem

1. The trader doesn't follow the rules
2. The trader's success results in special treatment, with no consequences

Nature of any resulting contract problems

1. Has a contract been created?
2. What are the terms of that contract?
3. What evidence of the contract will be admissible in court?

Allowing and sometimes encouraging uncontrolled business practices can create the same three contractual problems as indicated in the last example, not to mention the emotional toll this type of favorable treatment takes on other traders who are required to follow the rules.

The path to follow for problem resolution
1. The trader's supervisor must require compliance with internal procedures
2. Consequences should attach for failure to follow the procedures

Human error and routine business practices that are liability time bombs should be identified and corrected. Many organizations just struggling to make a profit in a world of shrinking margins may focus all their efforts on the sale or the purchase. This is myopic thinking when one considers that default on just one deal will affect the bottom line.

Applying Practice #3
Your origination group[25] typically spends several months negotiating potential multi-year sales to cogeneration facilities, industrial consumers, and LDCs where gas restructuring has not occurred. During the negotiation process, it is common to utilize letters of intent[26] that specify the nature of preliminary discussions and specifically state that no contract will be formed between the parties until final management approval of both organizations has been given and a formal contract document signed.

Pre-contract writings
Up to this point, the bargaining process of deal-makers has been explored with an emphasis on routine transactions that may have delivery periods of anywhere from several hours, one day, or perhaps as long as one year. Most gas sale organizations differentiate between their short-term business and their long-term business, and those in long-term sales or purchases follow some fairly distinctive patterns in their negotiations. An initial contact is made, and if the other party is interested, negotiations for a sale or purchase will proceed. Sometimes these negotiations will take many months, since much long-term business in the current and future gas mar-

ket is for new gas-fired electric generation or cogeneration facilities being planned and constructed. *Distributed generation*[22] is also a growing industry, and industry forecasters, as well as the Energy Information Administration (EIA)[23] have estimated that the demand for natural gas to fuel various forms of electric generation will increase to 23% of total natural gas consumption by the year 2010.

Understandably, the ability to make gas sales to these highly specialized industries requires a good knowledge of electric generation. When an organization has an established origination group, it is expected that the negotiations may be few in number at any given time—but intensive—and will result in excellent long-term benefits.

On the gas supply side, long-term sales by producers may also result in drawn-out negotiations. As you have already learned, every producer has its own set of issues because of gas quality, the expected producing life of the well, the vicissitudes of gas production, the unique characteristics of a required gathering system, and so on. Organizations that are successful in buying long-term gas from producers at the wellhead, or at the terminus of a gathering system, will employ supply representatives that have solid knowledge of the gas production industry.

Because the negotiations on both the supply side and market side take so long, neither party is willing to allocate its time and effort to negotiations if the other party hasn't exhibited some commitment to negotiating the proposed deal. The problem is that contracts are absolute—either you have one, or you don't—and before knowing all the circumstances of the transaction, neither party wishes to be contractually bound. While the parties negotiating a complicated long-term transaction want some commitment from the other party, they are usually unwilling to make too great a commitment themselves, so the two parties will agree to negotiate in good faith toward the goal of "doing the deal." Any negotiation is in part a fact-finding mission, and either party may realize during negotiations that the deal won't work for it, in which case they want to just shake hands, go home, and start over with some other organization.

Request for proposal (RFP)

The first document in a lengthy negotiation may be an RFP, to which receiving parties may respond. Sometimes LDCs and industrial consumers are just looking for the lowest price, but more commonly this device is used to find a supplier that will provide the *best* proposal for a particular gas supply. These RFPs will usually contain well-defined parameters limiting the receiving party's areas of response and will customarily be issued well in advance of the anticipated performance. If an industrial consumer, for example, wants to receive bids for its gas supply for the upcoming year, it may issue the RFP in May or June of the previous year. The standard elements of an RFP are the following:

- Delivery period(s)
- Gas quantity—may be tiered, so an explanation will usually be included
- Alternate fuel needs
- Acceptable delivery points and potential alternate points
- Any limitations on seller's rights or abilities that will be required by the contract
- Any proof of deliverability required by the issuer
- Contract price (optional—depends on the specific RFP and what the issuer wants)
- Time period that RFP is open for proposals
- Deadline for response

Depending on the business of the issuing organization and its gas needs, the RFP may include elements in addition to or different from the above list, but the essential elements remain fairly constant. The RFP issuer will distribute the proposal to organizations that it believes to be reliable suppliers. In response to the RFP, suppliers will submit proposals. Individual proposals may vary from the RFP as a potential supplier will try to highlight the *things* that differentiate it from other potential suppliers. These *things* may include imbedded services, provision of hedging alternatives, delivery options, and so forth.

The RFP issuer will, after the deadline for response has passed, review the proposals and select the lowest bid or best bid as the case may be.

From the RFP issuer's point of view, the RFP is just that—a request for proposals, and not a contract offer. But, depending on how the RFP is worded, the request may indeed by construed to be an offer requiring an acceptance. The RFP will typically state that it is not an offer, but merely a distribution for informational purposes.

When potential suppliers respond to the RFP though, those responses could quite easily be interpreted as offers, requiring acceptance by the issuer. Proposals must be carefully worded to protect the potential supplier from adverse market movement between the time the proposal is submitted and the award date. Proposals usually contain the following elements:

- Direct response to all direct requests
- Variations from the RFP when the supplier can add value (if allowed)
- Statement that the proposal is not an offer
- Statement that no contract will be formed without final management approval and contract execution
- Deadline for response (optional—depends on the pricing)

Both parties to the RFP process must carefully choose the words being used in writing. As I've noted elsewhere in this guide, *Merchants* under the UCC are held to a different standard whenever any offer is put into writing.[24] In essence, UCC Section 2-205 states that whenever a Merchant offers to buy or sell goods in a writing in which it is stated that the offer will be held open, that offer cannot be revoked within the stated period. If no response deadline is given in the offer, it must remain open for a "reasonable period of time" that in any event cannot exceed three months.

As you learned in chapter 4, the concept of Merchant under the UCC is important in differentiating the obligations of contracting parties under certain circumstances. It merits restatement that gas Merchants deal in natural gas, or hold themselves as having superior knowledge of gas sales. Industrial consumers sell widgets, or steel, or electricity, and they will be merchants of the manufactured product, but not necessarily Merchants of natural gas. Commercial consumers are Merchants of hamburgers or clothing, or pizzas—but not natural gas. So the gas supplier will usually be bound by the *Merchant rules* of the Code, while many consumers will not.

Letters of intent (LOI)

A LOI is usually in the form of a letter sent by one organization to the other, with signature lines for both parties. The LOI will list the topics being discussed by the two parties and will usually describe the nature of the obligation both parties have under the LOI. The obligation typically is one to negotiate in good faith toward completion of the deal. LOIs will carefully limit the effect of the parties' negotiations and will always emphatically specify that the LOI does not create an obligation to contract.

LOIs, mutual understanding agreements, and other pre-contract documents are the bane of the gas attorney's existence. Standard business practice tells us that some form of written commitment is necessary, but all of our legal instincts tell us, "Don't do it." With that natural conflict occurring, attorneys and others who draft these documents do so with great care and precision. Even the most thoughtfully crafted LOI will do your organization or client little good if the negotiation team acts hastily.

Negotiations

Not only the words used but actions may well affect the breadth of a pre-contract understanding. Let's assume that your organization's negotiators routinely maintain copious progress notes during the long negotiation process. In addition to written notes, we cannot overlook the importance of the words being used during negotiations. Since an offer can be accepted in any manner to indicate a willingness to be bound, negotiators must carefully balance their enthusiasm for the deal with common sense use of words and tangential writings.

Notes and other memoranda

Negotiators should carefully consider the depth and scope of notes they maintain during the negotiation process, because even seemingly innocent notes or other memoranda kept by the negotiators may come back to haunt your organization. What if, during the negotiation phase, one of your negotiators—in the interest of showing your organization's commitment to the process—says something like "I'm looking forward to this deal"? If the negotiations do not result in a contract because the numbers have gone against you and now favor the other party, do you have a deal? You might.

And, when the other party sues for breach of contract, the mere statement anticipating the "deal" may be enough to show your organization's intent to be bound, thus creating a contract. Negotiators must choose all their words carefully—both those put into writing and those spoken.

The progress notes kept by your negotiators, any notes to file, internal memoranda, e-mails, faxes, taped phone recordings, and other relevant evidence will be requested by the other party during the *discovery*[27] phase of the lawsuit against your organization. You will be required to produce the requested documents that may then be introduced into evidence in the lawsuit to show intent. The oral statements made by your negotiators may be introduced into evidence through witness testimony for the same purpose. The intent of the parties is an abiding concept in contracting, and if written documents don't clearly reveal the parties' intent, a court has the ability to look anywhere relevant to determine that intent. Any statement made in anticipation of the deal could be converted into evidence of an intention to be bound to the deed.

Contract Performance

Remember that a contract is, among other things, a way to force the other party to do something it doesn't want to do at the time. This usually comes into play when gas prices move against a party, and it no longer wants to perform. When that happens, you want the contract language to be clear in stating both parties' obligations. The problem in this regard is that gas prices could move either way and that price knife could be cutting *your* organization. You already learned that it isn't uncommon for the person who did the deal to argue for purposefully flexible contract language, hoping that if a dispute occurs, your organization will either have the better part of the argument, or the dispute will force resolution of the issue through negotiation. When the "flexible language" argument wins, the results will be vague or ambiguous contracts.

Assume for this section that the deal is done, a contract made, and performance is underway. Once the contract formation maze has been successfully navigated, the only thing you have to worry about is performance, right? Wrong. A myriad of issues transpire simply because of the way most gas sale transactions are recorded—some of these issues occur because the contract language isn't clear, and others occur during the ubiquitous confirmation process.

The routine use of fax confirmations during performance results in numerous questions like:

- "Whose confirmation controls?"
- "When the buyer and seller send simultaneous fax confirmations, which one controls?"
- "If the confirmation and a tape recording of the relevant transaction conversation don't agree, which controls?"
- "Can our internal gas management system handle the variety of deals being done?"
- "If gas flows without a contract in place, what are the *contract* terms?"

During performance, we routinely see essential contract questions and practical administrative questions develop. This section addresses the performance issues most frequently encountered.

Incomplete contract terms

What happens when an essential element of a gas transaction isn't included in the contract, or is unclear from actual contract terms? You already know that if the dispute winds up in legal proceedings, a court will try to determine the intent of the parties by looking beyond the contract language, limited only if the court cannot determine that there is a "reasonably certain basis for giving an appropriate remedy"[28] for breach. A court has a variety of options that it may explore under Article 2 of the UCC, and depending upon the parties' intent as determined by a court, it may even *fill in the blanks* for litigants that have not expressed their intentions in writing.

Courts must follow the UCC law

The commercial code was developed to enhance commerce and to encourage commercial dealings between citizens of the various states. In

Chapter 6 Essentials of Contracting

chapter 4, you learned that one of the three purposes of the Code is to "permit the continued expansion of commercial practices through custom, usage, and agreement of the parties".[29] To that end, courts feel compelled to view evidence in a way that would further the purposes of the Code, so the general attitude of courts hearing commercial disputes under the Code is to define and uphold a disputed contract, if possible, rather than to discard it.

Assume that a court is deciding the following contract case and the contract document is unclear as to the parties' intentions on the level of delivery obligation. Here are the facts.

PUMP has sold a one-year gas supply of 10,000 MMBtu/day to WIDGET at a price of $2.75/MMBtu. Under most gas contracts, a seller's delivery obligation will be either *firm* (an absolute obligation) or *interruptible* (neither party has an obligation to commit to any sales or purchases), but the disputed contract doesn't use either of the two industry-accepted terms. The undefined term *baseload* is used to describe PUMP's delivery obligation. When gas prices spike well above $2.75/MMBtu because of low temperatures, and PUMP's transportation being used for the transaction is curtailed, PUMP will not make gas deliveries until transportation is restored. This may all be done in good faith, with an innate understanding that the industry use of the term baseload means that gas will be delivered unless transportation is curtailed.

On the other hand, LDCs and industrial consumers look at the term *baseload* differently. When projecting gas supply needs for a future period, utilities and consumers alike divide their gas needs into various categories that were discussed in chapter 2. The basic gas supply needs are usually termed baseload, meaning that for this portion of the gas portfolio the same gas requirements exist every day of the year. Gas purchased as base load gas, from the consumer's perspective, is gas that must be delivered to the plant every day of the year and, in the illustration above, that was the basis upon which WIDGET bought its gas from PUMP.

When PUMP stops delivering gas, it will argue that the parties intentions all along were for baseload sales that allowed an excuse for transportation curtailment. At the same time, WIDGET will argue that the parties intended the sale to be firm, with no excuse for loss of transportation other than primary point firm transportation. What is a court to do when faced with vague contract language, and each party espousing a position in conflict with the other party's position?

Parol evidence rule

The UCC language assists a court as it tries to divine the real intentions of the parties. The general rule established by Article 2 Section 202 is that when a contract has been signed by both parties, the terms cannot be contradicted by evidence of a prior agreement or of any oral modification to the contract. This is generally referred to as the "parol evidence rule," and if a contract is clear on its face and evidences the parties' intentions, courts will not look beyond the written contract. But parol, or *extrinsic evidence*,[30] may be received as necessary to *explain* or *supplement* the contract language. And, this is the vehicle through which vague or ambiguous contract language may be further explored by a court to establish intent.

To determine the parties' intent, a court may look at *course of performance, course of dealing*, and *usage of trade*;[31] as well as evidence of additional consistent terms.[32] In fact, the three types of extrinsic evidence available are, in practice, so powerful that in some states they may be used to contradict, supersede, or confirm ordinary contract language[33] irrespective of UCC language seeming to limit the use of parol evidence. This is how those innocent notes and doodles can find their way into evidence if relevant to proving intent, and one of the reasons it is so important to understand the UCC as adopted and interpreted by courts in the governing law state.

Course of performance

Course of Performance is the tool courts use most frequently when digging around outside the contract to determine intent. As defined by the Code, course of performance is only applicable when the contract for sale involves repeated occasions for performance (standard in the gas sales industry), and one party has acquiesced in or not objected to the type of performance rendered by the other. In the situation described above, the fact that WIDGET never objected to the supplier's deliveries during the course of performance may be introduced to show that the understandings of the parties may have been for the previously reliable supply. If PUMP continually delivered the full contract quantity in prior months, that evidence may prove course of performance indicating intent. This isn't the only tool available, though, a court may look further, into *course of dealing*.

Course of dealing

The Code defines *course of dealing* as "*a sequence of previous conduct between the parties to a particular transaction which is fairly to be regarded as establishing a common basis of understanding for interpreting their expressions and other conduct.*[34]" Course of dealing, then, relates to the manner in which the parties have dealt with each other in the past, not under the present contract but in other transactions. So, if PUMP has sold gas to WIDGET previously, those prior transactions may be relevant to prove the intention of the parties under the currently disputed transaction. But, even this evidence may be further supplanted by reviewing standard industry practice, commonly called *usage of trade*.

Usage of trade

Usage of trade is defined by the Code as *any practice or method of dealing having such regularity of observance in a place, vocation or trade as to justify an expectation that it will be observed with respect to the transaction in question.*[35] Evidence of trade usage is factually presented to the court. Generally, usage of trade is evidence of what the standard industry practices regarding the matter in dispute may be, and a party may be chargeable with a usage of trade of which it is ignorant.[36]

In this lawsuit, each party will undoubtedly introduce evidence tending to prove that its definition of baseload is the correct definition, and then it is up to the trier of fact to determine from all evidence what the intentions must have been.

If a conflict exists between evidence used to prove any of the three, courts will give course of performance the most weight, followed by course of dealing, and finally trade usage. If, in this example, the court determines that PUMP did know or should have known that WIDGET was relying on it to supply firm gas, the court may resort to one of its additional powers to supply missing contract terms, in this case, the definition of baseload as a firm delivery obligation. This is done through the courts' *gap-filling* powers available to all courts deciding Article 2 cases.

Gap-filling powers

Unless it can be shown to a court that the parties clearly did not intend to agree and contract (which may have been the case if PUMP and WIDGET each mistakenly believed the other was defining baseload the same way it defined baseload), the court may utilize one of its extraordinary powers under the UCC. As briefly described in chapter 4, these powers are generally referred to as *gap-filling* powers where the court can fill in blanks or correct vague or ambiguous contract language agreed to by the disputing parties. The only issue that a court cannot decide if the parties were not clear in their intentions is quantity. Even then, if it can be inferred that the parties agreed to a "full output" or "full requirements" sale, the courts will fill that gap.

Delivery point

Some of the UCC rules on gap-filling could produce ridiculous results if applied literally to the sale of natural gas—the requirement for delivery at the seller's place of business or at his home as noted earlier.[37] If a contract or transaction confirmation is tightly written to define only one allowable delivery point the parties will not have contractual flexibility to switch points. But if the contract delivery point designation is extremely broad such as "any metered point into the field zone of PIPE Pipeline Company," either the seller or the buyer could, in any given circumstance, be forced to deliver or take delivery of gas at a delivery point where it would be losing money to do so, or be in breach of the contract, since the contract language in this case didn't require "mutually agreeable" points.

Contract duration

Unless otherwise provided for in the contract, contracts to sell goods are valid for a reasonable period of time unless terminated at any time by either party. The terminating party must provide reasonable notice and the parties may not contractually agree to dispense with that notice.[38] If an attempted contract termination by one party results in a legal dispute, the court will determine whether the notification time was *reasonable* in light of the facts of that case, but parties have an implicit right to terminate any master sales agreement in the absence of a notice of termination clause in the contract. For that reason, it is quite common to see a clause similar to the following in gas contracts:

"Except for Early Termination pursuant to Article __, this Contract will be in effect from the date first set forth above until terminated by either party by providing at least one (1) calendar month's written notice of termination to the other party; provided, however, the provisions of this Contract will remain in effect as long as any outstanding transactions are in place between the parties, whether performance has begun at the time of termination or is scheduled to begin at some future date, and for payment and indemnification purposes for transactions already completed or in progress as of the time of termination."

Payment terms

If the gas contract does not specify payment terms or omits any relevant portion of the payment terms,[39] the Code specifies that the buyer must pay with cash[40] or its equivalent. If a check is used, a seller's acceptance of the check is conditional until the check clears. When a contract is silent on payment terms, courts allow the buyers some minor concessions in time of payment since strict interpretation of the code for gas buyers would require payment at the delivery point meter while the gas is being measured!

Price

The Code's approach to pricing is found in Section 2-305 and establishes that if the price term has been left open, the price will generally be a reasonable price at the time for delivery.[41] Contracts for the sale of goods don't require a price clause to be complete and enforceable, as long as the element of intent can be ascertained and adequate remedy for damages can be found. A contract will be valid even if any of the following three conditions exist, unless it can be shown that the parties didn't intend to be bound until reaching agreement on price:[42]

- a contract does not specify a price

- the price is left to be agreed by the parties at a future time and they fail to agree, or
- the price is to be established by some market indicator (such as the NYMEX or any published index)

Price is usually addressed in gas contract, but sometimes in a manner that may be difficult to determine. Courts may be asked to determine price under any of the following circumstances:

1. The parties have specified how a price is to be determined, but the method fails. In this case, the court will determine a *reasonable* price. Keep in mind that a reasonable price may not always be the "fair market value of the goods."[43] It may be appropriate when determining the price for a court to look at course of dealing between the parties to see how prior transactions were priced
2. The parties agree to set the price at a later date. While some courts have found these clauses to be unenforceable because they are too indefinite, the UCC specifically provides that a court may apply its reasonableness test in this circumstance. The price will be determined as of the time and place for delivery under the contract
3. The parties agree to a "prevailing market price." In this instance, the court will receive evidence of the market price at the time of delivery and make its determination of "prevailing market price." Whenever vague price references are used in a contract, the court may look at all of the published indices (and other market determinants) to determine that price
4. The parties agree to *market quotations.* Market quotations are given by *market-makers,* those organizations large and liquid enough to determine pricing that will provide quotations of the price they would pay to assume the position of the party seeking the quote or the price the quote-seeker would have to pay the market maker to assume its position. Sometimes the market-makers in the gas industry are energy companies, and other times the market-makers are brokerage houses or banks that routinely trade in OTC derivatives products. For a gas contractor, the important point to remember about

market quotations is that quotations given to Company A may be different than quotations given for the same transaction to Company B. Assume that mega-producer and tiny marketer asked for a quotation on the same transaction. The market-maker may look to many variables when rendering the quote and may favor the bigger, more financially secure company.

Quantity

A court will not fill in the gaps on quantity unless it can be shown from evidence presented that the parties intended the contract to be *full requirements* or *full output* contract. An example of a full requirements gas contract is a sale to an industrial user by one supplier, agreeing to provide all of the consumer's fuel needs during the term of the contract. A full output contract is usually seen at the other end of the industry spectrum when producers are selling all their reserves to one buyer. Sometimes this type of contract is referred to as a *dedication of reserves* contract that dedicates all gas produces from a well, series or wells, field, or total production form all sources, to one buyer.

Gas buyers and sellers must be aware of the expansive powers of a court deciding an Article 2 case. Most commercial organizations really don't want a court interjecting its idea of reasonableness into the contract. To avoid that possibility, be thorough in contract drafting and try not to give in to the urging of others for flexible or ambiguous contract language.

Confirmations—do they help us or hurt us?

The most commonly utilized form of contract used for gas sales and purchases is a *master contract* or *base agreement* that contains all the general terms and conditions applicable to each gas sale and purchase transaction between two contracting parties. Because so many organizations both sell to and purchase from the same trading partners, the standard contract terms will generally apply equally to each party as either buyer or seller. Nearly every natural gas merchant has developed its house gas contract, and while there may be vast differences on individual points in these contracts, most contain the same elements. The result of this timesaving and very expedient practice is that gas contractors will negotiate the terms of the master contract and sales representatives will negotiate individual transactions that will be subject to the terms of the master contract and evidenced on individual transaction confirmations. This efficient way of contracting is used widely by organizations conducting multiple transactions with the same trading partner.

This structure is not ideal, however, and causes many contracting problems. To illustrate, let's assume that representatives from two organizations have reached an agreement by telephone that involves a sale of 5,000 MMBtu/day at $2.00/MMBtu to be delivered at a published delivery point from January 1 through December 31. This agreement represents one transaction, and the four major elements of quantity, price, delivery point and delivery period have been decided. The respective gas contractors are quite frequently given responsibility for negotiating the base contract terms after an initial agreement has been reached.

As discussed earlier in this chapter, if you wish to enforce your organization's contractual rights through a legal forum, courts will require some tangible proof that the other party agreed to the terms you are pleading. That is precisely when you want to have a well-written contract in place. Unfortunately, we can't foretell which transactions will be problematic and which will proceed without a hitch. In my legal practice, I have counseled my clients countless times against flowing gas before a written contract with the other party is in place. Every gas attorney reading this guide will probably smile and say, "sounds familiar," and undoubtedly will have had the same imperfect results I have experienced.

Gas flows without signed contracts all the time. The gas business simply doesn't accommodate prior written evidence of the agreement under all circumstances. Numerous transactions with numerous trading partners/suppliers/customers result in incredible burdens on the contract review, negotiation and administration process. Many times the contract terms are still being negotiated when gas starts to flow and contract negotiators, drafters, reviewers, analysts or administrators must adapt the written procedures to actual industry practice. We send confirmations to trading partners hoping that the conformations will protect us. But the questions resulting from confirmation use that surface during contract performance are not easily answered.

At the beginning of this section, you were given a number of questions routinely asked by gas buyers and sellers. Let's review them.

"Whose confirmation controls?"

This question can be asked in a number of scenarios:

1. Seller and buyer are negotiating the base contract, but it has not been signed. Deals are being done and gas is flowing
2. A base contract has been signed but the contract doesn't delegate authority to either party to send confirmations. Deals are being done and gas is flowing
3. A base contract has been signed, with one party responsible for sending confirmations by a deadline. It fails to meet the deadline and the other party sends its confirmation

Let's address these scenarios in the order they are presented. The first example is perhaps the most difficult because at this point the base contract has not been signed, so there is no agreed language. The confirmation language provided below may alleviate the uncertainty in this situation, but it's still a difficult situation. Part of the reason these issues are so tough is that the UCC doesn't really handle them too well, and gas contractors are frequently left with puzzles missing a few pieces.

You may have to go all the way back to the concepts of *offer* and *acceptance*, for your answers in this situation. Under the common law that governs other types of contracts, the *mirror image rule* applies and the acceptance of an offer must mirror the offer without change. Any alternation in an acceptance beyond the terms of the offer is considered to be a refusal and a counteroffer. The Code takes a different approach, not surprisingly, one that encourages parties to freely contract with each other. Section 2-206 provides that any offer will be construed to invite acceptance in any reasonable manner, including performance. When day traders are regularly phoning multiple suppliers and customers, does the fact that one party generated the phone call mean that it made the offer? If so, how would that rule be applied when the other party cannot do the original deal for which the phone call was made, but has another deal of its own to offer? Sometimes you just cannot ascertain which party offered and which party accepted, so the offer and acceptance rules may not be helpful *vis á vis* the trading floor.

Acknowledging that it may be difficult or impossible to ascertain which party is the *offeror* and which is the *offeree* (the accepting party), the time at which the original offer was made, and other critical elements, we take the next logical step and look to the documentation paper chain. It may be the best thing available—even if not perfect—particularly in the frenetic day trading atmosphere of many gas organizations.

If you approach the offer and acceptance issue only in terms of the confirmation, and rely on an admittedly imperfect approach (that the first confirmation sent is the formal offer), then that confirmation will establish the requirements for a valid acceptance from the other party. Because of the legal uncertainties, it's even more important that you, as a gas contractor, craft confirmation language to assure your confirmation controls. This is a rather fuzzy area of UCC law, though, and even the experts disagree on any "first to fix" preference.

Some of the elements you may want to include in confirmations used when the base contract is under negotiation include the following:

1. Limit the mode of acceptance to specific choices such as requiring the other party to sign and return your confirmation, to send a non-conflicting confirmation, or to nominate gas for transportation.

2. Set a deadline for acceptance commensurate with the type of business being done. For longer-term transactions, several days may be appropriate. For day-trades, one business day may be appropriate.

3. Specify that the offer may not be accepted if the other party changes or adds terms to the confirmation.

The offeror can set any reasonable parameters within which the acceptance must be made. Without this limitation, changes may be made either on your confirmation form or in the other party's confirmation, and the *additional* or *different* terms will become part of the contract. The importance of the UCC's

Chapter 6 Essentials of Contracting

Merchant designation is critical in this context because if either your organization or the other party is not a Merchant, any new language will become part of the contract unless the confirmation forbids additions or changes.

Between Merchants, additional terms automatically become a part of the contract unless the offer has limited that right, the additional terms materially alter the offer, or the offeror objects to the additional terms. The Code is silent on the effect of different terms between Merchants, but a majority of courts hearing contract disputes between Merchants have rejected terms different from those in the confirmation.

When additional terms *materially* alter the offer, they will not become a part of the contract. Courts have determined that *materiality* should be based upon what would *surprise* or *cause hardship* to the other party.[44] Examples of material changes include:

- disclaimer of an implied warranty
- seller's cancellation right if the buyer failed to pay an invoice
- changes to price, quantity, delivery period or delivery point
- limitation of remedies
- adding liquidated damages
- stating a choice of forum
- adding an indemnity clause

Non-material changes include:

- addition of a force majeure clause
- a clause providing for seller's reasonable credit terms
- a clause requiring the buyer to pay late interest

In all these cases, however, the facts of the particular situation controlled the ultimate decision.

4. Add a clause to the confirmation that addresses the base contract being negotiated, specifying that the terms and conditions for all transactions prior to final execution of the contract will be those in the standard contract, and that, once the contract has been signed the relevant

transaction will be subject to the terms and conditions of the negotiated and signed contract displacing the non transaction-specific elements of the confirmation.

"When the buyer and seller send simultaneous fax confirmations, which one controls?"

One of the most basic problems between Article 2 and gas marketing is the inherently different concept of a confirmation. The Code contemplates that a confirmation will serve as the response to an offer (true acceptance). In the business of buying and selling gas that could indeed be the case, but it is just as likely that confirmation may be viewed as the offer itself.

To further complicate matters, the rules may be different when one party (or both) is a Merchant.

For Merchants[45] sending confirmations, the Code specifies that *"between merchants if within a reasonable time a writing in confirmation of the contract and sufficient against the sender is received and the party receiving it has reason to know its contents, it satisfies the requirements . . . against such party unless written notice of objection to its contents is given within 10 days after it is received."*[46] What does this mean to your organization that commonly relies on fax confirmations as the only evidence of a deal? It depends on the circumstances.

1. If both companies are Merchants, the receiving party has 10 days to object to anything contained in the confirmation—unless the contract specifies a shorter period that is not unreasonable in the industry

2. If only one company is a Merchant, the Code does not specify any time limits for objection to the terms in a transaction confirmation

The original question may still go unanswered when faxes with differing terms simultaneously cross over the phone lines. The Code doesn't satisfactorily answer this question, and if the base contract doesn't address this issue, the answer may best be found in the rules regarding offer and acceptance for a particular transaction.

If the base contract specifies one party as the *confirming party*, then that party will have primary responsibility for sending the confirmation and will—absent any contradictory language in the contract—be the master of the offer having the ability to limit the form of response. The base contract may further address this question by providing that the only acceptable forms of acceptance are: (1) signing the confirmation sent by the confirming party; or (2) letting the response deadline pass without objection. When the non-confirming party does not have a contractual right to respond with its own form of confirmation, many of the problems associated with this question may simply disappear.

"If the confirmation and a tape recording of the relevant transaction conversation don't agree, which controls?"

When an organization's base contract resolves conflicts in favor of the taped record, that means that its traders have been well-trained, its records are maintained in a logical manner, and it has taken the necessary steps to move from a paper-driven transactions to data-driven transactions. The topic of trader and organizational preparedness has been covered in relation to admissibility of evidence earlier in this chapter.

The harder question arises when the contract doesn't specifically give priority over one or the other form of *writing*, thus opening the door for multiple problems that most certainly will occur. The Code gives very little guidance, if any, in this area. Careful contractors will include language in all contracts that deals with this and designates one form as the preferred form or proof.

The question above is typically asked when the contract has been negotiated and signed, without any contract language for resolving conflicts between two types of proof that the deal exists. At the time this guide was written, many industry participants had just begun using tape recordings. No formal procedures for use or maintenance of the recordings had been implemented by many organizations. Traders and others being taped either hadn't been provided the necessary training to ensure complete telephone records, or may have been careless in their phone habits. In states requiring prior consent of employees whose conversations are being taped, consent forms may not have been signed by affected employees.

The contract should specify which form of document will control, and during the transition phase from fax confirmations to tape recordings of trader conversations as evidence of the writing, most organizations still specify that the written confirmation will control. The industry is in a state of evolution from paper to electronic data and confirmations will undoubtedly become obsolete at some point in the future.

Applying Practice #4

The fourth practice given at the beginning of this chapter illustrates this situation:

"Your organization requires written confirmations of all transactions and also tapes telephone conversations between traders. The form contract you use states that in the event of a conflict between the written confirmation and a taped recording of the transaction conversation, the tape will prevail for evidentiary purposes."

The practical result may be that sometimes the confirmation will be the contract, but when the tape conflicts, then the tape itself will become evidence of the contract. This may be a major issue at some point and careful gas contractors should think through the evidentiary implications when crafting contract language that addresses both confirmations and tape recordings.

"Can we handle the variety and number of deals being done?"

This is an administrative issue that may become a contract problem. Let's say that your house contract provides only for *firm* and *interruptible* obligations and that the definitions in your contract are not unusual. Assume that the traders in your organization are buying and selling gas at obligation levels different from firm or interruptible. If your gas management or contract management system is electronic, you may be limited to the types of transactions you can identify in the system, so when a deal is entered into the system it can be classified in one of two ways—firm or interruptible.

When your electronic gas management system broadcasts information for nominations, the gas schedulers will rely on the designated level of delivery obligation when gas is re-routed, curtailed, or otherwise moved around.

Matching firm sales with firm transportation is routine, but what will the schedulers do when the deal was actually done for a level of obligation somewhere in between firm and interruptible—say, baseload? Does the scheduler have enough information to make decisions about curtailing deliveries she or he believes to be interruptible from the system designation when the deal was actually for a delivery obligation higher than interruptible, but less than firm? If that is the case, your organization may have to pay damages to the other party for failure to deliver when an unsuspecting gas scheduler cuts gas being delivered to the other party.

Since this is a practical guide to gas sales, you're going to get some very practical advice on this question. First, the gas management system upon which your organization relies must be structured to recognize all types of transactions being entered for disparate applications. Make sure your gas management system can accommodate all the types and levels of business being done, as illustrated below.

Applying Practice #5

"Your organization has an electronic energy management system, allowing multiple applications of transactional information entered into the system. Once entered, the information for a transaction is automatically converted into the appropriate form that will then be used by departments within the organization including accounts receivable or payable, volume administration, dispatching, contract administration, credit, etc."

A second component of this practical approach is currently seen in middle-office and back-office procedures. Many energy businesses have downsized to the point they've almost turned inside out. Fewer employees are being asked to do more work, and while shareholders may be giggling all the way to the bank, many of these businesses have fallen behind in the administration of transactions. This may include too few credit analysts, too few contract analysts; in fact, too few of almost every position but those who buy and sell gas. Gas contractors are being asked to do more with less support, so contract procedures in many cases have been streamlined to address business realities.

Applying Practice #6

"The number of transactions done on a daily basis by your organization has risen dramatically during times of downsizing and layoffs. You use fax confirmations as the record of those transactions, but the sheer number of confirmations coming across your fax machine has become overwhelming. You do not have enough employees to assure timely confirmation checks."

All of the problems that have been discussed relating to confirmations become magnified when too few employees are expected to review too many confirmations, particularly when a plethora of confirmation questions are raised repeatedly. If your organization regularly agrees to confirmation deadlines of one or two business days, with too few employees to handle the load, be prepared for problems down the line.

"If gas flows without a contract in place, what are the terms of that contract?"

This final question should nearly answer itself, since much of the information in this section relates to questions of performance and terms of performance. If there is no writing at all, a tape recording of the negotiations will establish the major points of quantity, price, delivery point, and delivery period. The rest of the terms will be those found in Article 2 of the UCC. Every issue of performance must be viewed in the context of the Code. If either party fails in its delivery obligation, the Code establishes the rules to follow. If a force majeure event occurs, the Code establishes the rules to follow. If delivery cannot be made at the agreed point, the Code establishes the rules to follow. Every issue of contract formation, performance, or remedy will be addressed in the Code.

REMEDIES FOR NON-PERFORMANCE

The UCC contains some of the most comprehensive set of seller's and buyer's remedies found in the law. Most of the remedial terms are found in Sections 6 and 7 of Article 2, and the remedies for specific failures differ, depending on the nature of the failure and the harm cause by the non-failing party.

Contracts establish a series of rights and obligations. Whenever one party has a contractual right, a corresponding obligation will attach. For example, sellers have the obligation to deliver gas and a right to receive payment for gas delivered. Buyers have the right to take delivery of the gas and the obligation to make payment for gas delivered. Because of this balancing of rights and obligations, when one party fails in an obligation, the other, perforce, has not received the benefit of the contractual right, and is given a remedy.

Sometimes, as indicated in the previous paragraph, the seller's obligations are different from the buyer's; other times the obligations are identical. For example, in the following gas contract clause, the rights and obligations are identical.

"To comply with transporter(s)' nomination deadlines and the routine business practices of transporter, seller and buyer shall communicate to each other the daily receipt and delivery quantity and all other information necessary for all third parties, including transporters, to schedule the gas. Telephone, facsimile, electronic data interchange, or any other mutually agreed means may be used by the parties to communicate nomination information to each other and to transporters. Both parties shall provide each other with timely notice of any changes to nominations prior to transporter(s)' nomination deadlines."

The word *shall* typically indicates contractual obligations, while use of the word *will* or other action word designates an action resulting usually from performance of an obligation of either party. In this nomination paragraph, both parties have an obligation to communicate nomination information and to provide timely notice of changes to nominations. Without this type of obligation language in a gas contract, the parties wouldn't have any duty under the gas sale contract other than the duty of good faith in communicating with each other about their respective transportation service agreements.

The UCC carefully describes the types of failures most common to sales of goods and the remedies available when failures occur. What can go wrong under a gas sale contract? A seller or buyer could, after the contract is made, decide not to go through with the deal. A seller could fail to deliver the gas or breach any of its warranties. A buyer could fail to take gas, wrongfully reject the tendered gas, or fail to pay for the gas.

Seller's or buyer's failure

In most cases, the remedies available to each party for the other's failure are not exactly the same. Since the seller is giving up title to and possession of the goods to a buyer, the seller's responsibilities are generally a bit more cumbersome than the buyer's since it was the party placing the goods in commerce. Consequently, the buyer's remedies reflect that implicit imbalance between its obligations and those of the seller, and the buyer's remedy for a seller's breach will usually be a bit more generous than those available to a seller for the buyer's breach. This is true even when the failure is one of repudiation. The discussion of this topic began in chapter 4, and will be expanded in this section to cover the applicable remedies.

One important distinction between remedies available to the seller and those available to the buyer under the Code is that buyers are allowed to recover both *incidental* and *consequential* damages in addition to the basic remedy. Sellers are entitled to recover only incidental damages. In proper situations, courts may award damages in addition to those meant to compensate the party for its direct loss, including incidental and consequential damages, and each of those will be discussed at the end of this chapter.

Repudiation

Either the seller or the buyer may fail in its obligations even before the gas has started to flow. This happens when one party indicates its intention, either expressly or implicitly, not to perform. Assume that a gas seller and buyer agree in September for a gas sale that will begin the following January and run the entire year. If between September and January the seller indicates that it will not deliver the gas in January, it will have repudiated the contract, giving rise to remedies available to the buyer. A failure of this type that occurs before the contract term begins is termed *anticipatory repudiation.*

Now assume that performance under the arrangement described above has begun, and the buyer notifies the seller in March that it will not continue to perform beyond that month. In this example, the buyer has clearly indicated that it will not perform, and this action constitutes contract *repudiation.* In either situation, the remedies available to the buyer for a seller's repudiation or to the seller for a buyer's repudiation are similar but not always the same. When one party explicitly states its intention not to perform, the other may pursue remedies available under the Code. The more difficult case occurs when the repudiation is implicit, rather than explicit.

As discussed in chapter 4, it may be difficult to prove whether an anticipatory repudiation has occurred, and the non-repudiating party must carefully proceed if it believes that the other has repudiated. Some situations are more difficult to pin down, but any repudiation must be unequivocal, and it must "substantially impair the value of the contract,"[47] although a repudiation may be for the entire contract or only for partial performance. Given those guidelines, if the seller declares that it will not make delivery of 5,000 MMBtu/day, but only 3,000 MMBtu/day during the term, that action will constitute a repudiation if "material inconvenience or injustice will result if the aggrieved party is forced to...receive an ultimate tender minus the part or aspect repudiated."[48]

This is an extremely vague area of UCC law. In the absence of a direct statement by the seller or buyer that it will not perform its contractual obligations, the other party is often left with a real dilemma on its hands. If it starts to enforce any of its remedies for the other's contract repudiation, and is successfully challenged in that action by the party, it may have breached the contract itself through its remedial activities.

To illustrate this dilemma in action, assume that in July PUMP and WIDGET negotiate a 2-year (January to January) gas sale for 5,000 MMBtu/day at an index price. In October, PUMP attends an industry trade show. He tells FLEX, one of his industry friends, that he is thinking twice about the upcoming sale to WIDGET because of some unanticipated additional transportation costs. Later that night, at one of the numerous conference parties, FLEX casually mentions the transportation situation and predicts this will effect every shipper on the system. A reporter for one of the gas industry publications hears the conversation and does some research on her own. She determines that the transportation issue is a big deal and authors a story in her organization's daily publication predicting doom and gloom for shippers on that pipeline.

PUMP becomes increasingly concerned after reading the story and calls WIDGET, asking to renegotiate the price. PUMP says that it will be very difficult to perform under the agreement because of the transportation difficulties. WIDGET replies that the deal has been made and no changes are possible. When PUMP responds, "I don't know if we can do this deal," has PUMP repudiated the contract?

Unless PUMP makes an unequivocal statement like "We won't do this deal," WIDGET must proceed carefully.

Anticipatory repudiation

If either party, before performance is scheduled to begin, indicates to the other in an unequivocal way that it will not perform, and if the failure to perform would substantially impair the value of the contract, then the other party may elect to do any of the following:

- wait for a commercially reasonable time for the other to begin performance
- resort to remedies for breach even if it has notified the repudiating party that it will await its performance, and
- suspend performance[49]

The paths taken by a buyer for the seller's anticipatory repudiation is different from the path taken by the seller under the same circumstances.

If the buyer has elected to treat the seller's anticipatory repudiation as a breach of contract, it may enforce any of its rights enumerated in Section 2-711. Those rights are:

1. To *cover*, by making a reasonable purchase of substitute goods, and collect the difference between the cover price and the contract price, together with any *incidental* or *consequential* damages, but less any expenses saved in consequence of the seller's breach[50], or
2. To collect damages equal to the difference between the market price for the goods at the time the buyer learned of the breach and the contract price, together with any incidental or consequential damages, but less any expenses saved in consequence of the sellers breach[51]

Repudiation—seller

In addition to the two remedies available for anticipatory repudiation, if performance has begun and goods have been delivered, the buyer may also entertain either of the following:

- as to any goods for which a prepayment has been made, recover the goods[52], or
- under some circumstances obtain specific performance, or *replevy*[53] the goods[54]

Either of these two remedies would be feasible when the buyer has prepaid for goods not yet delivered by the seller. Those circumstances would apply rarely in gas sales, so the first two remedies are by far the most frequently employed by aggrieved buyers.

Repudiation—buyer

Remembering that the seller has a right to wait for the buyer to perform for a commercially reasonable time and to suspend its own performance in the meantime, if a repudiation is for the whole contract, the seller has the right to pursue any of the following remedies for breach of the contract:

- stop delivery by any *bailee*,[55] which in the gas industry probably means delivery by the transporter[56]
- resell and recover damages measured as the difference between the contract price and the resale price, together with any *incidental* damages incurred by the seller, but less any damages saved in consequence of the buyer's breach[57]
- cancel the contract[58]

The UCC remedies available for repudiation to both aggrieved sellers and buyers were designed in large part for sales of widgets. Natural gas is a unique fungible product, and the inability to "identify the goods to the contract," a concept underlying some of the remedial measures available under the UCC, poses some very real problems for gas contractors. It isn't uncommon to be faced with a situation requiring a *stretch* of Code interpretation to make the enumerated remedies practical for gas sales. This is one of the reasons why routine gas sale contracts will contain a clause enumerating and sometimes limiting remedies available when one party fails. Contract repudiation is rarely addressed in gas contracts, so these provisions of the Code will apply when either party stops performing or unequivocally indicates that it will not perform.

Seller's failure

Sellers have an obligation to deliver the contract quantity. Failure to deliver all or any part of the agreed quantity may subject the seller to any of the damages provided in the Code. If the seller fails in an installment, as would be a typical situation in the month-to-month gas industry, any non-conforming delivery in one installment is considered to be a breach of that installment only unless the entire value of the contract is impaired.[59] If the entire value of the contract is impaired, the entire contract will have been breached.

Non-delivery. Failure to deliver gas is probably the most common reason for non-performance by sellers in the gas industry. A typical gas sale transaction means monthly nominations, delivery, and payment in monthly installments. Under the Code, if a seller makes a *non-conforming* delivery during any month, the buyer may reject that installment.[60] In natural gas sales, would a failure to deliver the entire contract quantity constitute a non-conformity giving rise to the buyer's remedial rights of rejection? There are no clear answers to this question as this section of the Code was undoubtedly meant to address quality of the goods. Gas sellers and buyers that have not specifically addressed installments in their gas contract may be left wondering how this section would apply. Even a general industry presumption that this section of the code applies to gas sales may not make it so—another reason that remedial limitations are specifically set forth in gas contracts.

If a seller fails in its delivery obligation, the buyer's damages are those enumerated in Section 2-711, the same remedies available for the seller's repudiation. Contract clauses that specifically describe and limit the remedies available to buyers usually mirror those available under the UCC.

Breach of Warranty

As discussed in chapter 4, only sellers make warranties under a routine

Chapter 6 Essentials of Contracting

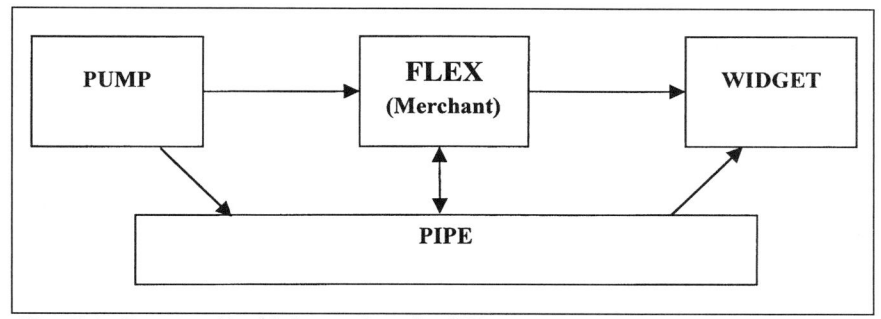

```
┌─────────┐      ┌─────────────┐      ┌─────────┐
│  PUMP   │─────▶│    FLEX     │─────▶│ WIDGET  │
│         │      │  (Merchant) │      │         │
└────┬────┘      └──────┬──────┘      └────▲────┘
     │                  │                   │
     │           ┌──────▼──────┐            │
     └──────────▶│    PIPE     │────────────┘
                 └─────────────┘
```

Fig. 6-1 *Situation Establishing a Breach of Warranty Action*

gas contract, and the Code specifies the types of damages that may be due a buyer if the seller fails to deliver goods that conform with contract specifications. Section 2-714 generally covers the damages a buyer may collect if the seller has failed to deliver goods in conformance with the contract *when the buyer has accepted the goods*. One of the major distinguishing factors of a breach of warranty action is that this type of action may be prosecuted when the seller has delivered goods and the buyer has taken delivery. In the gas industry, a breach of warranty based upon quality of the goods would be rare, but certainly not impossible as exhibited in Figure 6-1.

Assume that PUMP has sold gas out of the gathering system to FLEX, and that FLEX is reselling the gas at a field zone delivery point to WIDGET. The transporter, PIPE, may or may not take gas from PUMP, depending on whether the gas meets PIPE's quality specifications. Let's say that PIPE does accept some "non-spec" gas from PUMP, and that when the gas is delivered into WIDGET's facility, it causes harm to the plant.

A number of issues are raised in this example, but let's focus only on the obligation of FLEX as a seller to deliver merchantable gas to WIDGET. Because FLEX has delivered non-spec gas to WIDGET, the gas probably was not *merchantable* and not *fit for its intended purpose*, the two implied warranties given by Merchants that were discussed in chapter 4. PIPE accepted gas from PUMP (and perhaps other downstream producers) not meeting its own quality specifications. That is an issue between PIPE and PUMP under the TSA. Under the *gas sale contract* sellers have a contractual obligation to deliver goods that are at least of the same quality as other goods of the same nature, and FLEX has failed in its obligation to WIDGET.

WIDGET *must provide notice* to FLEX that the gas was non-conforming in accordance with Section 2-607 (3), but if it follows the procedures described in that section, it may sue FLEX if necessary for breach of warranty. The potential damages it may collect include "*the difference at the time and place of acceptance between the value of the goods accepted and the value they would have had if they had been as warranted, unless special circumstances show proximate damages of a different amount.*"[61]

In a proper case, *incidental* and *consequential* damages may also be recovered. In the illustration given above, the incidental and consequential damages are the topics probably keeping FLEX awake at night. (Although not addressed here, FLEX will have the same cause of action against PUMP for failure to deliver merchantable gas.)

Incidental damages

Incidental damages are defined in Section 2-715 (1) as: "*expenses reasonably incurred in inspection, receipt, transportation and care and custody of goods rightfully rejected, any commercially reasonable charges, expenses or commissions in connection with effecting cover and any other reasonable expense incident to the delay or other breach.*" Any costs incurred by WIDGET in identifying the extent of the losses, in inspecting damaged equipment, and in procuring any alternate gas or alternate fuel supply could be considered to be incidental to the breach.

Consequential damages

Consequential damages that are available only to the buyer under the current Code[62] include "*any loss resulting from general or particular requirements and needs of which the seller at the time of contracting had reason to know and which could not reasonably be prevented by cover or otherwise; and injury to person or property proximately resulting from any breach of warranty.*"[63] The element of a seller's previous knowledge is critical to a buyer's case in seeking consequential damages. Consequential damages could include lost profits,[64] liability to third persons,[65] and damages caused by delay.[66] Gas buyers that routinely waive their right to collect consequential damages in the event of a breach may be giving up the right to collect for damage to their plants, personal injury to employees, and any other damage that resulted directly because of the seller's failure.

In the illustration above, had WIDGET waived its right to receive incidental and consequential damages, without proving to the court that it would be unjust to uphold the waiver, WIDGET would be obliged to pay for any costs it incurred as a result of another's breach of warranty

Buyer's failure

The most frequent types of failure by buyers are:

- failure to take the gas, and
- failure to pay for the gas

Failure to take delivery of the gas

As was the case in the preceding discussion, a seller's remedy for the buyer's failure to take delivery of the product are the same as for a buyer's repudiation, as enumerated in Section 2-703. As was also the case with a seller's failure, the remedies available under the Code, that may be confusing and difficult to apply with natural gas sales, are typically made clearer through contract language that defines the seller's rights.

Failure to pay for gas delivered

The final type of remedy is that available to the seller when a buyer fails to pay for the goods. Failure to pay for goods delivered is a breach of the contract, and contract language will usually address that as a specific default for which the seller may cancel or terminate the contract and collect damages.

Without anything to the contrary in a contract, failure to pay under the Code is an event of breach for which the seller will have an action for the price. When the buyer fails to pay according to contract terms, the seller may recover, together with any incidental damages, the price of goods accepted and the price for any goods not taken by the buyer if the seller is unable to find an alternate market for the goods.[67]

Whenever a buyer has failed by wrongful rejection (failing to take deliv-

A Practical Guide to Gas Contracting

ery of the gas), or has failed to make payment when due, or has repudiated the contract, a seller that is not entitled to the price (because no gas was delivered) is nevertheless able to collect damages for non-acceptance of the goods. Those damages are as indicated above for are buyer's repudiation.[68]

---

1. The typical name given to the group that looks for new long-term opportunities to sell gas.

2. A letter of intent or LOI, is frequently written at the beginning of contract negotiations between two parties contemplating a long-term transaction. The LOI will state the understandings of the parties and usually specify that no contract will have been formed until a formal contract document is signed.

3. See chapter 4 discussion of UCC Merchants.

4. Direct links to many state secretary of state offices can be found at http//w3.uwyo.edu/~prosopect/secstate.html.

5. The term *redact* is used to describe the process of deleting (usually through bold marks obviously made to obliterate non-relevant terms) the language on an actual copy of the document. In a partnership agreement, for example, where contracting authority is placed with a managing general partner, all other contract terms on the page(s) would be redacted if the other party requested such proof.

6. *Ultra vires* literally means "beyond power."

7. UCC § 2-204 (1).

8. UCC § 2-204 (3).

9. UCC § 2-204 (2).

10. UCC § 2-201 (1).

11. UCC § 2-201 (3).

12. FRE, Rule 1001.

13. FRE, Rule 901.

14. FRE, Rule 803 (6).

15. FRE, Rule 801 (c).

16. FRE, Rule 803 (6).

17. FRE, Rule 901(a).

18. FRE, Rule 901(5).

19. FRE, Rule 901(6).

20. 28. USC § 1332 (a)

21. Id.

22. When large users of natural gas for electric generation construct and own their own generation facility, this is called *distributed generation.*

23. The Energy Information Administration, or EIA, is the information and publication branch of the Department of Energy. Its excellent web site is found at: http://www.eia.doe.gov/.

24. UCC § 2-205.

25. The typical name given to the group that looks for new long-term opportunities to sell gas.

26. A LOI is frequently written at the beginning of contract negotiations between two parties contemplating a long-term transaction. The LOI will state the understandings of the parties and usually specify that no contract will have been formed until a formal contract document is signed.

27. Discovery is the process whereby parties request non-oral relevant information from the other party. This is accomplished through interrogatories that typically contain a thorough list requiring production of anything that could be relevant. This information may then be used at trial.

28. UCC § 2-204 (3).

29. UCC § 1-102 (2) (b). See the discussion of Code purposes in chapter 4.

30. *Extrinsic evidence* is evidence other than that contained in the written contract. Any admissible evidence may serve this purpose, including letters, memoranda, notes to file, and witness testimony.

31. UCC § 2-202 (a).

32. UCC § 2-202 (b).

33. Id., p. 93.

34. UCC § 1-205 (1).

35. UCC § 1-205 (2).

36. White and Summers, *Uniform Commercial Code – Sales*, Fourth Edition, 1988, West Publishing Co., p.92.

37. UCC § 2-308.

38. UCC § 2-309.

39. UCC § 2-507 (1).

40. UCC § 2-511.

41. UCC § 2-305 (1).

42. UCC § 2-305.

43. Id.

44. UCC § 2-207, official copy, comment 3.

45. See the discussion of merchants in chapter 2.

46. UCC § 2-201 (2).

47. UCC § 2-610

48. *White and Summers*, infra. P. 185., quoting Comment 6 to UCC § 2-610.

49. UCC § 2-609.

50. UCC § 2-711 (a) and UCC § 2-712.

51. UCC § 2-711 (1) (b).

52. UCC § 2-711 (a) and UCC § 2-507.

53. Replevin is a personal legal action brought to recover possession of goods by the party entitled to the goods.

54. UCC § 2-711 (2) (b).

55. A bailee is one to whom goods are given for movement or delivery.

56. UCC § 2-703 (b).

57. UCC § 2-706 (1).

58. UCC § 2-703 (f).

59. UCC § 2-612.

60. UCC § 2-612 (1).

61. UCC § 2-714 (2).

62. In a proper case, a court might award consequential damages to a seller in some states.

63. UCC § 2-715 (2).

64. *Lewis v. Mobil Oil Corp.*, 438 F.2d 500, 510-11, 8 UCC 625 641 (8th Cir., 1971).

65. *Gambino v. United Fruit Co.*, 48 F.R.D. 28, 6 UCC 1056 (S.D.N.Y. 1969).

66. Strunk &White, infra. pp. 378-379.

67. UCC § 2-709 (1).

68. UCC § 2-708.

Chapter 7 | The Gas Contract

When pipelines bought their gas from producers, transported, and sold the gas to LDCs, gas contracts looked very different from the way they look today. As the gas industry became de-regulated after Order No. 436, most gas contracts retained the flavor and indeed the same language as their previously regulated contracts. Organizations that had pipe in the ground, such as LDCs and producers, continued to use the older contracts that typically contained detailed clauses on measurement, gas quality, and testing. But the *open-access* era meant that the gas contracts used for a majority of gas sale transactions became burdensome and much of the contract superfluous for gas marketing.

The traditional players (producers, pipeline companies, LDCs) were accustomed to operating under the same set of rules required when gas buyers and sellers also incorporated operational concepts into gas contracts. As larger numbers of marketing companies started buying and selling gas, they changed the landscape for spot gas marketing, and most of those changes have survived. Gas contracts today don't look much like those used in the early 1980s.

A number of similar contracts are customarily used to transact day-today gas marketing business. The types of gas sale contracts are as varied as one's imagination, but the format of a basic gas sale contract has become fairly standard. Gas contracts may be individual—*i.e.*, one contract for one deal. More commonly, a base contract is used for all types of gas sales and purchases to facilitate multiple transactions without the need to renegotiate basic terms and conditions for every transaction. Confirmations are then used to memorialize individual transactions. When a base (or master) contract is used, all individual transactions may be considered to be one *contract*, or each transaction may be a separate contract. Contracting parties should make a choice and state it in the base contract. Inadequate planning in this regard could be very costly when the other party defaults in a way that would effect all transactions or fails to perform on a single transaction.

Assume that producer PUMP and marketer GAS agree to a base contract with the following sentence:

"More than one transaction may be in place at any given time between the parties, and all transactions together will constitute one contract of the parties. A breach of one transaction will constitute a breach of the entire contract."

PUMP and GAS have seven transactions in place at one time. Let's assume that six of the transactions are for six-month transactions that will be settled at the end of each month—installment contracts—one is for a four-month delivery period where payment will not be due until the end of the period. If PUMP refuses to deliver gas under the four-month transaction, it will have repudiated that transaction. The provision referenced above may result in PUMP's repudiation of the other six transactions as well. GAS may be in a position where it cannot enforce its rights under that one transaction without placing PUMP in default of the entire contract.[1]

Let's change the facts a bit and assume that the same contract provision reads as follows:

"More than one transaction may be in place at any given time between the parties, and each transaction will be an individual contract between the parties."

In this case, if PUMP fails to deliver under the four-month transaction, it has clearly not automatically breached its obligations under the other transactions. Unfortunately, even this approach isn't necessarily fail-safe. If instead of breaching one transaction, PUMP breached the general terms and conditions of the base contract by declaring bankruptcy, for example, the bankruptcy court may consider all transactions in that case to have been breached.[2]

In this chapter the discussion will focus only on the base (or master) contract (or agreement)[3] and on the standard elements of that contract. Although the same form of contract may be used for a variety of transactions, the terms shouldn't be necessarily the same in base contracts used for relatively short-term, low-risk transactions as those for long-term, high-risk transactions. The first section of this chapter will analyze the differences between levels of risk in gas transactions and recommend some procedures to follow when making the determination of appropriate contract language in either circumstance.

You have already learned that contracts reflect the parties' rights and their obligations. Another way of saying this is contracts are a process of distributing *risks* and *rewards*. While fewer precautions may be taken in a short-term contract, there are risks present in every gas sale contract and those should be addressed—both as a matter of company policy and as a part of contract planning.

When you think about risks and rewards in terms of choosing the correct format for a contract, how do you come to your ultimate resolution? You must have an objective standard, and the ideas presented in this chapter will guide you through that process. The decision you make about which contract format to use should reflect decisions that have been made about risk acceptance.

After examining the differences between contracts developed for relatively low-risk deals and contracts developed for everything else, the remainder of this chapter will introduce you to a standard gas contract. By reviewing the standard elements of a contract, you will learn:

- the clauses that should be included in all gas contracts
- optional clauses that may be included in gas contracts
- some clauses that you will not want to include in your gas contracts

Short-Term Contracts

The terms *short-term* and *long-term* don't necessarily refer only to the delivery period of underlying transactions. As the two categories of contracts have come to be known, the short-term variety includes contracts with lower risk. When the delivery commitment is brief, the price is market-based—delivery points being used are seldom if ever constrained, the delivery obligation is not firm—the transaction naturally involves a lower risk to both parties.

Those classified as long-term are riskier because of long delivery periods, complicated formulas that determine quantity obligations and related payment rights, opportunities to re-price the deal based upon exchange prices or some other criteria, and others. So the designation, while used generically in the gas sale industry, isn't necessarily just a reflection of contract term or delivery period.

What is a short-term contract?

Short-term contracts should be different from long-term contracts. Contracts for purchase or sale of gas on the spot market typically have a relatively short delivery period. Fewer protective clauses may be needed, because this type of contract has lower risk.

Every gas organization has its own definition for short-term, by default putting all transactions not meeting that definition into the long-term (sometimes short-handed simply to *term*) category. It is not uncommon for one organization to define short-term as any transaction having a delivery period of less than one day; another might define it as any transaction having a delivery period of one week or less; yet another, one month or less. Some still define short-term as any transaction having a delivery period of less than one year. Let's assume that a house short-term contract is developed by an organization that informally defines short-term as one week or less. If that contract is used for transactions with a trading partner that defines short-term as a year or less, some contract language changes will probably be made during negotiations because risks willingly assumed for one week may be far broader than risks assumed for a year.

Short-term contracts that relate to low-risk transactions will have *fewer clauses,* will usually not include formal definitions, and may not contain clauses for early termination, confidentiality, assignment, alternate dispute resolution (ADR), or even notices.

Companies may use short-term contracts to *limit the scope* of particular transactions or the amount of business that may be done by inexperienced traders. The contract may be used for transactions having only a specified delivery period, for only interruptible transactions, or for transactions having limited dollar value.

Contracts for short-term transactions are an extremely useful part of any gas sale organization's gas contract portfolio, but there is always a danger that the contract might be used for business outside the defined parameters. When language has purposefully been abbreviated or deleted to suit the needs of one class of transactions, any transaction not meeting the definition of that class will be at risk. When that happens, the UCC may step in when a court invokes its *gap-filling* powers under Section 2, part 3.

With all that in mind, the bare bone elements of a short-term contract are the same as those found in any other gas contract—namely, that the parties have agreed to a transaction and have set forth the terms and conditions that will govern that transaction.

How short-term contracts came to be

Short-term gas contracts were first developed in response to Order No. 436. Gas prices in the closed and regulated environment were fairly stable. When open-access brought real competition into spot market sales and purchases, the industry experienced true gas price volatility for the first time. Long-term, take-or-pay contracts were a thing of the past, and while pipeline companies still used the same general format for buying gas to serve their LDC customers, new entrants to the industry brought with them an entrepreneurial attitude. Their competitive approach to a previously regulated business meant a new approach to contracting as well.

The concept of a *spot market* and of interruptible transportation service and gas sales was new in 1985, but nearly every company ultimately developed its own house contracts. At the beginning of the new gas industry, the concept of selling (or buying) both firm and interruptible gas *under one contract* hadn't yet been developed. There were separate contracts for sales and purchases and separate contracts for interruptible and firm business.

The first spot market contracts were a combination of the past and future, because the present hadn't been identified. The gas contracting process was in a transitional state from 1985 until the early 1990s when organizations finally began to hit their contracting stride. From those first basic one-sided short-term contracts, the gas contract has progressed to attunement with the commoditization of natural gas. Except for the most exotic types of transactions, a basic gas contract today provides maximum opportunities for buyers and sellers to utilize all the innovations available in transportation, intra-contract hedging, and electronic communications.

Short-term contracts and the future

Electronic, financial, and regulatory innovations made an enormous impact on natural gas contracting in the mid-to-late 1990s. Those innovations included electronic contracting (that will be discussed in chapter 9), increased use of hedging techniques as part of a company's risk management policy (discussed in chapter 1), and ongoing improvements to the nomination capabilities of transporters.

When shippers are in communication with transporters solely through electronic mechanisms, electronic contracting will become more a standard than a novelty. As risk management techniques are further integrated, gas contracts are poised to accommodate this future as well. As these developments contemporaneously occur, only the frailty of humans, the state of the law, government regulators, and incompatible electronic capabilities will limit the possibilities for transaction flexibility. Chapter 9 will propose that gas contracts change even more in the future to completely fold together the three elements of technology, regulation, and risk management.

LONG-TERM CONTRACTS

The contracts typically referred to as *long-term* or *term* contracts today may look like the regulated contracts in use before *open access* times or they may have a completely new look. Producers, for example, still require the same type of detailed measurement language that has always been a part of production sale contracts. LDCs still continue to be constrained by their regulators and by the operations of distributing gas, so contract language continues to address both of those issues. The word being used to describe these unique contracts, however, doesn't reflect in any measure the true nature of the transaction. For example, a very complex but short-term peaking sale may involve page upon page of necessary obligatory language, with various incentives for performance. This is not a routine gas contract, but some organizations would define this document memorializing the transaction as a short-term contract simply because of the delivery period, and by implication, a low-risk contract.

What is a long-term contract?

Term contracts usually do have a relatively lengthy delivery period. Although these words are used to identify any of a number of different contractual arrangements, the general understanding of *term* in the past, was that the relevant transaction had a delivery period of more than one year. During the latter part of the 1980s, everyone operated off the same definition for *term*, and other risk variables had little meaning. Prior to 1990 when NYMEX futures contracts began trading, the length of the deal was the biggest risk, so that one word served as the identifier for all such transactions. Today, as you will see in the information that follows, a *term* transaction is one characterized by a long delivery period (that may be any period from 30 days to many years), and one with relatively high-risks attached.

The risk may be quantified in a number of ways that may include the *price*, the *deliverability* of a seller, the continuing *creditworthiness* of the buyer and seller, the availability and cost of necessary *transportation*, the potential for *merger* or buyout of either party, *hedging* activities of either party (and associated risks), the *nature of the business* entities involved, and the *character of the transaction*. Each of these topics is discussed elsewhere in this guide and the list is given merely to illustrate the increased complexity of the gas sale industry today.

Another distinguishing feature of many *term* contracts is that the base contract format may not be an appropriate vehicle to memorialize this type of deal. A *term* contract is generally not one of multiple (and similar) transactions between the two parties, so the most commonly used format is a single contract for the transaction. The two parties may also have a base contract in place for all other conventional transactions.

Traditional long-term contracts

Three major elements dictated the look of gas sale and gas purchase contracts prior to 1985. The first was *regulation,* the second *operational control,* and the third *price*. Remember that prior to 1985, all interstate gas sales were subject to regulation by the FERC under Section 7 authority of the Natural Gas Act.[4] States regulated sales in intrastate commerce and also sales made by LDCs.

After Order No. 436 was issued, the interstate pipelines had the option to maintain their systems as they always had or to admit third-party shippers. All ultimately chose the latter, and one of the major benefits to open-access pipelines was the ability to transport gas under the *blanket transportation certificate*.[5] Once third-party shippers became an active part of the gas industry, gas marketers, LDCs, and producers had new opportunities to contract; thus new types of contracts.

Because both producers and LDCs were emerging from a previously regulated environment, they were accustomed to seeing the same types of clauses needed by regulators, even though regulation had changed. It wasn't uncommon for gas sale contracts to contain extensive language addressing only regulations imposed by any governmental entity having jurisdiction.

The LDCs and producers both owned equipment and facilities in the ground. In order to ensure freedom of operations, the old gas contracts also contained numerous clauses intended to clarify the necessity for control of measurements, gas flow, and other operational considerations. Many of these contracts today still contain necessary operations language, particularly when a producer is selling at the wellhead or an LDC is buying at the city gate.

The final distinguishing factors of these contracts were the price clauses. While many spot market transactions were for a fixed price or for a floating index price, longer term transactions required necessary price adjustment clauses discussed in Chapter 2. Pricing formulas could easily result in several pages of contract language.

Long-term contracts and the future

As noted earlier in this chapter, some of today's *term* gas contracts retain many of the same elements as those seen in contracts of bygone days. Let's focus on some of the new aspects of high-risk contracts.

Price risk. Even with the advent of hedging potential through regulated futures contract exchanges and OTC derivative transactions, price risk is a critical element in all term contracts. *Price risk* is the same as *market risk*, defined as the risk that the market will move against a party. If you are a gas seller, and your sales are at a fixed price, your risk is the market will rise. If you are selling gas priced at an index or other floating-price scale, your risk is the market will go down. On the other hand, buyers paying a fixed price will be at risk when the market goes down, and buyers paying a floating price will be at risk when gas prices rise.

To hedge a straight fixed price purchase in a declining market, a buyer may sell futures contracts (go *short* in the futures market) and liquidate the position by buying the same number of futures contracts later when the market is lower. Or the same buyer might find a willing swap counterparty and trade its fixed price for a floating price. In the simplest situation, these tools will work, but most term sales are also complicated sales that may involve different pricing for different delivery obligations—more complex hedging techniques are needed.

An additional price risk involves seemingly unrelated future events that make the effective sale price less or more than anticipated. For example, assume that FLEX is making a term sale to GASCO. In a *reasonableness review* by the PUC in the second year of a five-year sale, 10% of the gas cost is found to have been imprudently incurred. That 10% change over three years will vastly change the economics of that deal for FLEX, and GASCO may not be able to transfer that additional risk to a third party in subsequent corrective hedging.

Another future event that could dramatically change the effective price is a change in tax laws. On any given day, in any given year, the Congress could pass a law imposing a 5, 10, or 15% consumption tax. If that were to happen, (and absent contract language to the contrary) the buyer would probably be left holding the tax bag since any new tax will most likely be imposed on gas consumers. Most gas contractors negotiating or drafting tax language will account for a future tax in one of three ways:

- Tax-sharing
- Contract termination
- Price renegotiations

A Practical Guide to Gas Contracting

It would not be unusual to see one of these clauses in a *term* gas contract:

"If, at any time subsequent to execution of this Contract, any governmental authority with jurisdiction levies or otherwise imposes any new tax, surcharge, fee, or other charge (a "New Tax"), the effect of which is to (i) add a new tax burden or (ii) increase a current tax burden for either of the parties, the parties will equally share in any such New Tax payment and the party designated by the taxing authority as the party responsible for paying the tax shall assume all reporting functions."

"If, at any time subsequent to execution of this Contract, any governmental authority with jurisdiction levies or otherwise imposes any new tax, surcharge, fee, or other charge (a "New Tax"), the effect of which is to (i) add a new tax burden or (ii) increase a current tax burden for either of the parties shall negotiate in good faith to establish a new Contract Price for all gas subject to this Contract. The purpose for such price negotiation will be to distribute the financial impact of the New Tax to the parties in a manner that preserves the economic benefit to each at the time this Contract was executed."

"If, at any time subsequent to execution of this Contract, any governmental authority with jurisdiction levies or otherwise imposes any new tax, surcharge, fee, or other charge (a "New Tax"), the effect of which is to (i) add a new tax burden or (ii) increase a current tax burden for either of the parties, then the party to this Contract designated by the taxing authority as the party responsible for paying such New Tax (as distinguished from collecting or reporting the New Tax) may, upon no less than _____ notice to the other party, terminate this Contract as of the Termination Date, with no liability to the other party, other than any liability incurred prior to the Termination Date. For the purpose of clarification, the terminating party

will not be liable for payment of any early termination damages, whether provided for in this Contract or pursuant to the Uniform Commercial Code as adopted in the Governing Law state."

Seller's deliverability. During the discussion of the common interests of all gas buyers in chapter 2, the deliverability issue was discussed. In *term* contracts, if a buyer wants to be completely sure that its gas supplier will be able to deliver, it may include any of a number of contract provisions to give it that security, some of which were discussed in the previous chapter. An additional note should be added at this point, though differentiating the attitude of gas buyers during the 1980s and the attitude of gas buyers today.

Then, there weren't concerns about depletion of a natural resource. Today's world is different, and although the industry predicts a rosy future, we cannot escape the fact that it is getting harder and harder to find new gas reserves, and current exploration and drilling technology will not be able to sustain exploration and production activities in some parts of the world. Today's technology must be enhanced to provide more economical alternatives to exploration and drilling methods. Without continuing technological improvements in all sectors of the industry, a real prospect exists for gas supply depletion before the middle of the Twenty-first Century.

With all that in mind, gas buyers are more wary of their suppliers' ability to perform and are continuing to require their suppliers to provide proof of reserves or proof of adequate gas supply contracts. The issue of deliverability becomes critical in another area as well. Current transportation capacity cannot handle projected needs for gas in the future. A buyer's deliverability concern may also be evidenced by required proof of transportation dedicated to supply its gas needs during the entire term of the transaction.

Buyer's and seller's creditworthiness. This topic was also discussed in chapter 2, but you should understand how typical credit issues dovetail with deliverability as discussed in the preceding paragraph. Credit evaluations in the gas industry today are completely different from the evaluations that were done 10 years ago. During most of the 1980s and into the 1990s, natural gas was overproduced. Gas supplies were plentiful, and the buyer's only concern at that time was whether its supplier had access to natural gas. Today, with deliverability concerns, buyers are becoming increasingly more interested in the long-term prospects of their suppliers, both in terms of the supplier's creditworthiness, and because of deliverability.

Mergers, acquisitions, and so forth. This guide is being written in a time of intense acquisition activity in the energy industry. Cash-rich electric utilities that must compete in a deregulated environment are looking for desirable gas industry organizations that will assist them with their competitive endeavors in the future. Large oil and gas producers are becoming even larger through acquisition of other major gas operations. The fallout of this activity for companies involved in term deals is that the potential sale or acquisition of either company must be contemplated during contract negotiations. The contract clause most relevant to this concern is the *assignment* or *transfer* clause, and parties to a term deal must carefully consider the potential adverse results of any assignment clause because of a merger or buy-out. Assume that PUMP and WIDGET have a five-year firm deal for 5,000 MMBtu/day. A plain vanilla base contract is used for the transaction, and monthly confirmations are sent by PUMP to WIDGET. The contract contains a clause allowing either party to assign its rights and transfer the contract without consent only if that assignment is to a creditworthy affiliate or as part of a sale of assets. When WIDGET's parent company begins negotiations to sell WIDGET to WIDGET2, PUMP may have no right to object (other than for credit reasons) to the assignment and transfer—even if PUMP would not otherwise do business with WIDGET2. If PUMP had previously been involved in expensive litigation with WIDGET2 and has set a policy that under no circumstances would do business with WIDGET2, it may nonetheless be required to accept the assignment.

Hedging activities. Even though hedging is a way to transfer price risk to a third party, additional hedging considerations abound in term transactions. Once a hedge has been put on a transaction, sound risk management principles dictate constant monitoring of the relevant hedge to determine whether or not that original hedge continues to respond to the organization's goals. Goals change as vice presidents change, even though a solid risk management policy should tightly control and limit any changes to established policy. Hedges must be relevant to the transaction and to the market.

Organizations that implement hedging strategies consider a deal hedged to be a package, the physical gas delivery being one part of the package, and the hedge the other part. A term contract will frequently contain clauses that address the costs associated with liquidating a futures position or unwinding a derivatives position, if that "undoing" activity resulted from a failure of the other party. The rule seems to be that if one party causes another to incur hedge-related costs, the party causing the problem will be responsible for resulting costs.

Nature of the business of contracting parties. From a supplier's perspective, a *term* sale to an LDC is different from a *term* sale to a cogeneration plant, or a steel plant, or a paper mill, and all of those transactions are different from a *term* sale to a new business venture undergoing necessary financing and other start-up arrangements. The owners of such new ventures are routinely required by the financing parties to have numerous gas supply contracts in hand months or years before scheduled gas flow. What goes through the head of a gas supply representative when negotiating any of the different types of deals indicated above?

Every deal is different, and each will come with its own set of challenges and potential rewards. Savvy gas sellers will study and learn the business of the potential customer to better serve that customer's needs.

Character of the transaction. On one hand, let's consider a five-year sale by PUMP to WIDGET having three tiers of delivery obligation and three different price calculations. On the other hand, let's consider a five-year baseload sale by FLEX to GASCO. The businesses are all different, but so is the character of each transaction. Baseload sales (constant daily requirements sold on a firm basis) require arrangements of constancy—*i.e.*, gas reserves (or gas supply from contracts), that must be sufficient to meet a constant demand. Transportation must be constant to meet a constant demand as well. Only one type of delivery is required and only one price calculation is necessary. Contract language used for this *term* deal will probably be straightforward and contain few surprises, other than those necessitated by any special circumstances of either FLEX or GASCO.

The sale being made by PUMP to WIDGET is different. Tiers of performance obligations and corresponding tiers of prices will result in lengthy contract language to:

- describe each tier of performance obligation
- describe the price calculations for each price
- determine which tier will be considered to have been cut first if either party fails in its delivery obligation
- establish a balancing hierarchy if either party incurs imbalance charges on its transportation service agreement
- develop netting procedures to be used for routine payments

With more varied options in one contract, the contract language sometimes seems to expand exponentially, and it would be a mistake for a gas contractor to streamline this type of language. Whenever more than one delivery obligation exists in the same contract, special care should be taken to completely describe the obligations and any remedy that may apply if a failure occurs in one, but not all, levels of obligation.

Those are the major distinguishing marks of short-term and long-term gas contracts. Each has a purpose and was developed to meet a particular need. While larger numbers of long-term contracts were negotiated in the early years of open-access, the industry has moved to high numbers of low-risk transactions and low numbers of high-risk transactions.

ELEMENTS OF A STANDARD GAS CONTRACT

The standard elements that will be discussed in this section include the following:

Title
Heading
Definitions
Purposes and Procedures
Term

Chapter 7 The Gas Contract

Obligations
Transportation, Scheduling, and Balancing
Gas Measurement and Quality
Delivery Failure
Billing and Payment
Financial Responsibility
Default
Contract Cancellation/Early Termination
Force Majeure
Taxes
Warranties and Indemnities
Notices
Miscellaneous
Signatures

Gas contracts are always divided into sections differentiated by content. Those contract divisions are traditionally called "articles," (in which case, Roman numerals will usually be used to differentiate or number the articles) but may also be called "sections." Each of the articles is organized by paragraph, and every paragraph should be numbered. In Article I, the paragraphs will be numbered 1.1,1.2, etc. Most gas contracts follow a similar order as well. The *Purpose and Procedure* article, for example, is always found at the beginning and the *Miscellaneous* article is last. A natural progression of articles that roughly follows the progression of the transaction is typical. In practice, this is illustrated by:

- contract formation and administrative issues
- nominations
- flowing gas issues (including the *Remedies* article)
- billing and payment
- all clauses that are independent, usually grouped in the *Miscellaneous* article

In this section, we'll examine every gas contract clause, and you'll learn how to distinguish necessary contract language from dispensable language.

Base Contract for Gas Sales and Purchases

Title of the contract

Every written contract document has a name. Sometimes that name will just be "contract" or "agreement," but more commonly in gas contracting a document's title will reflect more about its contents. As you learned in chapters 4 and 6, the UCC differentiates the meaning of the two words, *contract* and *agreement*, and in usage, gas contractors must be aware of those differences. Industry usage, however, does not differentiate between the two, as one organization will call its house document a contract and another will call it an agreement.

It's usually better to more explicitly state the entire scope of the document by calling it a "Contract for the Sale and Purchase of Gas" or some similar title.

Heading

"This Contract is made and entered into this ___ day of _____, by and between _____, a Delaware corporation, having a place of business in _____ ([short name]), and _____, a Kansas corporation, having a place of business in _____, ([short name])."

This clause identifies both parties to the gas contract. Traditionally, the designation of "seller" and "buyer" was also made in this clause. When either party may be the buyer or the seller, the statement establishing that fact may be in the heading or in another part of the contract. Sometimes the heading will contain less information, other times more. Here are examples of each:

1. *"This Contract is made and entered into this ___ day of _____, by and between _____, a Delaware corporation, ([short name]), and _____, a Kansas corporation, ([short name])."*

2. "This Contract is made and entered into this ___ day of _____, by and between _____, a Delaware corporation, having [its principal/a] place of business in _____ ([short name 1]), and _____, a Kansas corporation, having [its principal/a] place of business in _____, ([short name 2]). [Short name 1]'s DUNS number is _____ and its identification number for tax purposes is _____. [Short name 2]'s DUNS number is _____ and its identification number for tax purposes is _____. Either [short name 1] or [short name 2] may be the "Buyer" or the "Seller" in a particular transaction. Buyer and Seller may be referred to as "Party" in the singular and "Parties" in the plural and each agrees as follows."

Each party wants to get as much information as possible about the other. Why? Because some day your company may have to sue the other party, and the heading provides you with the legal basis necessary to file that lawsuit and to effect *service of process*[6] on the other party. The locale of organization, the place of business, and the official name of each party are all relevant pieces of information for legal procedure purposes.

With FERC standards that require identification of parties by their DUNS number,[7] a more thorough identification is made in order to transport gas. The best identification includes the name, place of incorporation, the location of a place of business, the DUNS number and the federal tax ID number. The DUNS number and federal tax ID number may be included in the *Notice* article of the contract in lieu of the heading.

Either in the header or prior to the signatures at the end of the contract, the parties should include a statement of mutual agreement.

Definitions

Definitions will typically be the first contract article, but they are only used if the words being defined are:

- technical
- used in other than their ordinarily understood sense
- vague or ambiguous or
- required by context

Gas contracts typically contain some formal definitions. For the words that meet any of the four criteria indicated above, the first choice a drafter must make is whether the defined words will be contained in a separate article or defined as they appear in contract text. The usual reason for dedicating a separate article to definitions is in the numbers—if your contract requires many definitions, it is simply more efficient to place all definitions in one part of the contract.

Another school of thought when many definitions are required by context is to include the definitions in text for the contract reader's convenience—*i.e.*, to define the words as they are used in the contract. Either way, when definitions can clarify the parties' intent, they should be used. The importance of definitions? My husband, who is involved in lawmaking, is fond of saying, "I will let the opponents write the law if I can write the definitions."

Technical words. Many contracts for the sale of gas at the wellhead contain numerous references to measurement requirements, calibration, and meter testing. Since the terms used may be quite technical in nature, they are frequently defined in the contract. The same is true when gas is being purchased by an industrial consumer whose plant requires stringent quality specifications to prevent damage to facilities and equipment.

Another frequent use of technical words is description of the parties' respective obligations regarding tiered delivery obligations, prices, and remedies. Many term contracts define a seller's delivery obligations on a daily, monthly, or annual basis. It wouldn't be unusual to see defined terms such as: "maximum daily delivery quantity" or "minimum monthly delivery quantity." Corresponding price clauses and remedies for failure to comply with the obligation will frequently be stated in terms of the "maximum monthly deficiency quantity" if the seller is obligated to deliver a minimum quantity with a payment due the buyer for failure to maintain the minimum delivery for the stated period of time.

Words not used in their ordinarily understood sense. When you use the word "nice", don't you typically refer to something or someone who is agreeable and pleasant? Most of us probably would adopt that usage.

Would it surprise you to know that the first definition for the word *nice* in Webster's Dictionary is "difficult to please, very careful; fastidious; refined." The sixth definition for that word is "agreeable; pleasant; delightful; attractive; well-mannered; kind. . ." The English language is filled with words that sound alike—the same word may have disparate meanings, and sometimes the same word is pronounced differently, depending on the meaning. The language itself is a hurdle to be overcome, and it is dangerous to presume that the meaning you ascribe to a word is the same meaning used by your listener.

In any contract, a word that could be subject to different interpretations should be defined. Here are a few examples of the words that may have different meanings in the gas industry:

Baseload—As already discussed, this term is susceptible to multiple meanings in the gas industry. Is it a level of obligation or a delivery requirement?[8]

Swing—This word may have numerous meanings, but most commonly, *swing* is used to describe one of two things. First, it may be a delivery obligation level, similar to *interruptible*. But that same word might be used to describe the party responsible for the risk of overdeliveries and underdeliveries for balancing purposes.

Swap—This term gained acceptance in the gas industry when over-the-counter derivatives entered the picture. *Swaps* may be either financial (each party makes payments to the other based on the same notional quantity) or they may be physical, where gas is exchanged by two parties.

Schedule—After a pipeline confirms nominations to its shippers, it will *schedule* the gas to flow on its system. That is the usual understanding of the word. Some organizations use the same term to describe the nomination process itself. This term should always be clearly defined, either in the definitions article or by explanation in the relevant contract clause.

Confirm deadline—This term could be used in a contract to describe the deadline for a party sending the confirmation. It could also be used to describe the deadline for response by the receiving party. The GISB contract adopts the second definition, but the definition is not standardized throughout the gas industry.

Btu—GISB adopted a standard definition for Btu and the FERC later incorporated that definition in Order No. 587. Only interstate transporters and their shippers are required to follow this definition: *Btu is equal to the amount of heat required to raise one pound of water from 60 degrees Fahrenheit to 61 degrees Fahrenheit at a standard pressure of 14.73 psia.* The basic definition for heat content of the gas may vary between regions of the country. Pressure would be different in mountain states than in coastal states. Fahrenheit degree numbers may be different. All of these make a difference for measurement purposes. If the GISB definition is not used as the standard measurement, the contract must contain the applicable definition.

Gas—The standard definition for *gas* or *natural gas* is: *"All hydrocarbons and non-combustible gases, in a gaseous state, consisting primarily of methane."* Another typical definition for gas is: *"Any mixture of hydrocarbons and non-combustible gases meeting the quality specifications of buyer's transporter."* Still another definition combines the two as: *"Any mixture of hydrocarbons and non-combustible gases, in a gaseous state, consisting primarily of methane, which meets the quality specifications of buyer's transporter."*

The meanings could all be different. When a transporting pipeline accepts non-spec gas from a producer, does that mean the producer has not delivered "gas" to its customer under the second definition above?

Term—This word could describe the type of deal, *i.e.*, long-term, or it could be used to describe the duration of a contract.

Vague or ambiguous words. Vagueness and ambiguity have different meanings. A *vague* word is one that is so unclear in its usage that the reader does not know what it means. An *ambiguous* word is one that could have one of two (or multiple) meanings, and it isn't clear from the usage which definition applies. Examples of these words (or terms) in the gas industry include:

Imbalance charge—This is a vague term because the reader doesn't know from the words themselves exactly what an *imbalance charge* might be.

Baseload—In industry usage, the word could refer to a delivery obligation that excuses performance if the supplier's transportation is curtailed. It could also refer to a delivery obligation that is firm, with no excuse for curtailment other than for firm transportation using primary receipt and delivery points.

Words not used in the ordinarily understood sense. When you and I use the word *firm*, we are probably referring to something solid, hard, or not easily moved. When the same term is used in gas contracts, it refers to a level of transportation or a delivery obligation that is absolute. This term should always be defined in a contract that allows firm sales because, in addition to not being used in its ordinarily understood sense, a firm delivery obligation is subject to different meanings. To some industry participants, some "firm" deliveries are "firmer" than others.

Purposes and procedures

This article is always found in gas contracts and may alternatively be called the *Scope of Agreement* article. This article is included to describe the types of transactions allowed by the agreement, and to establish procedures for transactions and confirmations. If the parties are taping telephone conversations, the necessary authorization language is typically included in this article.

Article II
Purpose and Procedures

"*2.1 Agreement.* This Contract establishes mutually agreed and legally binding terms and conditions governing purchases, sales and exchanges of Gas (Transactions) *between [short name 1] and [short name 2] during the term hereof. Individual Transactions will be specified in Confirmations provided by the Confirming Party. More than one Transaction may be in effect at any time and all Confirmations together will be considered to be one contract between the Parties."*

Discussion of Paragraph 2.1. Several items should be noted about this clause. First, the words *Contract, contract,* and *Agreement* are all used, but they are not used in conflicting ways. Agreement is being used as a paragraph heading to reference the general understanding between the parties; in essence, that they agree to be bound by the terms of the contract—a broader concept than the word Contract, as used in the clause. Contract is being used to reference the title of the document, and the lower case contract is used to identify the total legal obligations between the parties. The clause written this way was included in this guide to illustrate different uses for these three easily confused words.

The clause concisely describes the nature of the relationship between the parties and the types of transactions that are allowed within the parameters of the contract. Sales, purchases, and *exchanges* are possible.

This contract specifies that all transactions as evidenced in confirmations will be considered to be one contract between the parties. In the worst-case situation, either party defaulting under one transaction will have defaulted under all transactions (because all transactions are *the contract*).

"2.2Procedures. The Parties may enter into Transactions in any manner allowed by law. If Transactions are made through oral communication, each Party hereby provides notice that it may be taping phone conversations between traders and other affected personnel. Each Party hereby consents to such recording of its own traders and other affected personnel, and agrees to inform all such affected employees if their conversations are being taped. Such recordings may be used for evidentiary purposes and the Parties intend that the taped evidence of such Transaction will satisfy any legal requirements for a "writing." When the Parties send confirmations, either Party's logo or letterhead will satisfy legal "signature" requirements. If the terms in a Confirmation and the tape recording identifying the same Transaction conflict, the terms of the Confirmation will control."

Discussion of Paragraph 2.2. This clause is intended to facilitate dealings between the two parties and allows agreements between the parties in any legal manner. In chapter 4, we discussed the code's preference for commercial transactions, and this contract reflects the Code's approach.

This clause also limits the ability of either party to rely upon tape recordings that conflict with the information in confirmations. You have already learned that as laws of evidence change, and the practice of taping phone conversations grows, the common election to prefer the written confirmation will undoubtedly change to prefer the tape as evidence of "the contract." Today, some contracts specify that the tape recording will prevail in a conflict with the Confirmation language and others take in the opposite approach.

In either event, though, typical gas contracts today contain some language regarding taped conversations. Since trader conversation taping is relatively new to the gas industry, most gas contracts address four issues relative to tapes:

- Each party notifies the other that it *may* be taping
- Each party consents to taping activities of the other party
- Each party agrees to notify its affected employees and obtain written consent
- Both parties agree that, in essence, they will not object to introduction of tape recordings into evidence based upon requirements of the UCC statute of frauds

Some contracts go further than the evidence issues presented in the fourth issue listed above, by stating neither party will be allowed to object to admission of tape recordings in a legal proceeding based solely upon any evidentiary exclusion rules in evidence system applicable to any court.

"2.3 Confirmations. When a Transaction has been made, the Confirming Party shall provide to the other Party a written Confirmation containing the elements of that Transaction and any additional Special Provisions relative to that Transaction by no later than the Confirmation Deadline. [Receiving Party] may provide [Confirming Party] with its own form of confirmation containing non-conflicting terms, or it may sign and return a faxed Confirmation to designate its acceptance of such Confirmation and the Transaction. If [Confirming Party]

has not received objections to its Confirmation by the close of business on the [no. of days] Business Day following [receiving party's] receipt of the Confirmation, then the Confirmation as written and the Transaction will be deemed to have been accepted and agreed to by both Parties."

Discussion of Paragraph 2.3. Enumeration of the process for sending confirmations and for agreeing to confirmations should be an important part of any gas contract. Without rules to follow, the parties will be left at the mercy of the UCC to decide which rules apply. This part of the Code isn't exactly a model of clarity, and court decisions on these matters often conflict. The best preventive measure a gas contractor can take in this regard is to put the confirmation procedures in the contract. An organization short on staff contract analysts or confirmation analysts should take careful note of any time or confirmation content limitations. Once the contract is signed, the obligation exists, whether or not it is practically feasible to respond negatively to a confirmation within the allotted time.

Some industry companies are routinely sending their own form of confirmation, even when they are not the confirming party, just to assure that the deadline for response is met. Some even place a standard clause in the confirmation that automatically objects to the confirming party's form. This objection is sometimes received before the confirming party has sent the official Confirmation.

"2.4 Transactions. The Parties may enter into Firm and Interruptible Transactions, with specifications for any particular Transaction indicated in a Confirmation."

Discussion of Paragraph 2.4. This contract limits the transactions to those that are *firm* and those that are *interruptible.* Some parties view "2-Transaction only" limitation as a hindrance on the ability of traders or sales/purchase representatives to do deals, but gas contractors can craft the contract or confirmation language to allow for time variations for either firm or interruptible transactions. Other variations within the two principle categories are also possible. A good example of this is seen in the GISB contract that allows for differences within the two major delivery categories to accommodate peak-need sales or transactions where the buyer has agreed to take firm delivery of gas from the seller at the seller's option. The GISB contract also allows for variable-quantity deliveries, yet the only two levels of delivery obligation available under the GISB contract are firm and interruptible.

Notice that the permissive word *may* is used to describe the parties' intentions. No obligation exists under this format for either party to enter into firm transaction, interruptible transactions, or any transactions whatsoever. The parties are completely free to contract when the right deal comes along for both.

Term

Most base gas contracts are formatted so that the term will be indefinite, subject to termination when either party provides written notice to the other party. Contracts for one transaction only, the *term* contracts with unique features, will almost always have a defined life, usually a number of years. These contracts may also contain an *evergreen* clause that continues performance beyond the primary term for stated periods of time unless one of the parties wishes to terminate.

Article III
Term

"*3.1 Term. Except as provided in Article __ (Early Termination), this Contract will be in effect from the date set forth above until terminated on thirty (30) days advance written notice by either Party; provided, however, the provisions of this Contract will remain and be in effect during Transactions agreed to prior to the date of termination.*"

Discussion of Paragraph 3.1. A few matters should be discussed regarding this simple and straightforward contract clause. First, an exception for continuing the indefinite contract term is carved out in the reference to the contract article that lists the reasons for early termination of the Contract. Second, the advance termination notice as indicated in this clause doesn't require a calendar month's notice, but rather 30 days. The obvious result is the notice could be given mid-month, effective the middle of the following month.

Note that this is a one-paragraph article. As a matter of contract form, some parties choose not to number the paragraph in clauses such as this, others do. For standardization of contract terms, and ease of reference, the latter method is preferred.

Finally, two words are underlined, "*provided, however.*" Many state statutes have very strict requirements for conspicuously noting any contract language that has the effect of limiting a statement just made or right just granted. Most gas contractors will, as a routine practice, conspicuously note any such language. Traditionally the words "provided, however" have been used to identify any such clause.

Obligations

Sometimes the indemnification and payment articles will be extended beyond termination until all obligations under those articles have been satisfied.

Obligations of a gas seller include one to sell the product and another to deliver the product. Similarly, the buyer's obligations are to purchase the product and to take delivery. These are separate and distinct obligations of the parties and the remedies for a failure of the sale obligation and failure of the delivery obligation are somewhat different.

Article IV
Obligations

"*4.1 Obligations. The Party designated as seller for a Transaction shall have the obligation to Schedule, to sell, and to deliver or cause to be delivered the Gas at the designated Delivery Point(s). The Party designated as buyer for a Transaction shall have the obligation to Schedule, to purchase, and to take delivery of the Gas at the designated Delivery Point(s).*"

Discussion of Paragraph 4.1. Even if a gas contract doesn't specifically state that the seller has an obligation to deliver the product and to transfer title or that the buyer has an obligation to take the goods and to pay for them, the obligation exists. The Code specifically sets forth those rules in

Section 2-301 (chapter 4). Many gas sellers and buyers now add another requirement to the basic obligations of the parties, placed conspicuously in the contract, and that is to *schedule* the gas.

When the obligation to schedule gas is included in a contract, failure to schedule is a breach of the gas sale contract. Without a statement explicitly agreeing that a contractual obligation to schedule the gas exists, failure to do so or to cooperate with the other party to do so may not be a contract breach. When the obligation to schedule is a part of the contract, the definition of schedule must be carefully considered and a careful gas contractor won't permit too broad a definition of the term used. As an example, consider the following definition:

"Schedule means that the nomination and confirmation procedures required by seller's and buyer's transporters have been completed and that gas has been scheduled to flow on the transporters' pipeline system."

When this broad definition is used, the seller and buyer are both obligated to assure that their transporters have scheduled the gas, a power far beyond the ability of any third-party shipper. If a transporter fails to schedule gas properly nominated and confirmed by the seller, for example, under the foregoing definition, the seller would be in breach of the gas sale contract. Now, consider this definition:

"Schedule or Scheduled means that seller and buyer have complied with all nomination and confirmation requirements of any transporter(s) for a particular Transaction."

This places a narrower meaning on the term since the parties are only agreeing contractually to do that which they can control—their own internal nomination process.

Finally, the obligation of sellers and buyers in this article is limited to the named delivery point(s). Under the UCC, as you will learn in chapter 8, when one party experiences a force majeure and is unable to deliver or take delivery of the goods at a designated point, it is obligated to find an alternate point to make delivery. With the language written as it is in this clause, the parties may not be required to find alternative points of delivery in the event of force majeure.

When the delivery point is not determined, such a delivery into a pipeline pooling point, the parties may wish to incorporate the use of "Title Transfer Point" rather than a delivery point.

Transportation, scheduling, and balancing

Transportation is an integral part of most gas sales, yet the contract for transportation service (the TSA) and the contract for gas sales are different. This is an area where an analogy with the manufacturing industry is apt. Gas buyers and sellers contract with pipeline companies, just like product manufacturers contract with carriers for the goods they sell. The transportation arrangement is separate from the gas sale arrangement, but the two transactions are interdependent. Most gas contracts will contain a clause incorporating the necessary transportation communications as part of the gas sale contract.

Because of *privity of contract* (chapter 4), no mechanism exists for either party to a gas sale contract to interfere with the other's performance under its TSA. Because of lack of priority with the other party's transporter under its TSA, every gas contract will contain a clause describing mutual obligations of cooperation so that neither will suffer adverse consequences as a shipper because of the other party's action or inaction. Without a clause requiring cooperation and communication for transporters' nomination requirements, no such obligation would exist under the gas contract.

Article V
Transportation, Scheduling, and Balancing

"5.1 Transportation. Seller shall be responsible for arranging transportation to the Delivery Point(s) and buyer shall be responsible for arranging transportation away from the Delivery Point(s). Title to Gas and physical possession will pass from seller to buyer at the Delivery/Title Transfer Point(s)."

Discussion of Paragraph 5.1. The seller has responsibility for procuring transportation while the gas is in its possession, and the buyer has responsibility for procuring transportation while the gas is in its possession.

This standard contract clause defines the physical point at which each party has the obligation for transportation. Without such a statement, the Code provisions would apply. This part of the UCC is very widget-oriented, and many of the UCC terms would be difficult to apply to gas sales and transportation. Careful gas contractors will always include a clause of this type in all gas contracts.

To expand upon the potential concern regarding transportation pooling.[9] Shippers that use *paper pooling* sometimes cannot identify the geographic point at which gas was delivered into or out of the pipeline system. For tax and title transfer purposes, paper pooling has not proven to be a complete success story. As you will learn in the discussion of taxes, the right to impose a sales tax attaches to the transaction itself at the title transfer point. If parties cannot determine exactly where a sale transaction took place, then to what taxing authority should such transactions be reported for tax purposes?

"5.2 Communications. To comply with Transporters' nomination deadlines and the routine business practices of Transporters, seller and buyer shall communicate to each other the daily receipt and delivery quantity and all other information necessary for all third parties, including Transporters, to schedule the Gas. Telephone, facsimile, electronic data interchange, or any mutually agreed means may be used by the Parties to communicate nominations to each other and to Transporters. The Parties shall provide each other timely notice of any changes to nominations prior to Transporters' nomination deadlines."

Discussion of Paragraph 5.2. Since gas sales and gas transportation are separate activities, the seller and buyer must contract separately with transportation service providers. These TSP's will transport gas for a shipper, if the shipper has sufficient credit with the pipeline company, holds title to (or has the right to transport) gas on the pipeline system, and has signed a TSA with the pipeline company.

Gas sellers and buyers must cooperate with each other in complying with their respective transporter's nomination process, because each relies upon the other for necessary nomination information. Here's how the nomination process works on many pipelines. When the midstream seller and buyer are both using third-party transportation, the relevant seller's TSA is numbered—so is the buyer's. During the nomination process, the seller and buyer communicate to each other and to their respective transporters the quantities that will flow, the TSA numbers, upstream and downstream operators, ranking, the quantity to be delivered and received at the designated delivery point, and any other information required by the transporter.

Once the transporters have received nomination information from their shippers, the confirmation process begins and the flow numbers are checked upstream and downstream to confirm that the quantities nominated by their shippers agree with immediate upstream and downstream nominations. Once the pipeline operators have agreed on the flow amounts, each pipeline company will notify its shipper that the quantities are confirmed. After confirmation to shippers, gas is scheduled on the transporter's own system.

During the month of flow, some transporters have ready access to measurement so they can communicate with their shippers, letting them know if their accounts are out of balance beyond acceptable limits. A shipper receiving such mid-month information may then take corrective action to bring its TSA (or TSAs) back into balance. Cooperation of the other party to the gas sale contract is usually necessary to resolve imbalances.

Without accurate and timely communication with the buyer, a seller would have incomplete nomination information and the same holds true for a buyer relying on information from its seller. Every gas contract should contain a clause such as this to require cooperation.

This clause also allows the parties to use any form of agreeable communications for routine communications. While not technically necessary, as future communications become more electronic, this clause paves the way for movement into more sophisticated spheres of communication between the parties to the gas contract.

Finally, the last sentence of paragraph 5.2 requires the parties to communicate when either learns of a situation that would require the other to change a nomination to its transporter. Failure to do so is a breach of the contract. Assume that PUMP is selling natural gas to FLEX and the principal well from which the sale is being made requires immediate maintenance. Under this contract clause, PUMP has an obligation to notify FLEX of any lower nomination resulting from the lower gas flow that FLEX must make to its transporter to meet its nomination deadlines.

"5.3 Imbalance Charges. The Parties shall cooperate to avoid imbalances and the imposition of Imbalance Charges resulting from performance under any Transaction.

(a) Each Party shall provide nomination information to the other in order to effect timely notification to Transporter(s).

(b) Each Party shall notify the other promptly if it receives notice at any time from a Transporter that imbalances are occurring or have occurred. Buyer and seller will work to resolve any imbalances that do occur and will use [commercially reasonable/diligent] means to avoid imposition of any Imbalance Charges on either Party.

(c) The Party responsible for causing any Imbalance Charge shall also be responsible for paying the relevant Transporter(s), or reimbursing the other Party if it has paid its Transporter, for any such Imbalance Charge. Billing and payment for any such charges will be in accordance with the provisions of Article ___."

Discussion of Paragraph 5.3. As you learned in the previous discussions of the frailties of this industry, sometimes the actual measured quantity of gas does not equal the quantity that was nominated and confirmed. If this happens, one result may be that the seller or the buyer incurs charges from its TSP for failure to maintain balance in its gas account. Pipeline companies must control system integrity, and they can only do that with a tariff-based method for resolving overages and underages. Transporters assess various types of charges against shippers who violate balance requirements. One party to the gas sale contract may cause the other to incur these penalties.

Without contract language requiring cooperation to avoid imbalances, no such obligation would attach to a particular gas sale. This is a "must-have" contract clause.

Also note that various levels of cooperation may be required in a contract. Under the UCC, Merchants already must utilize standards of commercial reasonableness in their gas transactions, but parties *may* wish to specifically state the level of cooperation that must be followed by *both* parties by electing one of the three choices given in this clause.

Gas measurement and quality

Article VI
Gas Measurement and Quality

"6.1 Measurement. Final measurements under this Contract will be those made by buyer's Transporters, and such measurements will be determinative for payment and balancing purposes for all Gas sold and delivered under this Contract; provided, however, that when buyer does not utilize transportation, the measurements of seller's Transporters will be determinative for payment and balancing purposes.

6.2 Quality. Quality and heat specifications for all Gas delivered under this Contract will be those of buyer's Transporter, and seller shall be responsible for any damages caused by Gas not meeting such specifications."

Discussion of Article VI. In many instances, both the seller's transporter and the buyer's transporter will measure gas flowing through interconnection meters, but typically only one transporter is responsible under the gas contract for determining the flow amount. If contracting parties don't elect one party's transporter for measurement purposes, the process of reconciliation (reconciling measurements between seller's and buyer's transporters) may take months, if not years.

I've already discussed pipeline quality and heat specifications. Pipelines also maintain temperature requirements for gas entering the system, and gas typically cannot enter a pipeline system if it has a temperature in excess of 120 degrees Fahrenheit.

In midstream transactions, the seller is delivering gas that has already moved through another pipeline, so the trend in contracting is to require that quality and heat content requirements be those of the buyer's transporter. As with measurement, if the seller is delivering gas directly into the manufacturing facility of its customer, there's no "Buyer's Transporter," and so paragraph 6.2 may more appropriately read as follows:

"Quality. Quality and heat specifications for all Gas delivered under this Contract will be those of buyer's Transporter, and seller shall be responsible for any damages caused by Gas not meeting such specifications; provided, however, that when buyer does not utilize transportation, the quality specifications will be those of seller's Transporter."

How important is this clause? From both the buyer's and the seller's viewpoint, the quality issue is critical to a gas contract and should be carefully considered by any producer, LDC, or consumer. Of course, the measurement requirement is important as well, but the potential liability for a breach of the duty to supply gas meeting quality specifications is far higher than a breach for failure to maintain measurement equipment or even to measure correctly. The distinction is made because of devastating property damage and personal injury that could result from non-spec gas deliveries. First, let's consider the measurement question and some of the considerations a gas contractor should take into account in this regard.

Measurement. If a seller cannot determine how much gas it sold, there is no finite basis for billing and payment. You learned in chapter 1 that measurement issues can begin as the gas is coming out of the ground when the first sale is made. Some wells are equipped with orifice meters, others have a variety of electronic measuring devices, and still others have turbine meters. Some gas meters may be read on a real-time basis, while others are only reviewed at the end of a delivery period. Calibrations of meters are a factor as well, and differences in meter calibrations will result in different measurements. The amount of water, sediment, non-combustible gases, and liquid hydrocarbons in the gas system will affect measurements.

Gas may be measured on a volumetric basis (cubic feet in the U.S., cubic meters in Canada) or on a heat content basis (British thermal units in the U.S., gigajoules in Canada). Gas transporters care about volume measurement because gas movement is determined in part by the volume of gas in the pipeline. Gas sellers and buyers, on the other hand, are more concerned with the thermal capabilities of the gas stream.

Gas buyers purchasing gas supply from a producer or transporting gas on a volumetric basis (Mcf) and selling that gas on a heat content basis (MMBtu) will sell more than they purchased when conversions are done. Here's an example:

Let's say that FLEX is purchasing gas from PUMP at the wellhead on a volumetric basis (Mcf) and reselling the gas to REFLEX based upon heat content of the gas (MMBtu). The usual conversion from Mcf to MMBtu is based upon an agreement that 1,000 cubic feet of gas is equal to 1,012 MMBtus. If FLEX is purchasing 5,000,000 cubic feet (5,000 Mcf) of gas per day, it can resell 5,000 MMBtu/day *plus* the difference in measurements, or 5,060 MMBtu. From a gas buyer's perspective, most measurement issues occur when gas is being purchased on a heat content basis and sold on a volumetric measurement basis because fewer units will be sold than were purchased without a commensurate price adjustment. Most gas organizations that buy and sell gas in the interstate market routinely price gas according to the heat content. This issue is more relevant in the U.S. intrastate market and in cross-border transactions between North American contracting parties.

The many differences in measurement capabilities may also mean that two operators at the same point of interconnection could easily make different measurements for the same gas. The transporter of one of the parties should be the designated measuring party to avoid lengthy negotiations on the amount of gas sold. Interstate transporters in the U.S. must limit resolution time for any invoice to six months, with a three-month rebuttal period.[10]

A gas contract should specify the basis of measurement and the measuring party.

Quality specifications. While the first paragraph in this article is the source of more frequent gas contract issues, the second paragraph is the source of the costliest contract disputes. Assume that PUMP is selling gas to WIDGET for use in its cogeneration facility. The gas is being transported by PIPE under a TSA that PUMP has signed with PIPE. Now assume further that PIPE *has accepted* non-spec gas from PUMP, because the total commingled gas stream in PIPE's facilities met quality specifications. Another downstream producer, PUMP2, delivers extremely *rich*[11] gas into

PIPE's system. When WIDGET receives the enriched mixture extreme damage is caused to the cogeneration facility, WIDGET's only recourse will be against PUMP under the gas sale contract.

If you think back to the discussion on the warranties given by every seller that is a Merchant under the UCC, you already know that PUMP may have breached its implied warranties of *merchantability* and *fitness for a particular purpose*, for which the UCC provides remedies in Section 7. In the discussion of damage limitation clauses later in this chapter, you'll see how complicated this issue can be.

Without going into the various liabilities the effected parties may have, the easy questions raised by the example above include:

- What liability may PUMP have to WIDGET?
- What liability may PUMP2 have to PIPE?
- What liability, may PIPE have to PUMP?
- May WIDGET be precluded from receiving damages?

Gas buyers and gas sellers obviously view the issues presented in Article VI differently. Sellers wish to be protected from liability once the gas is out of their possession and has been accepted by a transporter. Buyers want an avenue of recourse if sellers fail in their quality obligations. The methods utilized by a seller wishing to limit its liability were discussed in chapter 4, as were the methods that might be used by a buyer wishing to expand a seller's liability for damage or injury caused by non-spec gas. Gas contractors should carefully weigh the advantages and disadvantages inherent in the choice of words used in this clause.

Delivery failure

One of the most common failures during performance of a gas contract occurs when sellers fail to deliver the contracted quantity and when buyers fail to take the contracted quantity. The failure may be entire or partial, but the clause always relates to insufficient delivery, and usually not to overdelivery. This typical contract provision is directed only to delivery failures and the remedies apply only for that failure. Any other breach of the contract will be handled in accordance with other contractual agreements or according to the remedies available under the UCC.

Sellers may fail to deliver gas for a number of reasons, some beyond their control such as unexcused failure of a well or losing a necessary gas supply. Buyers may have lower or greater gas needs than the nominated amount, so they may also fail to take the contract quantity. Sometimes, price volatility is a factor in unexcused failures as well. As noted previously, most regular gas transactions are installment-based, and even *term* transactions will typically be monetarily settled on a month by month basis. Nearly every gas contract will contain a clause that establishes the mechanisms to be used by the parties when a failure occurs in any given month.

Article VII
Delivery Failure

"7.1 Remedy for Failure. If either Party on any day fails, in accordance with Article IV, to Schedule and deliver or take delivery of any part of the Contract Quantity, then, in addition to any Imbalance Charges, that Party shall be liable to the other for any damages caused by such failure. The Parties agree that, unless otherwise provided in a Confirmation, the Market Price of Gas at the Delivery Point is a fair and reasonable calculation for damages caused by a Party for the other's failure in its delivery obligation.

> *(a) Seller's Failure. Seller shall be liable to buyer and shall pay buyer an amount that reflects the difference, if any, between the Market Price and the lower Contract Price for the total quantity not delivered.*
> *(b) Buyer's Failure. Buyer shall be liable to seller and shall pay seller an amount that reflects the difference, if any, between the Market Price and the higher Contract Price for the total quantity not taken.*
>
> *(c) Cover and Resale. If the non-failing seller procures an alternate market for sale or the non-failing buyer covers its loses by producing an alternate supply of fuel, then the actual*

amount received from or paid to the alternate market will replace the Market Price for the calculations in subparagraphs (a) and (b).

(d) Litigation. Each Party shall mitigate any damages incurred because of the other Party's failure, and in no event will either Party have the right to collect both Imbalance Charges and the remedy provided in this article to the extent that any portion of Imbalance Charges relates to supplied gas rather than to Transportation.

(e) Exclusive Remedy. The remedies set forth in this Article VII will be the sole and exclusive remedy available to a Party for the Other Party's failure in its delivery obligation.

Discussion of Paragraph 7.1. Important elements in this clause include the following:

- Failure will be calculated on a daily basis
- Liability attaches and damages will be paid for any unexcused failure
- Imbalance charges are calculated separately (but not redundantly)
- Liability is for *any* damages
- *Market price*[12] is used to calculate direct damages unless a Party covers its loss
- The parties may agree to an alternate market price in a confirmation
- The Parties have an obligation to mitigate damages
- Exclusive remedy statement

Use of the word, day. In the clause used for this example, please note a couple of important points relating to the eight topics raised above. Gas may be nominated more than once during a day on the interstate transportation system, so gas contractors should carefully review any definitions that incorporate use of the word *day* to assure inclusion of any failure on any transaction, regardless of length of the nomination period.

Imbalance charges. Imbalance charges are calculated separately under this contract, but in certain cases, an imbalance charge could actually be the cost used to determine a buyer's damages resulting from the seller's failure to perform. Assume that the definition of *imbalance charge* is as follows:

> *"Imbalance charge(s)" means any scheduling, imbalance, or similar penalties, fees, forfeitures, cashouts, or charges (in cash or in kind) assessed by a transporter for failure to satisfy the transporter's balance requirements or nomination requirements.*"

Most of the issues raised by this definition could easily be calculated separately from the damages caused by a failure to deliver because those charges will be addressed by:

- the relevant TSA as between the pipeline and its shipper and
- the gas contract that typically assigns responsibility for imbalance charges to the party that caused the imbalance

One situation is not covered by this particular clause and is most common to the case where a buyer is receiving gas behind the city gate. LDCs have various means to maintain balance on their distribution systems and to render an accounting when a gas buyer behind the city gate takes more or less gas than was delivered by its seller. When a gas seller does not deliver the contract quantity to its buyer (on an LDC system) and the buyer takes that gas anyway, the LDC and not the seller is the entity supplying the gas. Many LDCs will impose high charges for any such *unauthorized gas*. The charges for unauthorized gas may be extremely high and represent, in essence, the cost of gas sold by the LDC to the buyer when a seller does not deliver gas. A gas buyer facing huge and unanticipated costs for taking unauthorized gas will argue to its seller that the unauthorized gas charge has become the market price, and reflective of the damage caused by the seller. That argument appears to be fairly sound, so some gas contracts approach the issue through contract language like paragraph (1) (d) above. Others—predominantly sellers that are aware of the severity of many LDC *unauthorized* gas charges—will take a different approach with a contract clause, like the following: *In no event will either Party be liable to the other under the Contract for payment of any unauthorized gas charges, however called by the Transporter(s) assessing such charges.*

Liability for any damage/limitation of remedy. Contract language that allows a remedy for *any* damage caused by a failure of either party may be of particular concern to contracting parties. Most gas contracts clearly delineate the extent of damages allowed. One way to address this problem would be to insert the word, *direct* in the allowable damages, so the relevant paragraph would read as follows:

7.2 *The Parties agree that, unless otherwise provided in a Confirmation, the Market Price of Gas at the Delivery Point is a fair and reasonable calculation for direct damages caused by a Party for the other's failure in its delivery obligation. The remedy provided herein constitutes the exclusive remedy available to either party when the other fails in its delivery obligations. All other damages—including special, incidental, consequential, indirect, exemplary, or punitive—whether arising in contract, in tort, or otherwise, are hereby waived by the Parties."*

Under the UCC, parties may contractually agree to reasonably *limit* remedies and to provide that the contract remedies are *exclusive*.[13] Without a statement of agreement in the contract specifically limiting available remedies to those indicated in the contract to the exclusion of other remedies, the parties could be subject to a court's interpretation that "if they wanted it in the contract, they should have put it in the contract."

Damages. The remedies picture that I began painting in chapters 4 and 6 is not complete without the final touches—a discussion of damages. The remedies article of any contract usually limits recovery to certain types of damages as indicated in paragraphs 7(2).

Direct Damages—that is, the losses incurred by one party because of the other's failure—are always collectable by the prevailing party. In an installment contract, this would most likely be evidenced in the difference between the contract price and the market price at the time of failure. But after direct damages are awarded, the issue loses its clarity.

Indirect damages—damages that are not directly caused by the failure—is a category overlapping with other identifiable damages.

Incidental damages and damages incurred because of the other party's failure, such as additional transportation or storage costs as discussed in chapter 6.

Consequential damages—also discussed in the previous chapter—are damages not directly caused by the failure, but rather as a result of the failure. This type of damage may be remote—damage to third party because a warranty is breached, for example.

Use of a market price indicator. It is becoming quite common to contractually specify that some current market indicator will be used to determine damages. Several issues occur when this singular approach is used to evaluate damages. Let's say that the market price definition specifies that *Gas Daily* will be the relevant publication and the actual price will be the price published at (or closest to) the delivery point. If *Gas Daily* is used as an index-based price for a particular transaction, and a failure occurs, the market price used is also the contract price. But what would be the result in organizations that buy and sell gas under *Gas Daily* prices, *IFGMR* prices, *Btu Daily* prices, and so forth? *Gas Daily's* price may not be reflective of the actual price if the transaction was based upon a *Btu Daily* price. Different industry publications publish different delivery points—they all have different methods for calculating the published prices, and the price published for the same point on the same day may differ from index to index. Gas contractors should use care when agreeing to a market price determination based on a named publication. An average of several index publication prices at the same delivery point might be a more appropriate determination of market price.

The other issue raised with market price determination is the point in time when the market price is calculated. Should it be as of the date of failure, when the non-failing party learned of the other party's failure, or when the non-failing party had an opportunity to cover by making an alternate sale or finding an alternate market? Under the UCC, the market price is determined as of the time of failure.[14] Contracting parties desiring to use another time for calculation must address that issue through contract language.

Cover. Under this contract, when a non-failing party finds an alternate market or supply of gas (or, in the buyer's case, an alternate fuel supply), it may apply the actual cover cost instead of the market price calculation specified in the contract. Approaches differ on this issue. Sometimes a contract will require the non-failing party to cover its losses, other times the market price approach will be used. One of the major issues related to this particular topic occurs when the contract does not provide whether the failing party or the non-failing party will have the right to decide whether market price or the cover price applies if those prices are different. Many contracting parties agree that the non-failing party will make the election, but for one reason or another, most gas contracts don't address the issue.

Administrative or overhead fees. Some gas contracts have a clause requiring the failing party to pay an additional fee to cover the other party's administrative or overhead expenses resulting from the failure. The fee is typically based upon a per-unit cost from five cents to very large fees. Two issues result from using this type of clause.

What happens when a seller or buyer fails in its delivery obligation and the other suffers no resulting harm? Let's say that FLEX has an obligation to deliver 2,000 MMBtu/day to GASCO at $2.00/MMBtu and fails to deliver 15,000 MMBtus during the month. If GASCO can find a replacement gas supply for less than $2.00/MMBtu, it has not been harmed (rather it has been helped) by FLEX's failure. Unless the administrative fee clause is worded to apply only when damages have been assessed, FLEX may be required to make the administrative fee payment for the 15,000 MMBtus not delivered, even though no damage was caused by its failure.

This second issue relates to the amount of the fee. As you learned in chapter 4, the UCC won't enforce a penalty, so contracting parties agreeing to an administrative fee clause must carefully consider whether the five, fifteen, fifty-cent, or one dollar administrative fee per MMBtu may be considered to be a penalty, and as such unenforceable.

Billing and payment

Article VIII
Billing and Payment

"8.1 Invoice. Seller shall provide buyer with an invoice for all Gas sold and delivered under this Contract during the prior Month. If buyer has not provided seller with necessary measurement and balancing information from buyer's Transporter, seller's invoice will be based on Scheduled quantities, subject to adjustment when buyer receives its statement(s) from buyer's Transporter(s)."

Discussion of Paragraph 8.1. Under the UCC, payment is due at the time and place of delivery, unless otherwise specified in a contract. That is why all gas contractors include a billing and payment clause. Sellers usually send invoices to their buyers between the 10th and 15th day of the month following the delivery. Sometimes the seller has an obligation to render the invoice on or before a certain date, but by structuring the payment date to correspond with the date by which the seller has provided the invoice, no deadline is required. A seller that is slow to invoice will be paid later.

Interstate transporters' statements aren't usually available to their shippers until after the invoice for gas sales has been prepared by the gas seller. If that is the case, billing and payment will be based upon confirmed or scheduled quantities, with adjustment by either the buyer (when it received its transportation statement) or the seller on the next month's billing, and that option (buyer's to adjust payment or seller to adjust the following month's invoice) should be clearly included as part of this clause.

"8.2 Payment. Buyer shall wire transfer its payment to seller according to instructions on seller's invoice no later than the Payment Date or ten (10) days after receipt of seller's invoice, whichever is later. If buyer disputes any portion of seller's invoice, then, subject to paragraph 8.3, such portion may be withheld if buyer provides seller with documentation substantiating the reason for such dispute."

Discussion of Paragraph 8.2. Through the years, a rather common practice in the gas industry has evolved in which the payment deadline is later than the deadline for sales of other goods where it isn't uncommon to see terms like "net 10" for the payment date, meaning that payment made within 10 days of receipt of the seller's invoice is accepted without any interest charge. Payment accepted after 10 days will be subject to interest. In the gas industry, because of reliance on transportation statements for billing purposes, a different approach has been developed. The current approach is to bill and pay on monthly cycles, with payment due for deliveries in one month being due on or about the 25th day of the following month. This typical 60-day cycle has been the traditional basis for evaluating credit of the other party as well.

Sometimes buyers must pay the entire invoice, subject to changes when the transporter provides its statement. Other times the contract language isn't specific about whether the buyer must pay anything at all until it receives measurement information from its transporter. Regarding this issue, the relevant sentence from paragraph 8.2 reads: *"If buyer disputes any portion of seller's invoice, then, subject to paragraph 8.3, such portion may be withheld if buyer provides seller with documentation substantiating the reason for such dispute."* The way this sentence is written the buyer may adjust the seller's invoice and include a copy of relevant parts of its transportation statement to substantiate payment different from the amount billed. A buyer could conceivably withhold the entire payment as well, but interest at the contract rate will apply to any amount withheld.

Some gas contractors will prefer having the obligation to invoice and pay clearly identified in the contract. In that instance, an appropriate payment clause might look like the following:

"1. Seller shall submit to buyer on or before the 10th day of the month an invoice setting forth the amounts owed by buyer for gas deliveries during the preceding month. The invoice may be transmitted in any form acceptable to Seller and buyer. If the actual total quantities as measured by [the measuring party's] Transporter are not available by the contractual billing date, billing will be based on quantities confirmed by buyer's transporter for the month of deliveries, subject to correction by [seller on the following month's invoice/buyer] when transportation statements are received."

"2. Payment by [wire transfer/Automated Clearinghouse] of immediately

available funds shall be made by buyer to seller to an account designated by seller to buyer in writing on or before the Payment Date. If the Payment Date falls on any day other than a Business Day, payment shall be due the [preceding/succeeding] Business Day."

"3. If buyer in good faith disagrees with or disputes any information set forth on the invoice, it nonetheless shall pay to seller such amount as it concedes to be correct and provide documentation to seller specifying the reason for the dispute."

In this use of contract language, each party should clearly understand its obligations regarding the invoice and the payment. Additional clauses specifying the amounts to which interest will attach (discussed below) will also be helpful to grasp and identify the entire process. In any event, billing and payment language should be clear, not ambiguous or vague.

"8.3 Interest. Interest will accrue on all late payments owed by either buyer or seller at a rate equal to the lesser of (i) the posted Prime Rate in the Wall Street Journal as listed under "Money Rates", plus two percent (2%), or (ii) the maximum legal rate, from the date payment is due until the date payment is made."

Discussion of Paragraph 8.3. A gas buyer wants to assure that any amounts owed to it by the seller will be subject to the same interest that it is being charged if it makes a late payment. Interest clauses should be bilateral since either party could owe payment to the other. Sometimes contracting parties have some difficulty in agreeing to the types of payments that should be subject to interest. For example, if a buyer makes a good faith payment that it believes is correct, should that party then be penalized later when it or the other party discovers that the payment was insufficient? Should late interest attach to that remaining unpaid amount? The answer to that question is found in the way interest is characterized.

If interest is viewed by the parties as a reflection of the time value of money, then the party that in reality benefited from holding funds belonging to the other party should have to return interest in addition to the other party's funds.

If interest is viewed as an incentive for making timely payments, the parties may simply decide that all late payments, whether due to malfeasance or simple lack of accurate information, should be subject to interest.

Interest may also be viewed by the parties as a tool to discourage certain types of behavior, such as routine late payment. In this instance, interest might be waived when a late payment was not the fault of the party owing the payment. Whatever the reality of the parties' relationship, the interest clause should reflect that relationship.

The second aspect of interest relates to *usury* laws in all 50 states. Usury laws are limits on the amount of interest that can be charged by parties in that state for various types of transactions. The usury rate is tied in part to the risk undertaken by the creditor, so a relatively high-risk transaction such as credit card purchases or loans to high-risk customers will carry a relatively high usury rate. The usury rate for contracts is specified in every state's statutes and is more closely tied to the value of the U.S. dollar and inflation than to a specific level of risk. Any contract clause allowing interest in excess of the relevant usury limits will not be enforceable.

The interest rate used is typically tied to a floating rate, usually the *prime rate*.[15] The most common reference is to the *Wall Street Journal*, but it is not uncommon for contracting parties to reference another publication, to select a rate charged by a specific bank, or to contractually establish a stated rate of interest.

Gas contractors should carefully read all interest clauses because contract language isn't always clear as to which party may receive interest, the amount of interest, or the time during which interest will accrue.

"8.4 Audit Rights. Each Party will have the right, during regular business hours and at its own expense, to audit the books and records of the other Party for the purpose of verifying the accuracy of any invoice, statement, charge, demand, payment, or computation made under any Transaction(s). For auditing purposes, all such books and records will be presumed to be correct at a time two years after such original invoice, statement, charge, demand, payment, or computation was rendered."

Discussion of Paragraph 8.4. Records, statements, invoices, and related accounting data are usually not audited by parties to a routine gas transaction, but every gas seller and buyer should want to have an audit clause in its gas contracts. Without this permissive authority, neither party has any contractual right to verify the correctness of any document it receives from the other party. As gas organizations are performing fewer functions on paper and more functions electronically, the audit clause may be more frequently used.

One audit issue that has already been addressed in this guide is that most clauses will clearly limit the time for auditing rights. A contract can be drafted to make the audit right an effective statute of limitations, thus precluding any legal proceedings when the audit right has expired.

---

1. Even though repudiation may be partial and needn't be for the entire contract, the provision in this contract may be so unclear as to render the intention of the parties with regard to repudiation unclear.

2. In a bankruptcy ruling by the Federal Bankruptcy Court in Houston, Texas, the court would not allow the debtor to "pick and choose" between individual transactions done under a GISB contract, stating that all were part of the same contract. *In RE: Kimball Trading Company L.L.C., Kimball Trading Canada, Inc* Jointly Administered Under Cause No. 99-32383-H2-11 "Findings and Conclusions Concerning Debtor's Motions to Assume and Reject Contracts with Quicksilver," CMS and PG&E March 31, 1999

3. See the discussion of UCC definitions of *contract* and *agreement* in chapter 1.

4. See the discussion of federal regulation of the gas industry in chapters 1 and 3.

5. Chapter 1, supra., fn. 16.

6. When one party sues another, the procedural rules of the relevant jurisdiction must be followed. One of those rules deals with serving the other party with a copy of the complaint or filing. This is termed *service of process*.

7. The Gas Industry Standards Board and Dun & Bradstreet jointly developed a process for designating a separate DUNS identifying number for every organization that ships gas. The DUNS number is used for a number of identification purposes in the gas industry.

8. See the example of different usages for *baseload* in chapter 2.

9. Pooling service is offered by many natural gas transporters. Sometimes this service is simply called "paper pooling" because the type of pool used in this circumstance is not physical, but rather a number used to designate gas deliveries into or out of a designated pool area. Pools have alleviated many nomination burdens, because the gas scheduler, prior to the time pipelines established paper pools, had to nominate every package of gas at its particular receipt point and delivery point. When pooling is allowed it nominates only into or out of a pool.

10. GISB Business Practice Standard No. 3.3.15 (Version 1.3), as adopted by the FERC in Order No. 587-K).

11. Gas high in liquid hydrocarbons.

12. When a term such as this is used in a contract, it must be defined. The basis for Market Price may be one of the industry publications that publishes index prices, the average price reported by several such publications, or other objective criteria used to determine the cash or current price of the gas.

13. U.C.C. § 2-719 (1).

14. U.C.C. § 2-723.

15. The prime rate is the interest rate large banks charge each other for loans.

16. 11 U.S.C. § 101 *et. seq.*

17. U.C.C. § 2-609 (2).

Chapter 8: The Gas Contract — Part 2

Financial Responsibility

Even though the UCC states that each party has the ability to question the other's creditworthiness during performance and to request assurances of performance, no further guidelines are provided so parties are often left wondering what rights, if any, they might validly have to request and what credit information they must maintain during performance. The article serving as an example for this issue obligates each party to maintain its creditworthiness to the satisfaction of the other party. If at any time either party has a good faith reason to suspect that the other party may be in a diminishing credit situation relative to exposure, the other party must provide annual reports and quarterly unaudited statements *plus any other* documentation or information that is requested by the inquiring party.

Article IX
Creditworthiness

"9.1 Creditworthiness. Each Party shall at all times satisfy the credit requirements of the other Party. If a Party (X) at any time during the term of this

Contract makes a reasonable request, the other Party (Y) must assist in the reasonable financial review, including delivery as promptly as possible of an annual report including audited financial statements plus the most recent unaudited quarterly financial statement prepared in accordance with generally accepted accounting principles in effect as of the date of the request, plus any other reasonable documentation X might request."

Discussion of Paragraph 9.1. The hidden strength of this article is found in its last line—"plus any other reasonable documentation X might request." When the financials of subsidiary organizations are included in a parent company's consolidated financial reports, the subsidiary companies may be unwilling to make information about their own financial status available to their trading partners, opting instead to provide only the financial statements of parent companies. This unwillingness may stem from a variety of valid reasons to resist providing the requested information.

But from the perspective of a contracting party evaluating the other party's credit, only a foolhardy risk management program would allow sales to be made without any knowledge of the credit risk being assumed. Absent proof of creditworthiness, the party with insufficient credit will be required to provide credit enhancements, which may include (among others) letter of credit, affiliate guaranty, prepayment, or granting a security interest in property. Creditworthiness, as illustrated in the clause above, and adequate assurance of a party's ability to perform are two different things.

As you already know from the discussion of the UCC approach to adequate assurance of a party's continuing ability to perform in Section 2-609, in a proper circumstance either party may at any time make a demand for adequate assurances. Between Merchants, commercial standards will be used to determine the reasonableness of any such demand.[1] Since the parties may contractually change UCC terms, it isn't uncommon for a contract to specify that *only the seller* will have rights to adequate assurance protection. Gas contracts will frequently contain language limiting the types of assurances that may be demanded by either the buyer or the seller as well, so a creditworthiness clause that includes the elements provided in the example above provides each party with a *continuing* protection, as opposed to a right that may be *triggered* by some circumstance experienced by the other party.

With this contract clause, the information must be supplied at any time during contract performance if reasonably requested by the other party. As a party's exposure increases, credit requirements placed on the other party may increase, as well. Both parties are under a good faith obligation,[2] so the request for information should only be made when appropriate under the circumstances. Failure to comply with the provisions of a creditworthiness clause may be a breach of the contract. Without such a clause in the contract, the parties may only have adequate assurance protection from the Code, and that protection is available only under limited situations.

Parties releasing financial information may require some confidentiality statement in the contract under which the requesting party agrees to keep any credit information of the other party confidential and to use it only for the purpose of credit evaluation.

Default

Article X
Default

"*10.1 Default. If either Party*
(i) makes a general assignment or other general arrangement for the benefit of creditors;

(ii) files a petition or otherwise commences, authorizes, or acquiesces in the commencement of a proceeding or cause under any bankruptcy or similar law for the protection of creditors, or has such petition filed against it and such proceeding remains undismissed for thirty (30) days;

(iii) otherwise becomes bankrupt or insolvent (however evidenced);

(iv) is unable to pay its debts as they fall due;

(iv) fails to pay or perform when due any payment obligation to the other Party, whether under this Contract or in connections with credit obligations of a third party on its

> behalf or otherwise if such failure it not remedied by three (3) Business Days after notice of such failure is given to the failing Party;
>
> (v) fails to give adequate security for or assurance of its ability to perform its further obligations under the Contract within two (2) Business Days of a reasonable request by the other Party;
>
> (vi) breaches any other provision of this Contract, if such breach is not cured within thirty (30) days; or
>
> (vii) makes any misrepresentation under this Contract;

then that Party will be in default of this Contract after the applicable cure period has passed. In addition to all other remedies available hereunder and at law, the non-defaulting Party shall have the right upon notice to the Party in default to cancel this Contract and proceed in accordance with this Article X, or to withhold deliveries, suspend payments, or both."

10.2 Early Termination Date. If An Event of Default has occurred the non-defaulting Party may declare the Contract canceled and at any time from the default date until twenty (20) days thereafter, exercise its rights to name an Early Termination Date in respect of Transactions upon providing notice to the defaulting Party. Such Early Termination Date may be the day such notice is given or any prospective day within the twenty-day period.

> (a) The non-defaulting Party may, but is not obligated to, name such date as to all outstanding Transactions. Each Transaction will be liquidated to its current market value, using information reasonably indicative of future market prices, including quotes from market makers.
>
> (b) The value of all such liquidated Transactions will be calculated, as nearly as possible, as of the same date and will take into account the period between the Early Termination Date

and the date on which such amount would have otherwise been due under relevant Transactions. Any such calculations shall be done in a commercially reasonable manner.

10.3 Settlement Amount. The non-defaulting Party may set off or aggregate, as appropriate, any or all payments due under this Contract and any or all other amounts owing between the Parties under any other agreements, and net to a single liquidated amount payable by one Party to the other. The net amount due will be paid by the close of business no later than three (3) Business Days following the Early Termination Date.

Discussion of Paragraph 10.1. Most organizations that buy and sell gas try to do business only with creditworthy parties. During the course of performance, it is not unprecedented for one of the parties to face financial calamity. This calamity may be evidenced in a number of ways. It may begin subtly with a credit watch posting by one of the financial rating agencies. Creditors may become suspicious when payments are consistently late by more than 30 or 60 days. Dun & Bradstreet reporting may indicate that a party is consistently late in paying as well, so a creditor consulting Dun & Bradstreet may find that it isn't alone and other creditors are receiving late payments too. Articles about a particular organization may begin appearing in the industry press. These may include stories of financial woes, bickering among the board of directors, or searches for a white knight.

A party undergoing severe financial difficulties usually tries to put the best "spin" on its situation, so a gas seller may learn of a buyer's deteriorating financial condition only when a bankruptcy action is filed. This article deals with bankruptcy, failure to pay, failure to provide adequate assurance of the ability to perform, other contract breaches, and misrepresentation.

Bankruptcy. Bankruptcy is treated separately from other financial inability because of the Federal Bankruptcy Code[3] and its strict rules covering the property of the bankrupt debtor and rights of creditors. Two broad areas of concern exist in the bankruptcy context. The first deals with actions that may be taken by the court-appointed bankruptcy *trustee*,[4] and the second involves actions that may be taken by the non-bankrupt creditor. A gas contractor trying to anticipate potential financial calamity of either party may easily get mired in the minutia of the bankruptcy code. While the *bankruptcy trustee* does have broad powers, the bankruptcy code is filled with numerous exclusions and limitations on those powers.

Once a voluntary or involuntary bankruptcy petition has been filed, generally all collection activities that may have been pursued by the debtor's creditors must cease. The bankruptcy court obtains exclusive jurisdiction over the specific issue of debts. The so-called *automatic stay* provisions of the bankruptcy code prohibit creditors from commencing any judicial, administrative, or other action; enforcing any judgement obtained against property of the debtor before commencement of the bankruptcy case; obtaining possession or exercising any control over property of the debtor; undertaking any act to create, perfect, or enforce any lien against the debtor's property; collecting, assessing, or recovering any claim against the debtor that arose before the case was commenced; setting off any debt owed by the debtor; and commencing any proceeding before the U.S. Tax Court.[5] This precludes creditors, except in specific situations set forth in the bankruptcy code, from taking any action that would be detrimental to the interests of the bankruptcy estate and other creditors. Given this, how will a bankruptcy court deal with contracts that are still in effect?

Bankruptcy trustee.

Executory contract—no default. Contracts for sale under which the parties have not fully performed are termed *executory contracts*, and a bankruptcy trustee has the ability to assume or reject such contracts.[6] This is true when there has been *no default* of the contract and if the following conditions are met:

1. The trustee cures or provides adequate assurance that the trustee will promptly cure the default,
2. The trustee compensates the creditor (or other party) for its actual pecuniary less resulting from the default, and (in either case)
3. The trustee provides adequate assurance of future performance under the contract[7]

Executory contract—default. The bankruptcy trustee will not be required to comply with the procedures above, and may assume or reject an executory contract if a *default* has occurred under the contract and the default resulted from any of the following conditions:

1. the financial condition or insolvency of the debtor prior to the closing of the case,
2. commencement of the bankruptcy proceeding,
3. appointment of a trustee after the bankruptcy filing, or a custodian prior to filing, or
4. failure of the debtor to perform nonmonetary obligations under the executory contract.

If an executory gas sale contract contains a clause allowing one party to place the other in default for any of the conditions indicated above, that non-bankrupt party must carefully weigh the benefits and disadvantages of any actions it might have the right to take under the contract, including contract cancellation or termination. If the creditor chooses to cancel the contract, it must exercise that right before the bankruptcy case has been filed, or that right will be lost (unless the trustee gives permission). After the filing date, a creditor may not terminate or modify the contract or any of its terms.[8]

Assignment rights. Bankruptcy trustees have the additional power to *assign* executory contracts (regardless of restrictive assignment language in the gas contract), but the assignment right is limited by the willingness of the non-assigning creditor to accept such assignment.[9] The trustee may assign an executory contract only if adequate assurance of future performance is provided by the assignee of the executory contract,[10] and once assigned any contract language that would terminate or modify performance because of the assignment will not have any effect.[11]

Creditors. Many contracts used for long term purchases and sales contain a clause stating that the contract is a *forward contract* as defined by the U.S. Bankruptcy Code. A forward contract is defined as "*a contract (other* than a commodity contract) for the purchase, sale, or transfer of a commodity, as defined in section 761(8) of this title, or any similar good, article, service, right, or interest which is presently or in the future becomes the subject of dealing in the forward contract trade, or product or byproduct thereof, with a maturity date more than two days after the date the contract is entered into, including, but not limited to, a repurchase transaction, reverse repurchase transaction, consignment, lease, swap, hedge transaction, deposit, loan, option, allocated transaction, unallocated transaction, or any combination thereof or option thereon."[12]

The relevance of a forward contract designation in bankruptcy is that a "forward contract merchant" has the contractual right to *liquidate* relevant forward contracts[13] after the bankruptcy case has commenced. Forward contract merchants are defined as businesses that enter into forward contracts as or with merchants in a commodity.[14]

Any gas contractor drafting default language that incorporates bankruptcy issues should be familiar with the applicable bankruptcy or insolvency laws and carefully craft the relevant language. The words *terminate, liquidate,* and *forward contract* have specific meanings under the U.S. Bankruptcy Code, and bankruptcy language should be drafted in accordance with those meanings. Then, if bankruptcy or insolvency of one of the contracting parties becomes an issue, the response of the non-defaulting party will more likely be appropriate for the situation. Bankruptcy law is intricate and gas contractors new to the area should consult bankruptcy counsel.

Failure to pay. If a party to the gas contract fails to make payments when due, it has breached the contract. Although this reason for contract breach is most commonly thought of as a buyer's failure, in a given situation, the failure to pay could be the seller's. If a failure to pay occurs, the non-failing party must give notice that it has not received payment to the other party. If the failure is not *cured* and payment made within three business days in the example provided, the non-paying party will be in default and the other party may proceed with remedies available in the contract, under the UCC, and under other law.

Failure to provide adequate assurance of performance. As you learned in chapter 4, the UCC allows either party to demand additional assurances of continued performance if it has any reasonable expectation that the other party will not continue performing. A failure to give the assurances demanded could lead to a declaration that the party failing to give assurances has repudiated the contract. That failure could also give rise to contract cancellation in the default clause used as an example above. Either party facing the uncertainty of the other's continued performance may require assurances, and the failure by either party to provide that assurance within two business days is a default circumstance under the sample default clause.

If parties do not address the adequate assurance issue contractually, the party receiving the demand is given a reasonable period of time to comply with the demand, and the reasonable period cannot exceed thirty days.[15]

Other breaches of the contract. Any failure of an obligation under the gas contract may be grounds for breach and the consequence for failure to perform an obligation is typically found in the default article. The cure period for miscellaneous contract breaches is usually 30 days, a much longer period than that allocated whenever a failure occurs that involves payment because such a failure may not have such an immanent effect on the other party. Any failure to perform in accordance with a contract's provisions could be a breach of the contract if the other party is harmed by that failure.

Misrepresentation. Consequences also attach when a party makes a contractual representation both expressed and implied, then breaches that representation. Sometimes no representations are formally made in contracts, but it isn't uncommon for parties to make representations as to their legal power and authority. Let's say a party represents that it has the legal power to enter into the contract and that anyone signing confirmations on behalf of it has the authority to do so. That a remedy is available for a misrepresentation of legal capacity may be little comfort to a plaintiff attempting to sue, particularly when it's lawyers later learn that the other party misrepresented its legal status.

Notice that the potential actions that may be taken by the non-defaulting party include cancellation of the contract and suspension of performance or payment (or both). One point to be made about the use of the words, *termination* and *cancellation*, is that the UCC differentiates between those two words such that cancellation refers to the action a non-breaching party may take after contract breach by the other, while termination refers to an early conclusion of contract performance not associated with a contract breach. The U.S. bankruptcy code uses only the word *termination*, seemingly to include both instances discussed above. Gas contractors should choose the word used in this context carefully and also consider defining termination in the contract itself, so that the word may have a meaning common to both the UCC and the U.S. Bankruptcy Code.

Another point to keep in mind is applicability of relevant state laws and court rulings concerning strict notice requirements that a party may be obliged to use when declaring the other party in default of its contractual duties.

Contract cancellation/early termination. In the early years of open access, most gas contracts didn't contain clauses that ordered the procedures to be followed when a substantial breach of a forward contract occurred. Business was typically developed on a monthly basis, and little thought was given to the "what ifs" revolving around breaches of future contract obligations. Under the UCC, it is clear that a substantial breach of the entire contract will result in a damage award for the lost value of the entire contract, but that concept didn't find its way into most gas contracts until the advent of OTC hedging. The contract most frequently used for all non-regulated future financial transactions is the ISDA Master Agreement, a pro forma agreement developed by the International Swap Dealers Association in 1987.[16]

The first "ISDA Agreement" was developed in response to the explosion of international OTC derivative transactions that had begun in the 1970s in England. The 1987 agreement was subsequently re-drafted five years later, and now the 1992 ISDA Agreement is used by virtually all counterparties issuing OTC derivative transactions. In the next chapter, you will learn more about the current ISDA Agreement, but the only aspects of that document relevant for the purpose of this discussion are those relating to early termination.

Discussion of Paragraphs 10.2 and 10.3. The manner in which the early settlement calculations are derived closely follows (but with less complexity) the basic patterns seen in the ISDA Agreement. Under that agreement, the non-defaulting party has a right to liquidate transactions and cancel (or terminate) the contract. This process is only available to the non-defaulting party as long as the default continues. In other words, for as long as the failure leading to the default continues, the non-defaulting party has the rights similar to those enumerated in Paragraph 10.2 above.

In a break from ISDA Agreement language, this clause places a limit on the amount of time the non-defaulting party has to set an early termination date and the clock begins running on the day the default becomes effective. Some parties may prefer this approach because it provides some certainty to the process, while others might adopt the position that even if the default lasts for weeks, months, or even years, the non-defaulting party ought to retain its right to liquidate.

Once the date for early termination has been set, all affected transactions will be liquidated and reduced to present value facilitating a one-time settlement payment by the defaulting party. After the calculations have been made, the party in default will owe the settlement amount to the other party by no later that the third business day following the early termination date. If the non-defaulting party has been damaged beyond the direct damages indicated in this paragraph, it may elect to sue the defaulting party for additional damages.

Many gas contracts contain clauses of this type but they are relatively new to the industry. The decision that any organization makes about the way early termination calculations are structured (if at all) should include advice from risk management, tax, credit, and legal personnel.

Let's put this paragraph into practice with an example. Say FLEX and REFLEX sign a base contract that allows either party to be the seller or buyer in any transaction. The parties have agreed to nine transactions as follows (Fig. 8-1):

Deal No.	Buyer	Seller	Delivery Period
1	FLEX	REFLEX	August-December (year 1)
2	FLEX	REFLEX	May-June (year 1)
3	REFLEX	FLEX	January-February (year 2)
4	FLEX	REFLEX	October-March (years 1 & 2)
5	REFLEX	FLEX	December (year 1)
6	REFLEX	FLEX	January (year 2)
7	FLEX	REFLEX	January-December (years 1-3)
8	FLEX	REFLEX	April-September (year 2)
9	REFLEX	FLEX	February (year 2)

Fig. 8-1 *Transactions Under the Base Contract*

A Practical Guide to Gas Contracting

By this illustration, you can see that at any given time, multiple transactions between trading partners may be in various stages of performance, that performance may not have begun under certain transactions, and that either party could be the seller or buyer. In this case, FLEX and REFLEX have signed a base contract and all transactions together constitute the *contract* so that if either party defaults on one transaction, it will have also defaulted under the others, and all will be liquidated. In other words, the individual deals cannot be carved out for any purpose.

Let's say that in October of year 1, FLEX fails to make payment for gas delivered to it under Deal No. 1, above. If the failure is not cured and the default results, REFLEX will have the ability to liquidate all outstanding transactions between the parties. This will include the future transactions as well. If an agreement was reached by FLEX and REFLEX to any transaction prior to the time of the default, the fact that performance has not begun is irrelevant—the deal is done.

Assuming a Monday, October 25 payment date and notice of default dated October 26, FLEX will then have three business days to make its payment or place itself in a position of default. If payment isn't made by Friday, October 29, REFLEX may name an early termination date as of any date beginning with October 29 (the day the default became effective) until 20 days thereafter. As each day passes without notice of the early termination date, the time available to REFLEX becomes one day shorter.

If REFLEX notifies FLEX that the early termination date for all effected transactions is to be November 10, then as of November 10 the termination calculations will be made. For the deals that require future pricing, the NYMEX strip, market quotations, a forward-pricing curve used by the non-defaulting party, or other indicia of future gas prices will be used to determine the value of all transactions to REFLEX. Once an amount of damages has been determined, REFLEX will apply the appropriate percentage factor to discount that amount to its present value. Payment will then be due three business days after November 10th.

Force majeure

Article XI
Force Majeure

"11.1 Definition of Force Majeure. "Force Majeure" means any specific event or occurrence not contemplated by the Parties at the time they entered into a Transaction that prevents performance either partially or entirely, and is not within the reasonable control or ability of the Party claiming Force Majeure to prevent or overcome.

11.2 Notice. If a Force Majeure event has occurred, the Party experiencing Force Majeure shall provide immediate notice of such event to the other Party using the most expeditious means. Initial notice may be in any reasonable form, but written notice with full particulars and documentation necessary to prove the Force Majeure must be provided prior to excuse of any obligations under this Agreement. Once such notice and documentation has been received by the other Party, the delivery obligations of the Party experiencing the Force Majeure, to the extent those obligations have been prevented, will be excused for the entire duration of the Force Majeure event or occurrence.

11.3 Inclusions. Force Majeure will include acts of God, civil disturbances, riots, acts of terrorism, weather-related events such as storms and storm warnings, hurricanes, blizzards, floods, acts of regulatory and governmental bodies, explosions, necessary maintenance or repair of machinery, lines of pipe, and necessary facilities, freezing of wells or lines of pipe, acts of third party service providers essential for transportation or storage of the Gas, necessity for compliance with any court order, law, statute, ordinance, or regulation promulgated by a governmental authority having jurisdiction, strikes and lockouts, civil disorder, and any other event or occurrence not listed herein that satisfies the definition of Force Majeure in this Agreement.

11.4 Exclusions. Force Majeure protection will not extend to the obligation to make payments when due, loss of firm capacity released because of recall rights or other reservations and curtailment or interruption of interruptible or secondary firm transportation unless primary firm transportation of other shippers is also curtailed at the physical point of the Force Majeure event; allocations or reallocations by well operators, pipelines or other parties; changes in the market; loss of Gas supply; loss of market; or regulatory disallowance of the passthrough of costs of Gas or other related costs.

11.5 Settlement of Strikes. The settlement of strikes, lockouts or other industrial disturbances will be entirely within the sole discretion of the Party experiencing such disturbance.

11.6 Obligations of the Parties. The protection provided in this Article will not apply if the Parties do not cooperate to resolve the event or occurrence with the goal of resuming performance; <u>provided, however,</u> if non-performance caused by Force Majeure extends beyond _____ [days/weeks/months], the Party not directly experiencing the Force Majeure event or occurrence may, at its sole option, and upon providing notice to the other Party, terminate this Contract."

Discussion of Article XI. In the law of insurance, *force majeure* is defined as a "superior or irresistible force."[17] The term is not defined for other usages, but is a French term literally translated as a major force, essentially the same definition used above. Originally the term was used to apply only to "acts of God," but has been expanded—particularly in commercial usage—to include numerous events or occurrences that make either party's contract performance impracticable or impossible. The UCC addresses force majeure as follows:

"§ 2-615. **Excuse by Failure of Presupposed Conditions.** Except so far as a seller may have assumed greater obligation and subject to the preceding section on substituted performance:

(a) Delay in delivery or non-delivery in whole or in party by a seller who complies with paragraphs (b) and (c) is not a breach of his duty under a contract for sale if performance as agreed has been made impracticable by the occurrence of a contingency the non-occurrence of which was a basic assumption

> on which the contract was made or by compliance in good faith with any applicable foreign or domestic governmental regulation or order whether or not it later proves to be invalid
>
> *(b) Where the causes mentioned in paragraph (a) affect only a part of the seller's capacity to perform, he must allocate production and deliveries among his customers but may at his option include regular customers not then under contract as well as his own requirements for further manufacture. He may so allocate in any manner which is fair and reasonable*
>
> *(c) The seller must notify the buyer seasonably that there will be delay or non-delivery and, when allocation is required under paragraph (b), of the estimated quota made available for the buyer"*

Even a cursory comparison of the force majeure article above and Section 2-615 of the UCC shows the dramatic differences in the two. Under most force majeure clauses, the events and occurrences that will give a non-performing party protection are listed carefully. In the Code, a general concept is given, but little else. Courts in the U.S. have tended to prefer the list approach, where the parties agree during contract negotiations to the types of occurrences and events that will qualify for protection if performance is prevented. For that reason it is important for gas contractors to carefully anticipate the events that could prevent performance and to list those events in this clause.

Discussion of Paragraph 11.1. Force majeure should be defined somewhere in the contract, either in the definitions article, in context in the force majeure clause, or as a definition in the force majeure clause. The four traditional elements of force majeure include:

- an event or occurrence
- the non-occurrence of which was part of the basic understanding of the parties at the time they contracted
- that neither of the parties could have reasonably prevented or overcome
- that prevented performance, either entirely or partially

Event or Occurrence. Most force majeure clauses will include both *events* and *occurrences* as the underlying cause of the failure to perform. The word event is defined as "anything that happens or is regarded as happening" while the word occurrence is defined as "something that happens having no particular connection with or causation by, any antecedent happenings."[18] An event then, may involve a series of occurrences, while an occurrence stands alone. Typically, both will be included in a standard force majeure clause.

Basic Understandings. This element relates to the understanding of the parties at the time the contract was made. Originally, the contemplation element meant that the parties would not have entered into the agreement had they known that the particular event or occurrence was to happen. That original requirement has essentially disappeared from commercial transactions and frequently is not included in the definition of force majeure. Many gas-producing properties are located in areas prone to hurricanes or extremely cold weather. As noted in chapter two, platform abandonment due to hurricanes (or hurricane warnings) may be anticipated in the months of August through October in the Gulf of Mexico. Blizzards and other extremely cold weather conditions are common in the mid-continent gas fields, Alaska, and Canada. Since these events are within the anticipation of contracting parties, many gas contractors opt to omit the "reasonable contemplation" element of the definition. Others prefer to retain all four elements in the force majeure definition and rely upon the difference between anticipating an event of that kind and anticipation of that exact hurricane or blizzard, for example.

Inability to Prevent. The third element addresses the fault or negligence of the party experiencing the force majeure. If a party could have prevented or overcome the event itself—or the resulting inability to perform—and failed to do so, no force majeure protection will be granted under most gas contracts. Assume that a producer has had some difficulties with production facilities on one of its wells, but has failed to undertake proper maintenance because the well is ending its productive life, and the producer has concluded that it would not be cost-effective to undertake the maintenance. When the well fails because of the improperly maintained facilities, should the producer be protected by force majeure? You be the judge, but most buyers would not accept force majeure under those circumstances—that is, if the buyer had knowledge of the producer's decision made under the circumstances stated above.

Impossible Performance. The final and most important element of force majeure is also the most frequently overlooked. That is, the event or occurrence must have actually prevented performance, either entirely or partially. Most routine gas contracts do not require proof of the inability to perform, but in a long-term, high-risk gas contract, documentation of the inability to perform would almost certainly always be required.

Discussion of Paragraph 11.2. Gas contracts typically require formal written notice of a force majeure incident, but the force majeure clause will usually provide for immediate notice by the most practicable means, followed by notice in writing. There are a few cases when the notice requirement causes problems, first when written notice inadequately describes either the event or how that event prevented performance in which case the performing party may demand documentation or other proof that performance is actually prevented. The second notice problem occurs if the notice wording is such that the excuse or suspension of obligation will only be effective *after* the written notice has been received. Compare the two following clauses:

> 1. *"Notice of Force Majeure must be given as soon as practicable by any reasonable means, but Force Majeure protection will only be given to the party experiencing the event or occurrence after written notice with full particulars of the Force Majeure has been received by the other party"*

> 2. *"Notice of Force Majeure must be given as soon as practicable. Upon receipt of written notice with full particulars of the event preventing performance, the party experiencing the event or occurrence that prevented performance will be excused from its delivery or receipt obligations to the extent and for the entire duration of the event"*

Even though the intention in both of the clauses provided above may have been to provide force majeure protection for the entire event (the written notice having a retroactive effect), the first clause could easily be read to allow force majeure protection only prospectively.

According to the definition given in the example, the result of a successful force majeure claim is *excuse* from performance. Sometimes the result of force majeure is *suspension* of obligations. The two words may have different meanings and different connotations when used in contract language. To *excuse* generally means to release from an obligation or duty. To *suspend* means to defer or postpone, implying a continued obligation to make up that which was suspended. Carefully select the words you use to describe the result of a successful force majeure claim.

Gas contractors that want to clearly describe the effect of using the word suspend may wish to add clarifying language to the effect that a suspension for force majeure will or will not extend the term of the transaction.

Discussion of Paragraph 11.3. Why don't we construct force majeure clauses such that if anything prevents either party from performing its delivery obligations, force majeure protection will be extended, provided the event was not the fault of either party and could not have been prevented or avoided by either party—without bothering with the long list of included events typically included in the force majeure clause? The short answer is that courts in the U.S. prefer the list approach—*i.e.*, listing in the contract all of the events or occurrences that may result in force majeure protection. From the gas contractor's point of view, the challenge in listing events that should be considered excusable for force majeure is to make sure the list is complete. Most force majeure clauses will contain a trailer that adds the following language: "*or any event not listed that fulfills the definition of force majeure described in this Article XI.*" Even with this all-inclusive clause, a court trying to divine the intentions of the parties with respect to the event or occurrence at issue may disallow the disputed event not listed under the theory the parties wishing to include that event could have easily done so.

One way an organization may ensure the inclusion of all relative force majeure events is to undertake a force majeure audit that involves the gas sellers and buyers, transportation representatives, and gas contractors. As you learned in agonizing detail in chapter 2, a force majeure clause that fits the needs of an LDC may not fit the needs of a marketer.

Discussion of Paragraph 11.4. Many parties drafting force majeure clauses have a keen understanding of the events or occurrences that should not be given force majeure protection. That list is growing as the industry is becoming more complex and as transportation services are becoming more flexible to meet the needs of the changing industry. When Order No. 436 was issued, two types of transportation were available to shippers on interstate pipelines—firm and interruptible. There were few nuances within each of those categories. Now the interstate pipelines have shed their merchant function and are providing transportation services only. Shippers using interstate pipelines to transport gas now have a menu of services that are available to them, part of the growing flexibility in the transportation industry. Most gas contracts will allow force majeure protection if firm transportation transporting the gas subject to the sale is curtailed. With priority curtailment within firm transportation, most gas contracts will exclude curtailment of firm transportation using secondary receipt or delivery points. Similarly, while firm shippers may release excess firm capacity on all interstate pipelines in the U.S., they may also reserve the right to recall that capacity, thus making the reliability of capacity release transportation suspect. Many gas contracts will exclude loss of firm capacity release transportation if the releasing shipper has placed reservations on its released capacity, including the right to recall and has invoked that right.

A buyer of gas at or near the wellhead will generally allow force majeure protection for its producer supplier if affected wells freeze, but only rarely will the upstream buyer agree that force majeure protection will extend to the producer when only one well freezes. Usually a specific exclusion will be listed for freezing of individual wells or lines of pipe.

The final most common exclusion involves passthrough of gas supply costs when an LDC purchases gas at a cost later found by the state regulators to have been imprudently incurred. In the discussion of different interests among producers, marketers, LDCs, and industrial consumers in chapter 2, you learned about this.

Discussion of Paragraph 11.5. A paragraph of this type is always included in force majeure clauses to allow the party experiencing a strike or other industrial disturbance to settle the strike or disturbance in the manner it deems most appropriate. Without a clause such as this, the performing party could force settlement of the strike by threatening to place the other non-performing party in default of its performance obligation when the effected operation has either slowed or stopped altogether. This force majeure addition merely allows each party to the gas contract to handle any labor/management matters without interference from the other party.

Discussion of Paragraph 11.6. Some force majeure events and occurrences last only a few hours or a few days. Others may result in inability to perform for months or even years. Most gas contracts will require cooperation to resolve the situation with the goal of resuming performance. That is standard force majeure language—how should the parties address force majeure in the following situation? Let's say that a marketer has made a five-year sale to a start-up cogeneration facility. The first year's performance runs smoothly for both parties, but early in the second year, an explosion at the facility has rendered performance impossible for many months. What are the marketer's options in that event? It must stand ready to provide the daily contract quantity. In the meantime, it probably has committed to a long-term gas supply for the cogeneration facility customers and now it must find alternate markets for that gas.

Because of the uncertainty in the repair time, one concern of the marketer may be that it cannot enter into any alternative long-term sale commitments using the gas it has purchased primarily for the five-year sale. Similarly, the operator of the cogeneration facility may be concerned about losing its long-term gas supply if the marketer wants "out of the deal." Long-term force majeure is frequently treated differently from other force majeure situations. Negotiating parties should negotiate up front to find common ground for long-term force majeure resolution, where each is protected to an extent in the event of a catastrophic force majeure situation.

Taxes

Article XII
Taxes

"12.1 Responsibility for Payment. Seller shall pay or cause to be paid all

Taxes lawfully levied on the Gas, the Transaction or the seller prior to the Delivery Point. Buyer will pay or cause to be paid, all Taxes lawfully levied on the Gas, the Transaction or the buyer at and after the Delivery Point.

12.2 Responsibility for Reporting and Remittance. The Party designated by the taxing authority as the reporting and remitting entity shall be responsible for such reporting and remittance of Taxes unless otherwise agreed by the Parties. Either Party remitting or reporting Taxes on behalf of the other assumes full responsibility and liability for its acts and omissions and hereby indemnifies the other Party from all resulting liability.

12.3 Exemption Certificate. If any sale under this Contract is exempt from taxation, buyer shall provide Seller with a valid and properly completed resale or exemption certificate and if a blanket certificate expires during performance of any Transactions under this Contract, buyer will provide an updated certificate for seller."

Discussion of Article XII. While the tax clause is one of the most important in gas contracts, tax issues are frequently overlooked or misunderstood. The three most common types of taxes associated with natural gas in commerce are:

- severance or production taxes
- sales/use or gross receipts taxes
- consumption or utility taxes

Severance tax. Taxes charged on extraction of natural gas from the ground are in place in many producing states. In most of these states the producer is required to pay, report, and remit severance taxes to the appropriate taxing authority. In some states, though, the *first purchaser*[19] is the party responsible for remitting, reporting, or paying the severance tax.[20] Once gas enters the chain of commerce, each and every seller of the gas automatically warrants title to the gas. If severance taxes at the wellhead have not been paid according to the law, the title is not clear. If multiple sales are made of the same gas, each seller has potentially breached its title warranty, so correct payment of required severance taxes is an important issue that continues downstream. Gas producers know the severance tax laws in the states where they own production. The same knowledge is essential for the first purchaser of the gas as well.

Sales tax. A second area of concern in natural gas sales concerns the ability of taxing authorities to tax sales transactions. In this case, the gas is not being taxed, but the sales transaction is. The essence of the sales tax is that the taxing authority may collect sales tax on the transaction for sale of goods only once. This tax is imposed by the state in which title is transferred or the state of consumption and state laws on the subject vary wildly. Not all states impose a sales tax. The amount of sales tax differs from state to state. Counties and municipalities may also impose a sales tax, so the precise location of title transfer is important. In some states that allow an exemption from taxation on sales made for consumption in the manufacturing process, a county or city may nonetheless impose such a tax.

The buyer pays sales tax to the seller. The seller then reports the tax to the appropriate taxing authority and remits the tax. Hence, the seller is the responsible party. One must always consider every gas sale to be a taxable event. Only if a particular sale is exempted from imposition of the sales tax will the buyer be relieved of its obligation to pay the tax to the seller. The most common reasons for exemption from sales tax revolve around three basic principles. The first involves the *status of the buyer*, *i.e.*, not-for-profit organizations that have received an IRS Section 503(c) exemption from taxes. The second prevalent exemption relates to the *purpose of the sale*. Sales for resale (wholesale transactions) are exempted from taxation in virtually every state. The third reason for exemption is the *use* for which the product was sold. Gas consumed as fuel in the manufacturing process is exempted from the sales tax in many states, and some revenue statutes contain a specific exemption for fuel used in the cogeneration process.

The buyer must provide exemption certificates to the seller according to the contract requirements. The certificate states the reason for which the relevant sales are exempted, and is effective for a limited time. Some states have their own form of certificate and buyers are required to complete the correct form and submit that form to the seller. Since the seller is the responsible party for filing and remitting sales tax exemption certificates with the taxing authority, it may be in the seller's better interest to provide the correct form of exemption certificate to the buyer.

A majority of states now adhere to the Uniform Certificate for Sales and Use Tax Exemptions,[21] but only a few of those states don't require additional filing or reporting requirements. So, while the taxing authority requires proof of the exemption, the required proof varies from state to state. Taxing *nexus*[22] requirements, filing procedures, and the time at which the exemption certificate is filed may all be important. As a gas contractor, you should have an open dialog with designated tax personnel to assure that contract language protects your organization from unknown tax risks.

Gross receipts tax. Some states impose a gross receipts tax rather than a sales tax. In the simplest terms, a gross receipts tax is similar to the sales tax, but the seller pays, reports, and remits the tax to the appropriate taxing authority. Since the seller, in general, relies on its own information to file and pay the taxes, many of the problems inherent in the sales tax (where both the buyer and the seller are involved) are not seen in gross receipts tax states.

Utility or consumption tax. You've already seen that taxes may be imposed on the production and the sales of gas. Another set of taxes relates to the ultimate consumption of gas, and depending on the point at which these taxes are imposed, generally falls into one of two categories, either a utility tax or a consumption tax. These taxes are totally local creatures, and every one is different. In some forms, the municipality or the county will assess the tax. A continuing threat for the national gas industry is that the federal government may impose a consumption tax. President Bill Clinton, in office for less than one year, proposed a 14% federal *Btu tax* to be paid by the consumer. Under that proposal, the tax would not have been subject to any exemption. While that proposal was soundly defeated, anyone reviewing or negotiating a long-term gas sale or purchase should carefully construct contract language relating to new taxes, because in any given year the Congress (or any taxing jurisdiction) could impose such a tax.

Warranties and indemnities
Article XIII
Warranties and Indemnities

"*13.1 Seller's Warranties. Seller warrants that all royalties, Taxes and other sums due have been paid and that it has the right to convey title to and possession of the Gas. Seller further warrants this it is transferring good and merchantable title to the Gas and that all Gas is clear of any and all liens, encumbrances and Claims.*

13.2 Indemnity. Each Party assumes full responsibility and liability for its own acts or omissions and indemnifies the other Party from all liability and expense on account of any and all damages, Claims or actions, including injury to or death of persons arising when title to the Gas is vested in the indemnifying Party unless the act or accident was the result of the [negligence][willful misconduct or gross negligence] of the Indemnified Party, its agents or assigns."

Discussion of Paragraph 13.1. As you already know, a gas seller subject to the UCC automatically warrants title to all gas sold, even if the warranty is not specifically made in the contract. Most gas contractors add some additional warranties including the *right to convey* both the title and the possession of the gas. While sellers do have a contractual obligation to deliver the product and to deliver good title, many questions have arisen regarding the use of intermediaries or brokers to buy and sell gas. This issue typically arises in the agent/principal relationship where the broker (as agent) is given certain authority to act on behalf of the seller (the principal). Brokers do not acquire title to the gas, so while a broker as a seller's agent doesn't hold title to the gas itself, it will typically have been given the right to convey that title to the buyer. Sometimes a broker will be given authority to sign contracts as part of its agency relationship with the principal and other times, the principal will sign the contract. Basic knowledge of agency law is becoming more important in the new gas industry just because of the increased visibility of brokers.

Discussion of Paragraph 13.2. Each party to a gas sale contract takes responsibility for its own actions and indemnifies the other party against any damage or injury suffered as a result of something the indemnifying party did (or failed to do) under the contract. In the broadest sense, an indemnification is a statement that each party has liability insurance, or if it doesn't have such insurance, that it will nonetheless pay for direct harm suffered by the other party if that harm resulted from the indemnifying party's actions. The impact of indemnities may be felt most strongly at the point of consumption, because damage caused by the gas may involve damage to or loss of equipment, bodily injury, or loss of life. Indemnifications are usually given for title claims as well.

Third-party claims should be included in the indemnification. When a gas buyer receives notification, either formally or informally, that the gas it is purchasing does not come with clear title (as when royalties or upstream taxes have not been paid properly) it is faced with a dilemma. If a buyer first learns of the title issue when it receives legal notice, attorneys and courts will be a part of the process. Most attorneys don't provide free legal services, and neither do the courts. One of the specifically stated inclusions in an indemnification clause is usually the cost of attorney fees and court costs. State laws vary on the issue of whether or not attorney fees may be contractually assessed to the other party, though, so remember that even if attorney fees are part of the indemnification, in some states that right may not be enforceable.

When one party agrees to indemnify the other and a claim arises, one of the practical issues is which party's attorneys will provide the necessary legal services. From the indemnifying party's point of view, it will want to have control of any legal proceedings and keep costs at a minimum. But, from the indemnified party's vantage, it may feel that it will receive the best representation by its own attorneys and it will want to control the proceedings. Sometimes additional clarifying language is added to this clause to identify the indemnification process when the legal system becomes involved.

The final issue raised in this paragraph addresses the involvement of the indemnified party in causing its own damage. If the indemnified party contributed to its own injury, the indemnification sometimes will not apply so then the discussion turns to the level of negligence of the indemnified party. The determinant in this case is a combination of state law and contract specifications. Some state statutes specifically prohibit a party from contractually indemnifying the other for its (the other party's) own negligence. In such a state, the following situation illustrates one potential outcome. Assume that PUMP has sold and delivered gas to WIDGET's facility located 50 miles from the wellhead. The contract contains a clause similar to Paragraph 11.3. Gas is transported through an intrastate pipeline system and delivery is made directly to the facility. WIDGET's employees fail to properly read and record data indicating that the gas delivered is extremely high in contaminants. When the hot gas enters the facility, equipment is damaged and destroyed, the operation must be shut down, profits are lost, and personal injuries occur. As between the seller and the buyer, which do you think should be responsible for damage caused? On the one hand (unless disclaimed in the contract), the seller has given an implied warranty that the gas would be merchantable and an implied warranty that the gas would be fit for its intended purpose. So, from these facts, two potential breaches of warranty may be pertinent, and you have already learned the remedies available for breach of warranty under the UCC.[23]

Now, take that one step further and consider the applicability of the indemnification clause. Paragraph 13.2 above provides indemnification for *acts and omissions* of the indemnifying party—not directly for breaches of warranty. So, if a contractual obligation exists and the damage to the other party results from the action or inaction of the indemnifying party, the indemnification clause will apply. In states that prohibit indemnification for negligent acts, under the situation described above, the indemnification may not apply, if it can be proven that failure to read and record the relevant data was due to the negligence of WIDGET's employees. In this situation, WIDGET may have a claim for breach of warranty, but no right to indemnification because of its own negligence.

Gas contractors that include the concept of negligence in the indemnification clause (particularly if the governing law is a state that prohibits indemnification for negligent acts) will be silent on the issue elect to except *negligence* from the indemnification, or choose the higher standard of *gross negligence and willful misconduct.* A negligent act occurs when a person fails to do something that a reasonable and prudent person would do in the same circumstances. Gross negligence is the intentional failure to perform a duty in manifest disregard of the consequences. It is a failure to exercise even slight care. The legal standards are different and state laws differ on the issue, so care should be utilized when drafting the language that would render the indemnification unapplicable.

Notices

Article XIV
Notices

"14.1 All billings, payments, statements and routine communications pursuant to this Contract and any Transaction(s) shall be provided to the other Party in writing. Each Party shall provide the other with timely notification of any name or address changes for such communications.

14.2 Other than routine communications, all notices given under this Contract and any Transaction shall be in writing and addressed to the following:

Party A	Party B
Address 1	Address 1
Address 2	Address 2
City/State/Zip	City/State/Zip
Attention	Attention
Phone	Phone
Fax	Fax
E-mail	E-mail
DUNS #	DUNS #
Federal Tax ID #	Federal Tax ID #

14.3 Notices may be in any lawful written form and will be effective as of the time of receipt. Any notice received after 5p.m. in the receiving party's time zone will be considered received at the opening of business on the next business day. If time of receipt cannot be proven, the following presumptions will apply: (i) if by fax, the time and date designated by the receiving Party's fax machine;

(ii) if by courier, the designated delivery date; (iii) if by mail, five (5) days after the postmark; and (iv) if by electronic means, the time at which final successful transmission is recorded by the receiving computer."

Discussion of Article XIV. Most gas contracts now contain different notice requirements for formal notices and for routine communication purposes. An industry as dynamic as this one naturally involves employee transitions. Downsizing, mergers, acquisitions, and new job opportunities mean that long-time employees are in the minority in most organizations that sell or buy natural gas. A badly drafted notice clause could require a formal amendment to the contract whenever a new scheduler is hired.

Formal notices should probably be directed to a job title, rather than a particular individual, as well. Parties feeling uncomfortable with this approach may wish to add a requirement that the name of the individual to whom formal notices must be provided be updated as necessary without amending the contract.

Finally, the time at which notice is effective should be indicated. You'll see in Paragraph 14.3 that notices under this contract are effective upon receipt. Then, if a party cannot prove actual receipt, several presumptions are made depending upon the type of notice provided.

Miscellaneous

This article is typically used for everything that doesn't fit anywhere else in the contract. Some of the major miscellaneous clauses are considered by one party to be so important that those will be set apart in a separate article. Three common examples of this are assignment, confidentiality, and alternative dispute resolution. Attorneys care about the miscellaneous article, because this is where some of the most basic contract protections are given. While not inclusive of every single clause that could be contained in the miscellaneous article, the following list includes most of the typical clauses.

Alternate Dispute Resolution (ADR).

"15.1 Alternative Dispute Resolution.
(a) If a dispute concerning Contract Price, quantity obliga-

tions, Term, Delivery Point(s) or payments arises under this Contract, the Parties may agree to submit to final and unappealable arbitration utilizing the processes and procedures of the American Arbitration Association.

(b) Prior to instituting any litigation on any matter under this Agreement, either Party may require the other to submit to non-binding mediation conducted by a neutral third party agreed to by the Parties hereto. The purpose of such mediation is to bring decision-makers of the Parties to settlement of the dispute."

The two brief clauses provided above provide the backbone to any successful ADR clauses, but the following discussion will illustrate the need for a well-designed ADR clause, containing much more than the bare bones given above. Parties in many states may contractually agree to resolve legal disputes outside the court system. The two common forms of ADR are arbitration and mediation. Because nearly every state constitution provides the right to free access to state courts, some states have determined that parties may not contractually agree (in advance of the dispute) to resolve that dispute through binding arbitration. No such prohibitions exist regarding mediation because parties don't involuntarily lose their access to the courts after agreeing to mediation. Alternate dispute resolution methods are usually employed by contracting parties that want to avoid the time and cost involved in litigation. Parties wishing to really avoid time and cost constraints must be specific about those two factors in the gas contract.

Arbitration. Arbitration is a well-known alternate system for resolving disputes. The highway and building construction industries have successfully utilized arbitration for many years. Negotiations between labor and management also utilize arbitration to settle disputes. In fact, the type of arbitration known as "baseball" arbitration gains its name from a distinctive feature of arbitration used by the National and American Baseball Leagues in the U.S. where each party presents its solution for resolution of the disputed issue. The arbitrator must choose from the two options and may not reach any other decision. In the gas sales industry, though, arbitration as a form of dispute resolution is still a relative newcomer and contracting parties are usually careful to establish thorough rules for arbitration in the gas contract itself.

In arbitration, an arbitrator (or panel of arbitrators) receives evidence of the dispute from both parties and reaches a decision. Unless limited by contract terms, the arbitrator may receive any type of evidence and reach any decision. While there are winners and losers in the court system, that isn't necessarily always the case in arbitration, as the decision many times gives "something" to both parties—unless the parties have agreed to baseball arbitration.

Parties wishing to use arbitration to settle disputes may want to consider some of the following questions and incorporate any relevant procedures or limitations into the contract language:

1. How many arbitrators? Selection process?
- Will one arbitrator be used? If so, how will the arbitrator be chosen?
- If an arbitration panel is used, how many will be used and how will they be chosen?
- Must arbitrators be attorneys?
- What types of qualifications should arbitrators have?

2. What time limits will be established for arbitration?
- Deadline for arbitration demand
- Deadline for selecting arbitrator(s)
- Limited time for arbitration proceedings
- Deadline for arbitration decision

3. Should we limit the scope of arbitrable disputes?
- Price
- Delivery obligation
- Net Settlement calculations

4. **Should the Federal Arbitration Act[24] control, or ADR law of the governing state?**
   - Federal Arbitration Act may be preferable for interstate transactions
   - Some state laws will not allow parties to agree to arbitration in advance of the dispute
   - Statute of limitations of state law

5. **Should the American Arbitration Association[25] (AAA) commercial arbitration rules apply?**
   - Provide uniformity
   - AAA rules' applicability to the arbitrable disputes

6. **Should the arbitrator be limited in its review and decision-making powers?**
   - Federal Rules of Evidence or evidentiary rules of the governing law state
   - Federal Rules of Procedure or state procedural laws
   - Arbitrator confidentiality
   - *Ex parte*[26] communications
   - Limits on the decision (*e.g.*, baseball arbitration)
   - Arbitrator must follow relevant provisions of the underlying contract

7. **What type of relief will be available through arbitration?**
   - Any relief under the contract and at law and equity
   - Limitations on relief (punitive damages, consequential damages, etc.)
   - Conditions under which a party could appeal the arbitration decision
   - Attorney's fees to the prevailing party
   - Arbitration filing costs equally shared

In the gas sales industry, there appears to be a well-defined split on the appropriateness of arbitration. Some in the industry are great advocates of arbitration. Others hate it. One of the major reasons for opposition to arbitration seems to rest in the fact that the types of issues that are disputed in most gas sale transactions may not meld well into the arbitration structure. But even the most rabid opponent of arbitration can usually devise contract language to address the issues that caused the hostility.

Mediation. While arbitration is usually a well-structured and well-defined procedure, mediation may be just the opposite. The purpose for mediation is to resolve the disputed matter before costly and time-consuming formal proceedings such as arbitration or litigation occur. One person serves as the mediator and should receive training in mediation techniques prior to undertaking the difficult task. Many state bar associations have mediation specialists qualified for specific types of disputes. State law should also be consulted to determine if any limitations on mediation or mediation procedures have been formally enacted.

Once chosen, the singular goal of a mediator is to try to help the parties reach a mutually agreeable conclusion to the matter in dispute. A mediator will meet with both parties together and with each party separately. Potential outcomes of litigation and arbitration may be discussed frankly with the participants so they fully comprehend the possible adverse result of any third party decision. Mediation really puts the two parties in control of the dispute. One of the critical elements of any successful mediation is that the primary participants are the decision-makers. Mediation will fail if someone not authorized to make a binding decision on behalf of the company is the major participant.

If a mediation process is successful, both parties may sign a binding agreement that closes the matter in dispute, thus barring any formal legal proceedings. However, one of the best features of mediation allows the parties involved in negotiations to be forthcoming with their positions and to justify those positions without the constraints of evidentiary rules. Most parties will be counseled by their attorneys to be circumspect in the mediation process. Even though mediation is considered to be a settlement negotiation and not admissible into any subsequent legal proceeding, a fine balance exists between that which *could* be said to the other party and that which is *advisable*. Once a statement is made during mediation, even if evidence of that statement will not be admissible in subsequent legal proceedings, if the mediation is not successful, each party will have information that it may use to its strategic advantage during legal proceedings. While admittedly a double-edged sword, many industry participants prefer to float in murky mediation waters rather than do battle in the sea of litigation.

Governing Law and Venue

"15.2 Governing Law. This Contract will be governed in accordance with laws of the State of _____, except for such laws concerning the application of the laws of another jurisdiction. The Parties hereby stipulate that with respect to any and all disputes between the Parties arising from or relating to this Contract, venue will lie in the federal or state courts of _____, _____ County, State of _____, and each Party voluntarily submits to the jurisdiction of such court(s)."

Two separate issues are treated in this particular clause. The first is governing law, meaning simply that a court hearing the case would apply the law of the selected jurisdiction. The second issue is that of venue—where the court hearing the dispute is located.

Governing Law. Parties in the same state or the same province. The rules governing *choice of law*[27] should be divided into those that affect domestic parties and those that involve interstate or international parties. The governing law section of a contract seems simple. The parties merely agree to the law of a particular state or province that will govern any dispute under the contract. But the story only begins there. If a company in Calgary, Alberta and another Alberta firm have a gas contract, the laws of the Province of Alberta will generally apply. The same would be true of counterparties located in the same state. That is the simplest resolution to the governing law problem, but which law should apply if the seller is located in Houston, the buyer is located in Kansas, and the delivery point for all gas sold is in Oklahoma? Now, the dissimilarities involved in the transaction begin to take shape and you can see that some additional concerns may be relevant.

Parties in different states. In chapter 4, you learned about the two court systems in the U.S.—federal courts and state courts—and the challenges associated with litigation in either system. You already know that if the parties have complete diversity of citizenship and if the amount in controversy exceeds $75,000, or if the dispute involves a federal law, the federal courts will accept cases from litigants in different states. In either federal or state courts, the variety of issues will be reviewed to determine whether a particular state's laws would be appropriate for resolving the dispute.

Simply stating the law of a particular law or province in the contract as the governing law doesn't necessarily end the matter. A court's discretion could supercede the parties' agreement if the selected law has no relationship whatsoever to the disputed matter or the parties themselves. Simply put, then, contracting parties (if they agree at all) should select a law that has a connection with the parties, *e.g.*, location of the home offices or states of organization of either party, or a connection with the transaction itself, *e.g.*, the law of the state where gas deliveries are made. This is the best way to ensure that a court deciding the dispute will not interject its discretionary powers and overrule the parties' agreement. It's not a surefire method because a court may still determine that another set of laws is more appropriate, but it is the best insurance available to the parties.

Venue. Parties may agree to the exclusive or the non-exclusive location where any lawsuit resulting from or affecting the contract may be heard, although many gas contracts don't include a venue clause. Exclusive jurisdiction means that the lawsuit may be heard only in a court within the particular state or province. An example of exclusive venue is the following: *"With respect to any and all disputes between the parties arising from or relating to this Contract, venue shall exclusively lie in the federal or state courts of Houston, Harris County, State of Texas."*

Non-exclusive venue would result in voluntary submission to the jurisdiction of courts in one specific locale without contractually giving up the right to bring the lawsuit in another venue. In such a case, the above sentence would be altered to read as follows: *"With respect to any and all disputes between the parties arising from or relating to this Contract, the parties voluntarily submit to the non-exclusive jurisdiction of the federal or state courts of Houston, Harris County, State of Texas."*

The reason for this type of language, as discussed previously, is to in essence tell a court determining appropriate jurisdiction that the parties agree in advance that among all potential venues, Texas is one to which they would both agree. If litigation does occur, one of the parties may argue that another venue is more appropriate, but for the Texas litigant, the non-exclusive venue clause provides a certain level of comfort.

Parties should use some care in selecting the governing law and agreeing to venue. In the U.S., the UCC makes Article 2 in all states but Louisiana *somewhat* uniform. Remember though that courts in many of those states have, by decision, interpreted sections of the UCC in non-uniform ways. Another point for the gas seller to remember is that third-party warranty laws are different from state to state[28] and that implied warranties may be given to different remote parties, depending on which version of Article 2-318 the governing law state has adopted.

Specifically regarding the venue issue, parties should be aware that many courts, particularly if a jury is used to resolve the dispute, tend to give some level of "home court advantage" to the local litigant. If you are agreeing to the venue in the other party's home state or province, be prepared to face this subtle challenge.

Assignment or transfer. As indicated in chapter 4, unless agreed otherwise in the contract, either party may assign its contractual rights except where the assignment would:

- materially change the duty of the other party
- increase materially the burden or risk imposed on the other party by the contract
- materially impair the other party's chance of obtaining return performance[29]

The non-assigning party has the rights to demand adequate assurance of the assignee's ability to perform under the contract in any event.[30] The general view is that contract rights are freely assignable, so most parties to a gas contract will include an assignment clause that in some way limits rights that parties would otherwise have under Article 2 of the Code.

The words *assignment* and *transfer* mean nearly the same thing, but nuances of understanding separate the two words. Both refer to a type of conveyance, but an assignment refers usually to conveyance of a right, while a transfer refers to a conveyance of something tangible, such as property. When both assignment and transfer are used in a gas contract, the reference is both to rights and to the document itself.

"15.3 Assignment and Transfer. This Contract will be binding upon and inure to the benefit of the successors, assigns, personal representatives and heirs of the respective Parties. Other than to a parent or affiliate or pursuant to a sale of all or substantially all the assets of the assigning Party, no assignment or transfer of this Contract, in whole or in part, will be made without the prior written consent of the non-assigning or non-transferring Party, which consent will not be unreasonably withheld or delayed. Upon any assignment or transfer and assumption, the assigning or transferring Party will not be relieved of or discharged from any obligations hereunder unless specifically provided for in the executed assignment or transfer agreement and upon written consent of the non-assigning Party."

Discussion of Paragraph 15.3. Paragraph 15.3 represents the most common way of addressing the two conflicting issues involved in any assignment or transfer clause. The first issue related to whether or not the non-assigning party should be forced to do business with another party with which it might not otherwise choose to do business. The second issue that frequently conflicts with the first is the need by both parties for unimpeded business activities—*i.e.*, the ability to undergo business reorganization changes without obtaining prior written consent from all of the parties with which it has active gas contracts. Consequently, most assignment clauses in gas contracts reflect a few standard compromises.

For most assignment clauses, prior written consent of the non-assigning party will be required. The required consent must not be *unreasonably* withheld and it cannot be unreasonably delayed. Sometimes a fine distinction is found in the wording of any clause that inserts a standard of reasonableness. For instance, do you think the following two sentences mean the same thing?

- "The (actor) shall not act in an *unreasonable* manner"
- "The (actor) shall act in a *reasonable* manner"

Is *not unreasonable* the same thing as *reasonable*? Not necessarily. Another fine distinction in the use of these two words is whether the action must have been reasonable (or not unreasonable) in light of the circumstances or whether the actor's motivation must have been reasonable (or may not have been unreasonable). When reviewing an assignment clause, be careful not to overlook some of these fine points.

Prior written consent is not required under paragraph 15.3 in those circumstances where the balance of interests favors unimpeded business activities. When a company is sold or is merged into another entity and all or substantially all of the assets of the assigning party are transferred to the assignee, rights under this gas contract may be transferred without prior written consent of the other party. Only if the assignment of duties is explicitly stated in the assignment or transfer document, and if the non-assigning party has consented in writing, may the assignor be relieved of its duties under the contract.

Confidentiality. When open access transportation became a reality in the gas industry, most gas contracts contained confidentiality clauses. Price discovery in the 1980s wasn't efficient and organizations tried to keep all information about gas pricing confidential. With the advent of NYMEX gas futures contract trading, though, gas prices became transparent and today, current pricing information is available on a daily basis, and the futures price is indicated for up to 36 months beyond the current date. Organizations that choose to maintain confidentiality clauses in their gas contracts today do so to control business information that becomes public. When it is the corporate policy to manage public image and public information, confidentiality clauses will still be seen in gas contracts.

"15.4 Confidentiality. The terms of this Contract and any Transaction under this Contract, including price of Gas and transportation costs or arrangements and any other confidential or proprietary information provided to the other Party prior to and during the term of this Contract, including business and credit information together constitute the "Proprietary Information." Proprietary Information shall be kept confidential by the Parties, their agents, employees and assigns, and will not be disclosed to any third party except (i) as necessary to required third party auditors, accountants and attorneys, provided such third parties have agreed to be bound to this Paragraph 15.4, and (ii) to the extent necessary to effect transportation or other necessary services. If either Party is required to disclose information to any governmental authority or court having jurisdiction over that Party, that Party will notify the other in writing in

advance of disclosure specifying the nature of the request and the information to be disclosed. Each party agrees to limit any such required disclosure, both as to information and to distribution, to the maximum extent feasible."

Discussion of Paragraph 15.4. This clause prevents either party from disclosing information (other than to outside auditors, accountants and attorneys and as required for transportation and other regulatory purposes) regarding this contract. If a party is required to disclose information in a court or regulatory proceeding, it must notify the other party in advance. If a contract contains information that would be detrimental to the other party's commercial position, the other party may institute appropriate proceedings to protect the confidentiality of such information.

If financials of a subsidiary organizations are part of a parent's consolidated financial statement, the release of credit information may not be allowed by the parent company. In such a situation, the parties may nonetheless agree to the release of credit-related information if the information is used only for the purpose of evaluating credit and if a separate confidentiality agreement is signed. In that case, a separate clause could be included in the creditworthiness article of the contract (as was discussed in Chapter 7).

Failure to comply with a confidentiality clause may be a breach of the contract resulting in contract termination if the breach substantially impairs the value of the contract to the other party, so parties should not take this clause lightly.

Conflict of terms. The contract should contain a conflict resolution procedure whenever the base contract format is used. Separate parts of the total base contract structure (which could include the base contract, confirmations, and tape recordings of trader conversations) could be in conflict, particularly when different individuals are involved in contract negotiation, confirmations, and deal-making.

"15.5 Conflict of Terms. If there is a conflict between the terms of this Contract and any Confirmation, the Confirmation will prevail. If there is a conflict between a Confirmation and any electronic or taped recording of the Transaction conversation, absent manifest error, the terms specified in the Confirmation will prevail."

Discussion of Paragraph 15.5. Paragraph 15.5 is straightforward and applies the usual approach, which is to resolve conflicts in favor of the more recent and more specific document. One of the tricky points still at issue in the gas industry is resolving conflicts between written confirmations and taped recordings of the deal being made. For all of the reasons discussed in the evidence discussion in chapter 4, most gas sale participants are currently resolving conflicts between written confirmations and tape recordings in favor of the written confirmation. In the paragraph above, this is true unless a blatant mistake is found on the confirmation.

This resolution process will work well in most circumstances, even when the same base contract is being used for longer-term transactions and for day trading activities (where no written confirmation will ever be forthcoming). For day trading activities, if no written confirmation is used for purposes of satisfying the statute of fraud, and only the taped conversations exist to prove the elements of a transaction, there will be no conflict, so the conflict resolution rules will not apply. If traders have been adequately trained to state the elements of each transaction in every phone conversation, even fewer problems should occur.

However, problems may occur with this approach if confirmations are required for all transactions and conflicts occur between a taped conversation and the written confirmation. Gas contractors should use special care when making the decision of whether the written confirmation or tape recordings should control for day trading activities that include both tapes and faxes.[31]

Entire agreement. Sometimes called a "merger" clause or an "integration" clause, this miscellaneous provision results directly from sections of UCC Article 2 that deal with the court's ability to look beyond the four corners of the written contract itself to determine the intent of the parties at the time they entered the agreement. Section 2-202 makes it clear that if the parties agree that the writing constitutes their entire agreement, no other evidence may be introduced in a court of law that would tend to dispute the contents of the writing. Other evidence may be used to explain or supplement the terms of the contract, but not to dispute what is clearly contained in the written document.

Assume that PUMP and FLEX agree to a 5-year sale of gas in which PUMP agrees to deliver 5,000 MMBtu/day delivered at the end of the gathering system with which PUMP has a gathering agreement. The contract states that the delivery point for all gas will be "at the interconnection of gathering system X into the facilities of PIPE," which was the only connection between the gathering system and the pipeline at the time of agreement. What if another pipeline company, REPIPE, subsequently adds an interconnection further upstream on gathering system X during the 5-year term, and PIPE now wants to deliver gas into REPIPE to lower its gathering fee? FLEX doesn't want to change the delivery point.

The contract language appears to be clear that PIPE and FLEX contemplate only one point of delivery. But if in their negotiations it was clearly understood by both parties that additional points could be added on the same gathering system and evidence of that agreement meets admissibility requirements, a court hearing the dispute may allow parol evidence of the additional agreement if it would tend to clarify, rather than dispute, the agreement of the parties. This is an excellent example of the UCC's different interpretation of the words *contract* and *agreement*, since the court in this case is looking to the intent of the parties in their agreement (or bargain).

"15.6 Entire Agreement. This Contract, the Confirmations, and (if relevant) the electronic or taped phone recording of Transaction conversations, set forth all understandings between the Parties respecting such Transactions, and any prior agreements, contracts, understandings and representations, whether oral or written, relating to such Transactions are merged into and superceded by this Contract. This Contract may be amended only in a writing signed by both Parties."

Discussion of Paragraph 15.6. The parties have indicated their preference that all negotiation information be excluded from any subsequent court proceeding. A clause such as this must be included in all gas contracts to preserve the right to indicate that preference.

Amendments in writing. You probably noticed in Paragraph 15.6 that the final sentence requires all amendments to the Agreement to be in writing and to be signed by both parties. This sentence is so important that it merited special attention in this guide. Unless this sentence (or one like it) is included in a gas contract, the parties may change or amend contract terms simply by their behavior. For gas Merchants, the implications are even more far-reaching since the Code specifically states that as between Merchants, any such requirement on a form supplied by one Merchant must be signed by the other party.[32] Thus, if the base contract does not contain a sentence requiring amendments to be in writing, and the confirming Merchant adds this requirement as a special provision to a confirmation, without the signature of the other Merchant, this added language would not be enforceable. Performance may be another issue.

Once a contract has been negotiated and the parties either scan it into their electronic contract system or lock the original in a fireproof file, the only evidence that traders, schedulers, transportation personnel, credit, accounts receivable or accounts payable, and contract administrators have is the information contained in their own gas management systems. These systems may be electronic or paper, but irrespective of the type of system used, few contract management systems contain all the elements of the deal. All systems will give the pertinent information relative to quantity, price, delivery point, and delivery period of a relevant transaction, the transporter or transporters initially used, level of sale and purchase obligations (firm or interruptible), and so forth. Some systems are quite sophisticated. Others are not.

Let's say that FLEX and GASCO agree to a one-year gas sale transaction that does not contain a requirement that all amendments must be in writing and signed by both parties. FLEX's electronic gas management system is state-of-the-art and enables quick access and accurate information to anyone using the system, but does not allow anyone other than authorized users to change items in the system. GASCO is a small entity, and its paper gas management system, while providing more flexibility in transactions, does not block any user of the system from making basic changes.

After FLEX has scanned the contract and GASCO has locked its copy in the fireproof file, performance begins. In the second month, FLEX has difficulty making deliveries to the designated delivery point and, with the trader's permission (who is not an authorized user of the system), FLEX's scheduler asks her counterpart at GASCO if a change in delivery points is possible. Receiving an affirmative response, FLEX's scheduler nominates gas to the new delivery point. GASCO's scheduler has changed the delivery point in the GASCO's gas management system. Three months later when FLEX can resume deliveries to the original point, GASCO objects on the grounds that the parties effectively changed the terms of the contract by their performance. Without the sentence requiring written, signed amendments, the parties have probably effectively changed contract terms—this is so even with the sophisticated gas management system designed in part to disallow transaction changes without approval.

Representations of the parties. A representation made by either party to a contract is a statement of fact that may be either express or implied, and that may be made either before or at the time of making the contract, but which in either case, is so important to the other party that it influenced the decision of that party to enter into the contract.[33] Most gas contracts don't contain representations, unless the contract is unique (*i.e.*, not a base contract) and the circumstances of either party would warrant addition of representations. But a common representation made by both parties in today's gas contracting environment is an expression of the legal capacity of each party to enter into the contract. In the discussion of legal capacity in chapter 4, the distinctive differences between the *power* to contract and the *authority* of a party signing the contract to sign on behalf of an organization were discussed. This clause simply makes the legal capacity of both parties a basic premise upon which the other party relied.

"15.7 Representations. Each Party represents to the other that it has the full, complete, and current power to enter into and perform this Contract for as long as this Contract is in effect, that the person signing this Contract has full and complete authority to do so, and that the person(s) entering into and/or signing any Transaction(s) have full and complete authority to do so."

Discussion of Paragraph 15.7. This clause requires each organization entering into the contract to have requisite power to do so. For a corporation, a statement will usually be made in the articles of incorporation or by-laws that the corporation may enter into contracts and undertake any lawful activity. A partnership's power will be found in the partnership agreement, and an individual must be of the age of majority and not otherwise mentally impaired.

The person signing the base contract must have authority from the organization to bind the company contractually, and the party signing confirmations must have the authority to undertake that endeavor—one reason many confirmations are not signed.

Any misrepresentation is an automatic default under most gas contracts, and any misrepresentation made generally must include intent to deceive or mislead. The appropriate remedy in the case of a material misrepresentation generally is rescission of the contract. The UCC indicates that for fraud or a material misrepresentation all remedies available for a non-fraudulent breach may be available, and that neither rescission nor a claim for rescission will bar any other remedy that may be available.[34]

Severability

"15.8 Severability. If and to the extent any court of competent jurisdiction determines that any term of this Contract is invalid or unenforceable, that determination will not affect the validity of the other provisions of this Contract, which will remain in full force and effect."

Discussion of Paragraph 15.8. If a contract is severable and a court determines that part of the contract is not enforceable, *i.e.*, unconscionable terms or penalty provisions, the unenforceable part will be excised from the contract and the rest of the contract terms may be unaffected. However, if a severability clause is not included in the contract, a court finding part of the contract to be unenforceable, may declare the entire contract unenforceable. The severability clause has two important applications in gas sale transactions today. The first relates to bankruptcy and the second to the administrative fee included in many remedy articles.

The federal bankruptcy laws allow creditors of the bankrupt debtor in certain circumstances to setoff, liquidate, and terminate transactions between the creditor and debtor. However, some exclusions in the bankruptcy code prohibit setoff, liquidation, and termination in certain circumstances. Section 365 (e) (1) of the bankruptcy code specifies that even if an executory contract allows termination or modification of rights and obligations by the non-bankrupt party, such actions may not be taken after the bankruptcy proceeding has been filed if the contract provision is conditioned upon any of the following:

- insolvency of the debtor prior to filing the case
- commencement of bankruptcy proceedings
- appointment of a trustee before the bankruptcy has been filed

Gas contracts usually list as events of termination the three circumstances listed above. If the non-bankrupt party terminates the contract after the bankruptcy has been filed, a court may find the clause to be unenforceable because it is contrary to bankruptcy code provisions.

As noted in the chapter 8 discussion of Article IX Default, setoff rights in the bankruptcy code are subject to the Section 365 automatic stay provisions. But liquidation rights for "forward contract merchants"[35] allow those merchants to liquidate any contract that satisfies the definition of "forward contract"[36] in the bankruptcy code, outside the automatic stay, avoidance, or other prohibitory provisions of the bankruptcy code. In many cases, the gas contract giveth and the bankruptcy code taketh away. The point relating to the severability clause is this. Gas contractors, and those who make any termination, liquidation, or setoff decisions must be attuned to applicable law to avoid inadvertant unenforceability.

The second major issue regarding severability has to do with the administrative fee provisions of many remedy clauses in routine gas contracts. No U.S. court has been faced with this particular issue, so whether these clauses are enforceable in the U.S. gas industry is still unknown. If an administrative fee is determined to be a penalty by a court of competent jurisdiction, the affected clause, the entire remedy section, or the contract in total may be found to be unenforceable. Most parties include severability clauses in gas contracts because of these two potentially explosive issues.

Third-party beneficiaries. *"15.9 Third-Party Beneficiaries. The Parties agree that this Contract was not entered into for the benefit of any third party, that there are no third party beneficiaries to this Contract, and that the Contract* *does not impart enforceable rights to anyone who is not a party or a successor or assignee of it."*

Discussion of Paragraph 15.9. The discussion in chapter 4 regarding warranties given by a seller to downstream parties relates directly to this clause. However, some potential upstream parties not subject to warranties given by the seller may also be deemed to be primary beneficiaries of the underlying gas contract. When the parties indicate that the contract was not entered into for the benefit of any third party, they add a type of insurance against third party warranty claims and also against the non-warranty claims of any third parties. This is not a warranty disclaimer, but rather an indication of the intention of the parties. Warranty disclaimers are included in most contracts developed by gas suppliers, and those operate separately from the provisions of this clause.

Waiver. *"15.10 Waiver. The failure of either Party to insist upon compliance with any term of this Contract, to exercise any option, enforce any right or seek any remedy shall not affect, nor constitute a waiver of that Party's right to insist upon strict compliance with that or any other term of this Contract, to exercise that or any other option, enforce that or any other right or to seek any other remedy."*

A waiver is the intentional relinquishment of a right. Waivers of rights and contract modification through amendment (as discussed in the "Written Amendment" clause) are closely related. The UCC does not require amendments to be in writing (unless the contract provides otherwise), but states that an amendment not in writing can serve as a waiver instead. Does this mean that the ability to waive a contract right, *i.e.*, the right to place the other party on notice that it has defaulted under the agreement, is superior to the agreement made by both parties that amendments must be in writing and signed by both parties? Let's take the example of the transaction between FLEX and GASCO indicated earlier in this chapter and add a few additional facts. If the gas contract contains neither a written amendment clause nor a waiver clause, the facts are fairly indicative that the parties have impliedly agreed that a noncontested variation from contract provisions may constitute a waiver of rights. Now assume that the contract between FLEX and GASCO does contain a written amendment clause but no waiver clause. Could FLEX's failure to take some prophylactic action, even as it acceded to the temporary change in contract performance, have limited its rights to demand a return to the performance specified in the contract? Maybe so.

Courts hearing disputes of this type have reached incompatible conclusions, and the state of the law in this area is not clear.[37] Careful gas contractors will always include both a waiver clause and a clause requiring all amendments to be in writing and signed by both parties.

Joint efforts of the parties. *"15.11 This Contract will be considered for all purposes as having been prepared and drafted through the joint efforts of the Parties, and will not be construed against one Party or the other as a result of the preparation, negotiation, drafting, or execution of this Contract."*

When courts are reviewing contracts, one of the rules of contract construction is that, whenever a question of particular contract language is at issue, the terms will be viewed in a light most favorable to the non-drafting party. "Take it or leave it" contract terms in contracts for necessary goods and services are being targeted by this rule.

The applicability in wholesale gas transactions is somewhat limited, but as the gas industry involves itself more in commercial and other retail transactions, the relevant knowledge of the parties, as well as their relative negotiation strength, may become an issue. For this reason, many gas contractors will add this clause to their house contract as protection should contract language ever be litigated.

Signatures. Finally, the parties must sign a contract. Notwithstanding the applicability of diverse types of "signatures" that will satisfy both statute of frauds concerns and evidentiary requirements of a court, most organizations that buy or sell gas will still require the written signature of a party having authority to bind the company. That party will also provide its title, and the date on which the contract was signed. Once this has been accomplished, the contract has been fully negotiated with terms that reflect the understanding of both parties.

While many gas contracts may contain additional terms other than indicated in this guide, these are the basic terms found in nearly every routine base gas contract. One special note should be added regarding unusual additions to gas contracts. As noted earlier, whenever you see something in a gas contract that just doesn't make sense to you, that particular clause may have been added to the other party's house contract following a particularly bad experience that occurred because the contract didn't specifically address that particular issue before.

---

1. U.C.C. § 2-609 (2).

2. U.C.C. § 1-208.

3. 11 U.S.C. § 101 *et. seq.*

4. The representative of the estate. 11 U.S.C. § 323 (a).

5. 11 U.S.C. § 365 (a).

6. Id.

7. 11 U.S.C. § 365 (b).

8. 11 U.S.C. § 365 (e) (1).

9. 11 U.S.C. § 365 (c) (B).

10. 11 U.S.C. § 365 (f) (2).

11. 11 U.S.C. § 365 (f) (3).

12. 11 U.S.C. § 101 (25).

13. 11 U.S.C. § 556.

14. 11 U.S.C. § 101 (26).

15. U.C.C. § 2-609 (4).

16. ISDA subsequently changed its name to International Swaps & DerivativesAssociation.

A Practical Guide to Gas Contracting

17. *Black's Law Dictionary*, Revised Fourth Edition, St Paul, Minn. (1968). The term is not defined in the context of contracts for sale of goods.

18. *Webster's Encyclopedic Unabridged Dictionary of the English Language*, Grammercy Books, New York, 1989.

19. A term used to designate the party purchasing gas for the first time. The original purchaser.

20. For a good review of current state severance tax laws, see the Energy Information Administration's website at http://www.eia.doe.gov/neic/press/pressrel.html.

21. The website of the Multistate Tax Commission contains several documents, including the Uniform Sales/Use Tax Exemption Certificate, at http://www.mtc.gov/txpyrsvs/services.htm.

22. *Nexus* is the word used to describe the level of business activity in a state that would warrant registration with that state. These regulations are frequently written very broadly, so the party anticipating registration with a state in which it does business must rely upon its best judgement in many circumstances when deciding whether its level of business in that particular state sufficiently meets the nexus requirement.

23. See discussion of warranties in chapter 4.

24. 9 USC 1-14, 43 Stat. 883, 61 Stat. 669, 68 Stat. 1233, P.L. 101-369.

25. The American Arbitration Association Commercial Arbitration Rules may be found at http://www.adr.org/rules/commercial_rules.html. Another excellent source of information about arbitration may be found at http://www.gsa.gov/regions/4k/legal/legal89.htm.

26. Communications by one party to a proceeding to the decision-maker that is outside formal procedures.

27. Every state has both substantive laws and procedural laws, and when parties elect a governing law for their contract, they have implicitly adopted both substantive and procedural laws. One aspect of procedural laws is that which addresses the issue of the appropriate law for the particular matter in dispute. For example, the laws of every state in matters relating to real estate determine that the law of the state where the realty is located should govern any disputes. The same issue is true with contracts, although state procedural laws vary on this topic. The choice of law provision in the governing law state could, in fact, direct that another state's laws prevail.

28. See the discussion of third party beneficiary law in chapter 3.

29. U.C.C. § 2-210 (1).

30. U.C.C. § 2-210 (5).

31. See the discussion of evidentiary rules and potential problems when one form over another is deemed to be the "contract" in chapter 4.

32. U.C.C. § 2-209 (2).

33. *Black's Law Dictionary*, Revised Fourth Edition, West Publishing Co., 1978.

34. U.C.C. § 2-721.

35. 11 U.S.C. § 101 (26).

36. 11 U.S.C. § 101 (25).

37. *White & Summers supra*, pp. 29-36.

Chapter 9 | New Issues: Over the Counter (OTC) Derivatives Documentation and Electronic Contracting

Bob Dylan sang it well back in 1964, ". . . and the times, they are a changin'." Thus goes the saga of the gas industry as well. The original U.S. integrated gas companies, huge conglomerates that included production, transportation, distribution, and other gas-related subsidiaries, are becoming even larger beasts with voracious acquisition appetites to fuel their growth. Vast, multinational energy companies are being formed through the process of merger and acquisition, and traditional competitors, seeing the advantages naturally associated with sheer size and strength, are beginning to eye each other lustfully. Large integrated gas companies are doubling or tripling their original size, and sometimes being merged into other large integrated companies. The gas industry is not doing this alone. The most visible players in energy merger mania are electric companies.

When restructuring of the electric industry began with Order No. 888,[1] a new type of electric utility, the power marketer, was created. Cash-rich electric utilities have new competitive opportunities and many are seeking out alliances with other energy service providers. Natural gas, electricity, coal, wind, solar, and other forms of energy (and energy-related products) are now being purchased, sold, and traded under shared organizational umbrellas.

While still tempered by some regulatory constraints, the new allied energy industry emerging is far greater than the sum of its original parts.

The FERC has indicated its preference for greater uniformity in the energy industries it regulates and for expanded competitiveness within its ambit of regulatory oversight. The FERC has mandated the use of electronic communications between natural gas transporters and shippers, and for dispatching communications in the electric industry. Once electronic communications became commonplace with those industries, advances were made to expand the use of electronic communications for other purposes as well.

The GISB activities also create a new level of consistency between interstate gas transporters and shippers who use the pipeline system. As other industry participants become *GISB compliant* through incorporation of GISB standards in day-to-day gas sale and transportation activities, even the most stubborn sectors have begun to cooperate and negotiate with each other.

Another major development, irrespective of the type of energy at issue, is the use of hedging products. OTC derivative transactions comprise a large part of the hedging portfolio of many energy organizations, and the contracts used for OTC transactions are becoming an influence in energy sale contracts as well. More and more, gas contracts are featuring language that originated in financial transaction documentation. The long-term contracts of yesterday that contained numerous clauses addressing the price of gas to be delivered in the future now barely address the price issue. Through contract language, the parties seem to assume that each will undertake its own form of hedging activity to realize its own goals. The old price renegotiation clauses are being replaced by language that requires a defaulting party to compensate the other for any losses associated with unwinding (OTC transactions) or liquidating (futures contracts) hedging positions taken. This new approach to contract language also includes some crossover from actual language found in the ERMA Agreement or the ISDA Master Agreement.

For example, the notion of *material adverse change* is included in the schedule of many OTC derivative hedging agreements. The term material adverse change sometimes attaches to a change in the creditworthiness of a party caused by a downgrade in rating by one of the major rating agencies, *e.g.*, Standard & Poor's[2] or Moody's.[3] If the change in creditworthiness would detrimentally affect a party's continuing ability to perform under the financial transactions, the agreement may be terminated.

The *Creditworthiness* sections of physical gas contracts are now showing much greater detail than in the past, which may be a total understatement. Gas contracts in the 1980s and 1990s frequently went without any language addressing creditworthiness. Today creditworthiness clauses may include descriptions of the events that give rise to a right to demand credit assurances or to terminate the contract because of a failure or diminution of the other party's creditworthiness. Not so long ago, most gas buyers didn't even check into the financial standing of their gas suppliers. Now many gas contracts contain language giving both parties contractual rights upon the occurrence of adverse financial circumstances of the other party.

Even though the situation of a brief failure to deliver or receive gas is typically addressed in the Failure to Perform section of a gas contract, the more difficult question of a long-term failure is left unanswered in many gas contracts today. All OTC derivative master agreements contain an *Early Termination* or *Liquidation* section that provides a mechanism for liquidating transactions, reducing the liquidated amount to present value, and terminating the contract in the event a party substantially defaults. These clauses are appearing in an increasing number of master gas contracts when effected transactions have a term longer than a few months. Some energy participants believe that more of the language found in the most common master agreements for OTC derivatives should be incorporated in physical gas contracts. I agree with this position for many of the reasons discussed below:

OVER-THE-COUNTER DERIVATIVES DOCUMENTATION

While some participants utilize their own form of agreement, the most commonly used master forms of derivative agreements in the energy industry are:

- the form developed by the International Swaps and Derivatives Association (ISDA)
- the form developed by the Energy Risk Management Association (ERMA)

When an organization contemplates multiple transactions with one swap counterparty (this may be for a number of reasons including risk management and ease of contract negotiation), selection of the master agreement used is entirely at the discretion of that organization.

ERMA Agreement

The ERMA Agreement is well established, particularly in the oil industry, and was developed by OTC energy derivative users as an alternative to the ISDA Master Agreement. It appears, however (at least in the natural gas industry), that the ISDA form is used more frequently than the ERMA Agreement. This is probably due to the routine use of the ISDA form by derivative dealers with which many energy industry participants conduct OTC derivatives business.

ISDA Master Agreement

When ISDA released the 1992 ISDA Master Agreement, it was intended for international use, primarily by dealers. The same agreement being used today by a gas firm for its OTC derivatives hedging is being used by banks in Tokyo, Copenhagen, Johannesburg, and London in their dealings with other multinational banks and securities houses. The 1992 ISDA Master Agreement is admittedly a difficult read for the first-time peruser, and it continues to be a difficult document, even for the accomplished specialist. It was developed by a committee—surely part of the reason for its verbosity, but also the reason for its thoroughness. The 1992 ISDA Master Agreement has addressed nearly every issue that could conceivably arise between OTC derivative counterparties.

This agreement is owned by ISDA and may be purchased for a small fee.[4] The user has a non-exclusive right to reproduce the agreement without infringing the copyright owned by ISDA. In fact, the appearance of an ISDA Master Agreement is so important that most contractors will refuse to review an agreement that purports to be the ISDA Master Agreement, but doesn't look like the form provided by ISDA.

There are two versions of the ISDA Agreement for non-specialized use, as noted in the upper left-hand corner of the document itself. Those are:

Chapter 9 New Issues: Over the Counter (OTC) Derivatives Documentation and Electronic Contracting

- the Local Currency—Single Jurisdiction version
- the Multicurrency—Cross Border version

Few differences separate the two documents, as they were both created from the same template. The Multicurrency—Cross Border version includes additional language that would be appropriate to international transactions, and addresses the issues that may arise when payments are made in different currencies. As you can imagine, the problems associated with cross-border transactions where payment currency may be other than the currency of the payee's jurisdiction can be substantial. Not only is the difference in the value of currencies (the exchange rate) an issue, but taxes may play a major role in international swap and other derivative transactions.

The ISDA Agreement is typically not altered on its face and is not negotiated. It is what it is. One of the great advantages of a master agreement is that it saves time. Once learned, the ISDA Agreement needn't be re-read for every negotiation because every Multicurrency—Cross Border ISDA Agreement with every counterparty will be exactly the same and every Local Currency—Single Jurisdiction ISDA Agreement will also look like all others. But negotiation does happen and the result of those negotiations will appear in the *Schedule*. A Schedule is the form provided by one negotiating party to the other that contains deletions, alternations, and additions to the ISDA Agreement terms.

One feature of the ISDA Agreement that is not known by all industry participants and gas contractors is that the same Master Agreement commonly used for financial OTC derivative transactions could also be used as a Master Agreement to sell peaches, natural gas, or electricity. The ISDA Agreement is a general set of terms and conditions that may govern many types of commercial transactions. If an organization were to sell natural gas, coal, or electric power under the ISDA Agreement, the format for the schedule would look somewhat different from the Schedules used for OTC derivative transactions. In addition to the information already required by a schedule, terms for the relevant non-OTC derivative transactions would be added. While expanding the uses for the ISDA Agreement to additional types of transactions is not without technical problems, it certainly is possible to accomplish.

When teaching my seminar on the ISDA Agreement documentation, I have heard from many class participants that the "ISDA language" is becoming more prevalent in physical gas contracts. I heartily approve. A discussion of the entire ISDA Agreement is beyond the scope of this guide, but in this section you will learn about the most common ISDA Agreement clauses being incorporated into gas contracts.

Representations

As noted in chapter 8, most gas contracts today will include the basic representation that each party has the legal capacity to enter into the transactions contemplated by the base contract. The ISDA Agreement contains the following language regarding representations of the parties. Some, or all, of the language below may look familiar to you even if you don't negotiate derivatives documentation:

> *(a) Basic Representations.*
> *(i)* Status. *It is duly organized and validly existing under the laws of the jurisdiction of its organization or incorporation and, if relevant under such laws, in good standing;*
>
> *(ii) Powers. It has the power to execute this Agreement and any other documentation relating to this Agreement to which it is a party, to deliver this Agreement and any other documentation relating to this Agreement that it is required by this Agreement to deliver and to perform its obligations under this Agreement and any obligations it has under any Credit Support Document to which it is a party and has taken all necessary action to authorize such execution, delivery and performance;*
>
> *(iii) No Violation or Conflict. Such execution, delivery and performance do not violate or conflict with any law applicable to it, any provision of its constitutional documents, any order or judgment of any court or other agency of government applic-*

Chapter 9 New Issues: Over the Counter (OTC) Derivatives Documentation and Electronic Contracting

able to it or any of its assets or any contractual restriction binding on or affecting it or any of its assets;

(iv) Consents. All governmental and other consents that are required to have been obtained by it with respect to this Agreement or any Credit Support Document to which it is a party have been obtained and are in full force and effect and all conditions of any such consents have been complied with; and

(v) Obligations Binding. Its obligations under this Agreement and any Credit Support Document to which it is a party constitute its legal, valid and binding obligations, enforceable in accordance with their respective terms [subject to applicable bankruptcy, reorganization, insolvency, moratorium or similar laws affecting creditors' rights generally and subject, as to enforceability, to equitable principles of general application (regardless of whether enforcement is sought in a proceeding in equity or at law)]."[5]

Discussion of Representations. First, we should examine the capitalized words in the highlighted contract clauses. The term *Agreement,* as used in the ISDA Agreement refers to the master agreement, all confirmations of individual transactions, the 1991 ISDA Definitions and 1993 Commodity Derivative Definitions (if applicable) incorporated by reference, and any included Credit Support Document(s). *Credit Support Documents* generally refer to any guarantees given by parent or affiliated companies of either party and also to the Credit Support Annex, if the parties wish to establish a system for collateral exchange. The ISDA Credit Support Annex (or other security agreement) establishes rules and procedures for exchanges of collateral between parties when either party's exposure to the other is beyond the monetary limit established in the credit support annex. It also establishes the basis for a *security interest*[6] held by the secured party.

The parties represent that they have the legal status and power to enter into the agreement; they have procured all necessary consents; execution, delivery, and performance under the agreement won't violate any laws; and

345

obligations under the agreement will be enforceable. These concepts and requirements are basic to any contract, but in the ISDA Agreement, they have been raised to the level of representations.[7]

Default and Early Termination

Many gas contracts now contain default clauses that specifically list the events that may result in default. In the following clause, the relevant excerpted portions of the ISDA Agreement are used to illustrate the origin of some new gas contract clauses relating to default.

"(a) Events of Default. The occurrence at any time with respect to a party or, if applicable, any Credit Support Provider[8] of such party or any Specified Entity[9] of such party of any of the following events constitutes an event of default (an "Event of Default") with respect to such party: —"

Failure to make Payment
"(i) Failure to Pay or Deliver. Failure by the party to make, when due, any payment under this Agreement . . . required to be made by it if such failure is not remedied on or before the third Local Business Day after notice of such failure is given to the party;"

"(ii) Breach of Agreement. Failure by the party to comply with or perform any agreement or obligation (other than an obligation to make any payment under this Agreement . . .) to be complied with or performed by the party in accordance with this Agreement if such failure is not remedied on or before the thirtieth day after notice of such failure is given to the party;"

Discussion of Failure to Pay and Breach of the Agreement

These two default events are contained in most gas contracts today although the language structure may be a bit different from that provided here. In the ISDA Agreement, the cure period for each default event is stat-

*Chapter 9 New Issues: Over the Counter (OTC)
Derivatives Documentation and Electronic Contracting*

ed in conjunction with the explanation of the default. When contracting parties wish to alter provisions of the law that place time limitations on payment obligations, they frequently will combine a risk avoidance view and a common sense approach to their choices. Failure to pay carries a high risk and may be cured relatively quickly, so only brief cure periods are usually designated for that default. On the other hand, failure to comply with a procedural matter in the contract, such as compliance with governmental requirements—while somewhat risky—will usually require a number of days or weeks to cure—consequently the 30-day allowance for other breaches of contract.

Credit Support Default

"(iii) Credit Support Default.
(1) Failure by the party or any Credit Support Provider of such party to comply with or perform any agreement or obligation to be complied with or performed by it in accordance with any Credit Support Document[10] if such failure is continuing after any applicable grace period has elapsed;

(2) The expiration or termination of such Credit Support Document or the failing or ceasing of such Credit Support Document to be in full force and effect for the purpose of this Agreement (in either case other than in accordance with its terms) prior to the satisfaction of all obligations of such party under each Transaction to which such Credit Support Document relates without the written consent of the other party; or

(3) The party or such Credit Support Provider disaffirms, disclaims, repudiates of rejects, in whole or in part, or challenges the validity of, such Credit Support Document;"

Discussion of Credit Support Default. Third-party activities are ordinarily not brought under gas contract language, except to the extent that the activities of gas transporters, hedging counterparties, and governmental regulators effect either party's ability to perform. The ISDA Agreement brings third-party Credit Support Providers (typically the guarantor of each party) within the parameters of the agreement itself, so that failure of a guaranty, or failure of the guarantor to make payment under the guaranty, will also be a breach of the gas contract. The guarantor is not a signatory party to the ISDA Agreement. Because the creditworthiness of each party is so crucial, the refusal or inability of a Credit Support Provider to maintain adequate creditworthiness or to make a guaranty payment due will result in a breach of the contract by the party on whose behalf the guaranty was given.

When a gas seller makes sales to a buyer totally dependent upon the creditworthiness of the buyer's parent or other affiliate, and requires credit enhancement in the form of a guaranty, that seller must rely on its ability to collect amounts due under the guaranty. Under the ISDA Agreement, guarantors that fail to make payment, or otherwise challenge their obligation to make payment, or that allow the credit support document to expire before obligations have been paid, cause a default of the agreement for which the guaranty has been given. It's as simple as that.

While that concept hasn't uniformly found its way into gas contract formats, the time will come when gas contracts will routinely make a failure by the guarantor into a default by the debtor.

Misrepresentation

"(iv) Misrepresentation. A representation made or repeated or deemed to have been made or repeated by the party or any Credit Support Provider of such party in this Agreement or any Credit Support Document proves to have been incorrect or misleading in any material respect when made or repeated or deemed to have been made or repeated;"

Discussion of Misrepresentation. Misrepresentation of a material fact is a breach of the contract, but it is sometimes not included as a contract clause. When it is addressed, misrepresentation is seen simply as an untrue statement of fact without any further elucidation, so that "misrepresenta-

Chapter 9 *New Issues: Over the Counter (OTC) Derivatives Documentation and Electronic Contracting*

tion" is an event of default. Contract language is trending toward further explanation of misrepresentation to put both parties on notice regarding the parameters of false or misleading statements made by either the contracting party or its guarantor.

Default under Specified Transaction

"(v) Default under Specified Transaction. The party, any Credit Support Provider of such party or any applicable Specified Entity of such party (1) defaults under a Specified Transaction[11] and, after giving effect to any applicable notice requirement or grace period, there occurs a liquidation of, an acceleration of obligations under, or an early termination of, that Specified Transaction, (2) defaults, after giving effect to any applicable notice requirement or grace period, in making any payment or delivery due on the last payment, delivery or exchange date of, or any payment on early termination of, a Specified Transaction (or such default continues for at least three Local Business Days if there is no applicable notice requirement or grace period) or (3) disaffirms, disclaims, repudiates or rejects, in whole or in part, a Specified Transaction (or such action is taken by any person or entity appointed or empowered to operate it or act on its behalf);"

Discussion of Default under a Specified Transaction. Most gas contracts don't incorporate any additional business between contracting parties in their default language. Some are beginning to see the advantages of maintaining control and management of risk by contractually integrating all of their business activities through contract language such that a default on one contract will automatically trigger default under other transactions between the same two parties. When the parties agree to such terms, a default under a nonrelated contract between the parties (or their guarantors) would also trigger a right of the non-defaulting party to declare other contracts between the parties in breach because of the default. If a clause of this type were used in a gas contract, other types of specified transactions may include electric power sales, OTC derivative transactions, and other transactional dealings. Depending on the nature of business between any two parties, the list of other contracts that could be included within the parameter of "specified transactions" may only be limited by the dealings between two particular parties.

349

Bankruptcy

"(vii) Bankruptcy. The party, any Credit Support Provider of such party or any applicable Specified Entity of such party:—

>*(1) is dissolved (other than pursuant to a consolidation, amalgamation or merger);*

>*(2) becomes insolvent or is unable to pay its debts or fails or admits in writing its inability generally to pay its debts as they become due;*

>*(3) makes a general assignment, arrangement or composition with or for the benefit of its creditors;*

>*(4) institutes or has instituted against it a proceeding seeking a judgment of insolvency or bankruptcy or any other relief under any bankruptcy or insolvency law or other similar law affecting creditors' rights, or a petition is presented for its winding-up or liquidation, and, in the case of any such proceeding or petition instituted or presented against it, such proceeding or petition (A) results in a judgment of insolvency or bankruptcy or the entry of an order for relief or the making of an order for its winding-up or liquidation or (B) is not dismissed, discharged, stayed or restrained in each case within 30 days of the institution or presentation thereof;*

>*(5) has a resolution passed for its winding-up, official management or liquidation (other than pursuant to a consolidation, amalgamation or merger);*

>*(6) seeks or becomes subject to the appointment of an administrator, provisional liquidator, conservator, receiver, trustee, custodian or other similar official for it or for all or substantially all its assets;*

Chapter 9 New Issues: Over the Counter (OTC) Derivatives Documentation and Electronic Contracting

(7) has a secured party take possession of all or substantially all its assets or has a distress, execution, attachment, sequestration or other legal process levied, enforced or sued on or against all or substantially all its assets and such secured party maintains possession, or any such process is not dismissed, discharged, stayed or restrained, in each case within 30 days thereafter;

(8) causes or is subject to any event with respect to it which, under the applicable laws of any jurisdiction, has an analogous effect to any of the events specified in clauses (1) to (7) (inclusive); or

(9) takes any action in furtherance of, or indicating its consent to, approval of, or acquiescence in, any of the foregoing acts; . . ."

Discussion of Bankruptcy. Most gas contracts list bankruptcy as an event of default, but few go into the detail utilized by the ISDA Agreement. This clause includes any type of proceeding that incorporates the *concept* of appointment of a receiver, conservator, or trustee for a party, as well as formal winding-up proceedings. Many gas contractors are beginning to see the advantages in a broadly-worded bankruptcy clause to assure that any event that could be a formal signal of either party's financial distress should be covered in the bankruptcy default clause (with the cautions provided in chapter 8).

Early Termination or Cancellation clauses

What happens when a default has occurred and the defaulting party must pay to the other party an amount that would compensate it for its losses? Under UCC law, when one party defaults, it must pay the other as *compensatory damages,* an amount that would place the non-defaulting party in the same position it would have been the default not occurred. Article 2 addresses this issue in Section 7 but many parties contractually agree to the procedures by which forward contracts will be liquidated and reduced to their present value to determine compensation for the non-defaulting party.

The paragraphs that follow have been derived from Section 6 of the ISDA Agreement and, while not identical to the ISDA language, these clauses give you an indication of the complexity of the ISDA Agreement calculation formula. Even though Section 6 could be used in gas contracts as written, few participants choose to include all of the ISDA language, partly because the language is complex, but in most part the reason is that not all of Section 6 is applicable to gas contracts.

One example of current inapplicability of the language to gas contracts lies in the ISDA distinction between:

- events leading to contract liquidation that are the fault of a party (Events of Default)
- events leading to contract liquidation that are not the fault of a party (Early Termination Events)

Under the ISDA documentation, the termination calculations are different when culpability of a party is involved and when the transactions are terminated because of the action of a third party, or for action of a party not involving some bad intentions or actions. In gas contracts, culpability of a party is almost always the reason for liquidation and cancellation, so only one calculation system is used.

(a) Right to Terminate Following Event of Default. *If at any time an Event of Default with respect to a party (the "Defaulting Party") has occurred and is then continuing, the other party (the "Non-defaulting Party") may, by not more than 20 days notice to the Defaulting Party specifying the relevant Event of Default, designate a day not earlier than the day such notice is effective as an Early Termination Date in respect of all outstanding Transactions. . ."*

Discussion of Right to Terminate. In most gas contracts, a non-defaulting party is limited in its remedy, insofar as the date on which calculation of damages will be made, to the price at the time of failure. Let's assume that FLEX and PUMP had a single-transaction contract and at some point FLEX failed to take delivery of gas. The calculation for resulting damages would take place according to Article 2 of the UCC as of the *time of failure*

Chapter 9 New Issues: Over the Counter (OTC) Derivatives Documentation and Electronic Contracting

(unless the contract specified otherwise, which the contract used in this example, did not). FLEX failed to take delivery of gas currently priced at $5.00/MMBtu, and the market moved down to $3.00/MMBtu by the time PUMP discovered the failure and had an opportunity to sell the gas to another party. Some might argue that PUMP had actually been damaged by the difference between the $3.00/MMBtu price and the $5.00/MMBtu price.

While the UCC does provide extended damages in certain cases, most gas contracts contain a damage limitation clause that excludes anything other than the price set forth in the contract. Unfortunately, the UCC does not generally take into account the market volatility of goods like natural gas. Gas prices may fluctuate wildly within a brief period of time, so the date on which the calculation of damages is made may make a dramatic difference in the resulting numbers just because of market volatility. The ISDA Agreement has taken an approach that accommodates market volatility by allowing the non-Defaulting Party to name an "Early Termination Date," which will be the date when all effected transactions will be liquidated.

Under the ISDA Agreement, the Early Termination Date may only be selected while the default is continuing, so if the Defaulting Party cures its default prior to the time the Early Termination Date has been named, the non-Defaulting Party loses its right to name the date. Once the Early Termination Date has been selected, the date named may be no more than 20 days after the notice is given. Ostensibly, then, a default could continue for months, and until cured, the non-Defaulting Party retains its right to name the Early Termination Date.

Effect of Designation

"(c) Effect of Designation. If notice designating an Early Termination Date is given under Section 6(a) or (b), the Early Termination Date will occur on the date so designated, whether or not the relevant Event of Default or Termination Event is then continuing."

Discussion of Effect of Designation. One of the unique characteristics of the ISDA Agreement is found in Section 6 (c), which simply states even if the default is later cured by the Defaulting Party, once the Early Termination Date has been set, the calculations will occur. Let's assume that

353

A Practical Guide to Gas Contracting

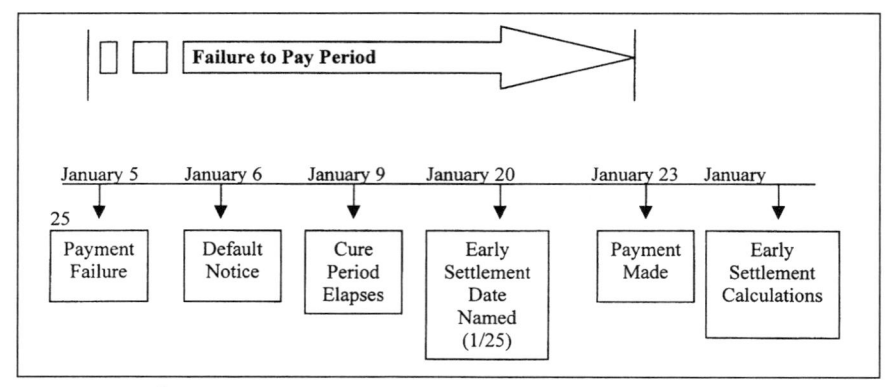

Fig. 9-1 Early Termination Process

WIDGET and PUMP are counterparties to an ISDA Master Agreement and WIDGET fails to make a routine payment on the Payment Date (Fig. 9-1). WIDGET owed the payment on January 5 and failed to pay, PUMP gave notice to WIDGET that if it did not pay within three business days, WIDGET would be in default. WIDGET did not pay until January 23, but in the meantime, on January 20th, PUMP designated January 25 as the Early Termination Date. In this example, the default calculations will proceed since the notice of Early Termination Date was given (1) while the default continued, and (2) WIDGET paid only after that date had been set.

Some gas contractors include the requirement that the default must be continuing to give the non-defaulting party a right to proceed with early termination calculations.

Calculations

> " *(d) Calculations.*
>
> *(i) Statement.* On or as soon as reasonably practicable following the occurrence of an Early Termination Date, the non-defaulting party will make the calculations specified in Paragraph _, and will provide to the defaulting party a statement (1) showing, in reasonable detail, such calculations and (2) giving details of the relevant account to which any amount payable to it is to be paid. In the absence of written confirma-

tion from the source of a quotation obtained in determining a Market Quotation, the records of the party obtaining such quotation will be conclusive evidence of the existence and accuracy of such quotation."

Discussion of Calculations. In a later part of this section, you will learn about the most common basis for determining the value of forward transactions as of a date certain and reduction of that amount to present value. This portion of the ISDA Agreement requires that the non-Defaulting Party make the value determination of the transactions being liquidated from its own position. Frequently this will involve seeking market quotations from dealers or market makers to determine future value of the involved transactions to the non-Defaulting Party.[12]

In one respect, the ISDA Agreement approach to early termination is quite different from that found in most gas contracts. As noted earlier in this section, most gas contracts don't differentiate between early fault-based terminations and terminations for reasons other than default. As a general rule, early termination (or cancellation) of a gas contract that results in liquidation of outstanding transactions occurs only through the mechanism of default. But the ISDA Agreement does make a distinction. The practical ramification of this is that the separation of "Events of Default" (those involving culpability) and "Early Termination Events" (non-culpable events) under the ISDA Agreement leads to different procedures for early settlement. Since those different procedures are not found in gas contracts, only one set of calculation rules applies.

The following language from the ISDA Agreement illustrates a second difference between early termination calculations for financial transactions and those for the sale of goods.

Payments on Early Termination

"(e) Payments on Early Termination. If an Early Termination Date occurs, the following provisions shall apply based on the parties' election in the Schedule of a payment measure, either "Market Quotation" or "Loss", and a payment method, either the "First Method" or the "Second Method". If the parties fail to designate a payment measure or payment method in the Schedule, it will be

355

A Practical Guide to Gas Contracting

deemed that "Market Quotation" or the "Second Method", as the case may be, shall apply. The amount, if any, payable in respect of an Early Termination Date and determined pursuant to this Section will be subject to any Set-off."

For the purpose of this guide, only two relevant issues come from the preceding paragraph. The first issue is whether or not the non-Defaulting Party would, under any circumstances, be required to make payments to the Defaulting Party upon an early termination (*Second Method*) or whether only the Defaulting Party would be liable for a settlement payment (*First Method*). The second issue is determination of the basis upon which the early termination calculations are made. The parties to an ISDA Agreement will elect either the *Market Quotation* method or the *Loss* method.

Discussion of First Method/Second Method. Under the *First Method*, only the Defaulting Party is required to make a settlement payment. In the ISDA context, U.S. counterparties will not agree to this method for several reasons. Banks cannot agree to the first method because the Office of the Comptroller of Currency (OCC) will not allow it. Potential penalty issues concern private industry, so counterparties elect the *Second Method* discussed below. In gas sales, the context is a little different, but not much.

Assume that PUMP has agreed to sell gas to GASCO under three gas sale contracts. In all three cases, since PUMP is always the seller and GASCO is always the buyer, GASCO will owe payments to PUMP and PUMP will owe gas to GASCO. Now change the facts a little so that FLEX and REFLEX have a number of gas transactions in place, such that either party could be the buyer or the seller under a particular deal. When REFLEX fails to perform under one of its transactions with FLEX, that action will place all transactions in potential default if all transactions are considered to part of the "Contract." Under the UCC Article 2 remedy terms, the non-defaulting party is to be placed in the position it would have been had the contract been completely performed. The default includes both purchases and sales, so calculations will be a collection of all affected transactions, both purchases and sales.

Under the Second Method of the ISDA Agreement, the non-Defaulting Party could in certain circumstances owe a final payment to the Defaulting

Chapter 9 New Issues: Over the Counter (OTC) Derivatives Documentation and Electronic Contracting

Party if the value of all transactions to the non-Defaulting Party were a negative number. Most gas contractors haven't been faced with this issue yet, since the concept is quite new to gas sales contracts, but it would be wise to fully understand the implications of making either choice (if at all) in a gas contract.

Market Quotation/Loss. If the parties elect that Market Quotations will be used to determine the liquidated value of terminated transactions, the non-Defaulting Party seeks four quotations from market makers for all transactions. In any given financial derivative transaction, either party could be in a favorable or an unfavorable position. So the market quotations would reflect the economic impact of those transactions to the non-Defaulting Party. Market makers willing to do so will quote a price to the non-Defaulting Party. For transactions favorable to the non-Defaulting Party, the quote is the price a market maker would pay the non-Defaulting Party to assume the non-Defaulting Party's position in the particular transaction or set of transactions. For transactions not favorable to the non-Defaulting Party the quote would be a price the non-Defaulting Party would have to pay the market maker to take its position in the effected transaction or set of transactions.

Under the ISDA Agreement, four market makers provide market quotations, the top quotation and low quotation are discarded, and the settlement amount is the arithmetic mean of the two remaining quotations. While the actual procedure is a little more complicated than presented above in the OTC derivatives context, this is the essence of a Market Quotation calculation.

If the parties to an ISDA Master Agreement wish to use the Loss calculation, they have essentially agreed to a general indemnity position. The Defaulting Party pays to the non-Defaulting Party the value of its actual losses (minus any gains) along with liquidation or unwinding costs, loss of profits, and other applicable costs incurred by the non-Defaulting Party. To determine loss, the non-defaulting party may seek market quotations, but the procedures detailed in the previous paragraphs regarding market quotations will not necessarily be followed.

When the language relating to First Method/Second Method and Market Quotation/Loss is part of a gas contract, contracting parties are making their own choices about which option to follow. Some contracts will require that the non-defaulting party seek market quotations (without further explicit language like that in the ISDA Agreement) and others will simply state that the non-defaulting party will determine its losses without rigid requirements as to the method that must be used to determine losses.

Regarding first method or second method, the implications of choosing one over the other are quite different because of the unique nature of the gas marketing industry. Some organizations that choose a method elect the first method since the procedures followed are basically the same as those detailed in the ISDA Agreement. Others will choose the second method because of concerns that the first method may result in an unenforceable penalty situation for the non-defaulting party.

Expenses

Many gas contracts don't contain provisions for the defaulting party to pay collection and litigation expenses incurred by the non-defaulting party. This particular clause is finding its way into more gas contracts to explicitly place the parties on notice that a defaulting party will be responsible for legal costs, including attorney fees and court costs, of the other party. With respect to this particular contract clause, you might recall in chapter 8 that some state statutes prohibit contracting parties from agreeing to a provision of this type, specifying instead that only fees actually provided for by statute may be recoverable by a prevailing party. With that caveat, the clause utilized in the ISDA Agreement is as follows:

"A Defaulting Party will, on demand, indemnify and hold harmless the other party for and against all reasonable out-of-pocket expenses, including legal fees, incurred by such other party by reason of the enforcement and protection of its rights under this Agreement or any Credit Support Document to which the Defaulting Party is a party or by reason of the early termination of any Transaction, including, but not limited to, costs of collection."

Jurisdiction

"Jurisdiction. With respect to any suit, action or proceedings relating to this Agreement ("Proceedings"), each party irrevocably:—
   (i) submits to the . . . non-exclusive jurisdiction of the courts of

Chapter 9 New Issues: Over the Counter (OTC) Derivatives Documentation and Electronic Contracting

the State of New York and the United States District Court located in the Borough of Manhattan in New York City, . . .; and (ii) waives any objection which it may have at anytime to the laying of venue of any Proceedings brought in any such court, waives any claim that such proceedings have been brought in an inconvenient forum and further waives the right to object, with respect to such Proceedings, that such court does not have any jurisdiction over such party."

Nothing in this Agreement precludes either party from bringing Proceedings in any other jurisdiction . . . nor will the bringing of Proceedings in any one or more jurisdictions preclude the bringing of Proceedings in any other jurisdiction."[13]

"(c) Waiver of Immunities. Each party irrevocably waives, to the fullest extent permitted by applicable law, with respect to itself and its revenues and assets (irrespective of their use or intended use), all immunity on the grounds of sovereignty or other similar grounds from (i) suit, (ii) jurisdiction of any court, (iii) relief by way of injunction, order for specific performance or for recovery of property, (iv) attachment of its assets (whether before or after judgment) and (v) execution or enforcement of any judgment to which it or its revenues or assets might otherwise be entitled in any Proceedings in the courts of any jurisdiction and irrevocably agrees, to the extent permitted by applicable law, that it will not claim any such immunity in any Proceedings."[14]

Discussion of jurisdiction. In the last chapter, the discussion of governing law and venue covered some of the important points that gas contractors must consider during contract negotiations. The jurisdiction clause above was carefully crafted to assure both parties' rights to file a lawsuit or institute other legal proceedings in the venue of their choice. This clause has been included to illustrate the plethora of issues that might spring from a seemingly simple statement as to jurisdiction or venue, and is recommended as a good starting point for anyone wanting to understand the types of issues that may be raised during the initial phases of litigation.

One noteworthy item with respect to jurisdiction is that the ISDA Agreement and related documents were drafted under New York law. Because New York is the financial center of the world, the New York legislature enacts many laws specifically targeted to dealings between finance centers. One of those laws passed in part of the *General Obligation Law* of New York makes it possible for litigants to gain entry to New York courts even if neither party has sufficient contacts with the State of New York to otherwise allow New York to claim venue.[15] As a consequence of the impressive changes to New York law (including a waiver from New York's strict statute of frauds provisions),[16] most counterparties to an ISDA Agreement will elect New York as the governing law state. If forward contracts meet the definition of *qualified financial* contracts, gas contract disputes may receive the same special treatment.

Set-off

Some fairly common gas contract terms are not found in the ISDA Agreement, but rather in the Schedule that supplements the master agreement. The *set-off* language provided by ISDA in the *User's Guide to the 1992 Master Agreement*[17] is frequently added by parties negotiating the ISDA Agreement Schedule. While most jurisdictions allow set-off by a non-defaulting party, the laws vary and many contracting parties wishing to clarify the set-off right will do so in the contract.[18]

A party having the right of set-off may subtract any amounts that it owes under the terminated derivative transactions from current obligations owed to it by the defaulting party under other agreements. Set-off typically comes into play only when a default has occurred and a date for termination of all transactions has been set. This remedy is usually not available to a party unless the other party has defaulted, although the ISDA Agreement allows set-off by non-Affected[19] Parties as well.

As energy transactions evolve, it is evident that sale contracts must evolve as well to meet market realities. This invasion into gas contracts by language typically used in other types of transactions is happening now and will only expand into the future. Gas contractors, even those not involved in ISDA-type transactions, would greatly expand their contract skills by reading through the ISDA Agreement or the ERMA Agreement to gain a fuller understanding of the state of energy contracting.

Chapter 9 *New Issues: Over the Counter (OTC) Derivatives Documentation and Electronic Contracting*

Electronic Contracting

When FERC issued Order No. 587, the natural gas industry was forced into the electronic future much to the dismay of many interstate pipelines and their shippers newly required to communicate electronically with the pipelines. While progress is inevitable, the extraordinary costs and time required to develop and maintain fail-safe electronic systems slowed the pace of progress. In addition to the requirement of Internet communication and contracting capabilities between interstate pipeline companies and their shippers, other aspects of the industry are being effected by electronic commerce as well.

The benefits of enhanced electronic communications are numerous. Electronic technologies permit searching and matching, bargaining, and exchanging payment and performance (the three functions of markets) to be performed with lower transaction costs. Conventional electronic contracting technologies allow parties to make arrangements and agreements to do business with each other. Enhanced electronic contracting technologies permit buyers and sellers to find each other and reduce transaction costs for both parties, whether the exchange of goods and services and payments occurs conventionally or electronically.

Internet commerce

Part of Order No. 587 was a requirement that the all interstate pipelines have operational home pages on the Internet.[20] Communications previously done through proprietary EBB systems were replaced by on-line systems so all nominations and other contractual matters would be communicated via a public electronic system. This was not accomplished without pain.

At the advent of Internet communications, a number of very real issues precluded many Internet users from using that mode of communication to transact business. Some of the more obvious issues were: privacy of the network, reliability of communications and communication providers, questions of proof and enforceability of contracts, the current UCC Article 2 and the revised UCC Article 2, and antitrust. The two most prevalent issues continue to be privacy and reliability.

Privacy. Today virtually anything can be purchased through the Internet by anyone willing to transact the business. Prior to refinement of security systems, transmittal of credit card information via the Internet has resulted in capture of that information by anyone with sufficient hacking skills and criminal motives. Although the security systems have been improved, no one can represent with 100% certainty that personal information (including capture of proprietary financial information) transmitted via electronic mechanisms is trustworthy.

On the other hand, when faxes are transmitted across telephone lines and left on fax machines for anyone to see, or when telephone conversations are overheard or even dishonestly tapped, the same security risks are present. Is it safer to send a credit card number via the electronic Internet or via fax? The security risks are probably higher when faxes are used. When we use the postal system, security may also be an issue. For example, postcards have absolutely no privacy, since anyone with even moderate curiosity handling the postcard may peruse it openly. No communication systems are failsafe, but electronic messaging systems, including the Internet and electronic data interchange, probably receive the greatest scrutiny regarding security, and thus have the highest level of security. Anyone with sufficient skills wanting to breach security systems, whatever the mode of communication, can probably do so as we have seen at various times.

Many organizations conducting business via the Internet will allow their customers to select a safe mode when conducting business, meaning that information transmitted must pass through a security system. In the gas industry, those conducting business online will give security the highest priority.

Although there will be multiple security attributes in any well-managed electronic messaging network, a level of control and security risk remains. Additional security protection ranges from agreed passwords incorporated in the text to much more elaborate security schemes.

Perhaps the best and most widely used mode of security is encryption. Under a public key encryption system, the sender has two keys—one public and one private. Using this method, the sender will calculate a special digest of the message text, then scramble it using the private key and the necessary algorithm.[21] The product of this computation is appended to the message

Chapter 9 New Issues: Over the Counter (OTC) Derivatives Documentation and Electronic Contracting

text. This is the digital signature referenced earlier in the discussion of statute of frauds requirements in the electronic age.

Upon receipt of the message, the computer unscrambles the digital signature using an appropriate algorithm and the sender's public key, then it recomputes the special digest from the message text. If the unscrambled digital signature and the special digest are the same, the message came from the sender and is unchanged.

Reliability. Even if you can satisfy yourself that the information you transmit via the Internet is safe, will it be reliable? The answer will vary, depending on the Internet service providers utilized and the reliability of the service provider's system. Reliability is really an issue relating to hardware and software used by Internet users and the service providers with which the Internet users contract. As with the security issue, reliability is becoming a more important issue with development of new technologies. Hopefully, these increase the level of comfort required by users who must communicate via the Internet.

You learned about the basic legal issues during the chapter 4 discussion of the statute of frauds and admissibility of electronic records into a legal proceeding. In this section, the focus will be on the other issues that will become more commonplace as the natural gas workplace becomes more "electronified." There isn't one set of rules for electronic commerce. In fact, depending on the nature of the electronic commerce, different concerns apply.

The Legal Issues

In addition to the evidentiary discussion in chapter 4, it might be useful to review some of those same issues in terms of electronic evidence since the only "document" proving a transaction may be one contained in the bowels of a computer. International commerce has virtually forsaken the primary mode of legal communication, which for centuries was paper. Laws have been passed under the notion that evidence would be preserved in various forms of writing, on paper. This is simply no longer the case in a majority of transactional matters, so litigation attorneys are venturing into the unknown in many instances. With the advent of electronic trading the available evidence has been presented, on a case by case basis, to "fit" rules of evi-

363

dence that were established for other forms of proof. Here are some of the issues that may arise in commercial litigation.

Evidence (business records). The three general types of records that might be offered at trial as evidence of electronic transactions are:

- a record of the contents of an electronic message at some stage in its existence
- a computer audit record
- a statistical or analytical report generated from a computer survey of a quantity of stored data

Computer *record reliability* depends on the controls around the data in the records, from the data's inception, to its recording, to its ultimate retrieval. These requirements are generally fulfilled on the operations side by proving the following:

- controls over data input
- controls over transmission
- control over record creation, indexing and storage, and
- security features throughout the system

As is true with other records, any given computer record offered in court could have some flaws. However, a careful analysis of the security and control features in a system can help a court evaluate record reliability. Courts today tend to admit general business computer records with relatively little foundation establishing the reliability of the record. This effectively creates a presumption of reliability for business computer records, but not everyone believes this is a good idea. The arguments forwarded by opponents to this position are:

- the proponent's superior access to information about system controls and security
- too much potential for error, and
- computer records may be easily altered

Chapter 9 New Issues: Over the Counter (OTC) Derivatives Documentation and Electronic Contracting

Contract formation. As with any form of errors or problems that might effect contract formation, the issues relating to contract formation in the electronic age raise some quite practical concerns.

Offer and acceptance. When is a deal a deal and when may one party rely upon the expression of another party? Many of the old principles and truths have been set back a bit with the advent of computer contracting. Many of the important legal issues of yesterday and today will in all likelihood become dusty from lack of use and in many cases, may disappear. Offer and acceptance hits home with two concerns pertinent to electronic contracting:

- Will both parties be aware of a proposed contract term?
- Will there be mistakes in communicating a party's intent?

Notwithstanding the continuing importance of assent in deciding contract formation, in some circumstances, one party may alter contract terms without the assent of the other party.[22]

The UCC has made it much easier to form a contract because of the desire to promote commerce, so, as with non-electronic modes of constructuring, an acceptance need not mirror the offer in order to be effective as long as acceptance is not made conditional on assent to the additional or different terms.[23]

Battle of the forms. The issue having its most immediate application in confirmations that disagree also finds its way into the electronic arena. While in theory, the battle of the forms works in electronic contracting the same as in paper trading, it may be substantially more difficult to communicate with another human being when the mode for communication of the contract terms is electronic.[24] This issue will undoubtedly take on new significance, as organizations become more and more electronic in their modes of contracting.

Parol evidence rule. You already know that when parties reduce their contract to an integrated writing, a court will not consider extrinsic evidence to determine what the agreed-upon terms were. It cuts off evidence of prior

365

negotiations and agreements on the theory that when the parties adopted the integrated writing, they intended it to supersede all prior understandings. The important issue in electronic contracting is determining what is a "written integration." An integrated writing does not necessarily require a signature. Moreover, in principle, the "writing" need not be an expression in any particular form—even an unrecorded oral statement could be a "writing" for parol evidence purposes,[25] so an electronically communicated message is inherently just as capable of being an integrated writing as a paper document.

Statute of frauds. The difficulty of applying writing and signing requirements to electronic transactions is only the latest manifestation of a continuing problem. Over time, in fact, as business practices and technologies have changed, courts have grown far more lenient in their interpretation of what constitutes a "writing" and a "signature." Courts seem to be adaptable to new technologies if the clear language if the law itself is also adaptable to the newest forms of writing and expressions of a signature. Courts have altered interpretation in other forms of electronic transactions, such as telegraph, telex, and mailgram. No case has directly addressed the issue of suitability of a fax to satisfy the requirement of a signed writing, but the issue has been more or less overlooked, with the consensus being that a fax does meet the requirement of a signed writing.[26] A few courts have addressed whether a tape recording satisfies the statute of frauds, and the trend is toward acceptance.[27]

The trend of businesses to utilize more and more modes of communication, coupled with the integration of the other forms of electronic contracting indicated above and the willingness of courts to adapt to new technology indicates a movement to formally accept any electronic record into evidence if it meets other evidentiary requirements. In the following paragraphs, we'll focus on the types of communications that are causing some jurists and contractors to reassess their way of looking at electronic business records.

Electronic Contracting and Electronic Data Interchange (EDI)

Electronic contracting is the exchange of messages between buyers and sellers, structured according to an agreed format so that message contents are

Chapter 9 New Issues: Over the Counter (OTC) Derivatives Documentation and Electronic Contracting

processed by the computer and automatically give rise to contractual obligations. Electronic contracting may include solicitation of bids, bids or proposals, or purchase order, invoices, and payment order.

EDI is the most developed form of electronic contracting and permits trading partners to exchange commercial documents electronically by means of *transaction sets*. Transaction sets are data structures agreed on privately or established by standards-setting organizations. The agreed EDI standard can be private or public. Companies tend to develop their own private codes—either individually or as part of a consortium—but public EDI standards have been developed by a number of large industry committees. The standard adopted by GISB is ANSI X12 (American National Standards Institute),[28] a public standard.

Before the FERC approved standards for the pipelines' electronic bulletin boards, a voluntary industry-wide group known as "GASFLOW" developed standards for the communication between pipelines and customers. In conjunction with the work done by the GASFLOW group, the FERC mandated an industry-wide EBB work group that initiated the direction of electronic communication standards in the industry. As an adjunct to the work undertaken by the FERC group, the natural gas industry formed GISB, which became the predominant force in electronic standards for the natural gas industry.

Electronic trading partner agreements. With EDI, computers communicate with computers and communications may only be read and deciphered by the receiving computer. The message itself is indecipherable to the human eye and appears as gibberish when printed. The computer receiving an EDI communication can automatically transfer that information into diverse application programs. In an energy setting the receiving computer may automatically transfer information in a confirmation immediately and directly to gas control, gas accounting, contract administration, and so forth. Since human eyes rarely see EDI communications, those using this form of communication require some common understandings regarding the rules for computer communications. A contract is necessary just to agree to communication parameters.

A Practical Guide to Gas Contracting

Gas contractors may now be asked to negotiate *electronic trading partner agreements, on-line trading agreements*, and a variety of other contracts that involve the use of computer-based communications. During negotiations with a non-regulated entity providing electronic services, there are several areas that may be of particular interest, both from the service provider's perspective and from the user's perspective.

International rules. To aid the negotiation of EDI rules, organizations have issued a code of conduct and several model trading partner agreements. The first product was the Uniform Rules of Conduct for Interchange of Trade Data by Teletransmission (UNCID), developed internationally and published by the International Chamber of Commerce (ICC). While these are voluntary rules, they may be considered to apply by default. These are considered to be the "Ten Commandments" of international EDI. Generally speaking, UNCID requires that EDI users:

- abide by their chosen EDI standard
- communicate with care
- instruct networks not to change messages without authority
- properly identify themselves in messages
- acknowledge or confirm good receipt when requested to do so
- take remedial action if a received message is not in good order or wrongly delivered
- maintain the security of protected data
- keep an unchanged log of exchanged data
- designate a person to certify the log
- Refer questions of UNCID interpretation to the ICC

Electronic trading partner agreements. In Order No. 587, the FERC has required all pipelines using EDI to provide a standard electronic trading partner agreement for all shippers to sign. Why are electronic trading partner agreements necessary? The answer can best be given through an analogy.

Electronic data interchange is a mode of communication used to transmit information from computer to computer, a practice new to the gas

Chapter 9 New Issues: Over the Counter (OTC) Derivatives Documentation and Electronic Contracting

industry. Another acceptable mode of communication is the postal service, but parties using the mail system to communicate with each other don't need a separate agreement establishing the parameters for communication by letter. We know how that system works, but not everyone understands how electronic messaging systems work, and furthermore, there is no "grand poobah" of EDI similar to the Postmaster General. We all must agree to adhere to the same communication parameters, and those are established in trading partner agreements. Once the industry has become standardized in EDI communications, the need for electronic trading partner agreements will probably disappear.

Before any two parties contract via the electronic medium, a trading partner agreement is to be signed by both. These are usually bilateral paper contracts that include the allocation of rights and responsibilities. A GISB workgroup adapted the first model trading partner agreement in 1996, which was later replaced by a more specialized agreement adaptable to gas industry needs in 1998. When it issued Order No. 587-K, adopting version 1.3 of the GISB standards, the FERC required all interstate gas pipelines to use the same standard trading partner agreement[29] that had been developed through the GISB standards process.

The standard trading partner agreement contains the following elements:

- A statement that the agreement addresses communications for transactions
- An agreement on the electronic standards that will be used by both parties
- A definition of the applicable types of "data communications"
- A list of the types of data communications available between the parties
- The procedures for sending, receiving, acknowledging, and verifying messages and
- The procedures to be followed when messages are garbled or otherwise vary from the procedures indicated above

While the electronic realm is still somewhat foreign to many industry participants, gas contractors must remain aware of new developments in technology that would effect the ability to prepare and negotiate the best contracts possible. Changes to the UCC (chapter 4) and to laws of evidence in many states; development of newer, better, and faster computer technology; and the evolving business of a united energy industry present challenges for the future that might require new types of knowledge.

Third party systems. Companies are also using third-party systems that bring buyers and sellers together electronically. These systems all require participants to sign a contract agreeing to the general terms and conditions of system usage.

The first broad category of issues relates to proprietary software systems developed by service *providers* and marketed to energy buyers and sellers who become the *users* of the system. System users are bound by the terms of the provider's contract that establishes parameters for buying and selling gas through the system. Both providers and users have unique concerns relative to the other party.

User issues.
- Many users are concerned that system providers may not have adequate subscriber numbers to ensure liquidity in the relevant market

- Security of the system is a major issue to users, particularly when sensitive information about the organization is at risk

- Telecommunications and electric backup capabilities should be addressed before relying too heavily on any one proprietary system

- The financial stability of the service provider is critical in two circumstances. The first arises when the provider is directly involved in transactions as a buyer or a seller. The second involves the stability of the system itself. Users that buy or sell on a long-term basis through a proprietary system need credit assurance from all relevant parties.

- The quality of software and hardware furnished by the provider

- The competence of the service provider's employees is important to a business that has entrusted its commerce system to the provider's ability

Chapter 9 *New Issues: Over the Counter (OTC) Derivatives Documentation and Electronic Contracting*

- Users need a system that is easy to navigate

- The responsiveness of a provider to system problems is usually an important component of contract negotiations. Even with the best system in the world and well-trained employees, if the provider fails to adequately respond to problems, the system may be a failure for the organization experiencing the problem

- Back-up capability of a service provider is also important. If the system crashes, the provider should be responsible for rescuing the data

- Users need to rely on information provided by the system.

Contractors negotiating any contract for provision of third party services should pay attention to the issues described above to assure that the 10 issues have been addressed in the contract.

Provider issues.
- Financial ability of system users

- As with system users, the need for sufficient subscribers is important to the provider to ensure that the system will be liquid and profitable

- Providers are concerned that users will misuse provided equipment, so contract language will include provisions specifically describing the user's responsibilities

- Users must be trained on the system to avoid the consequences of the previous paragraph, but also to protect the investment of the service provider

- Providers will never be liable for user mistakes

- Force Majeure that reflects the types of interruptions that would effect the ability to buy and sell gas will be included. Depending upon the nature of the provider's involvement, whether as a broker or a party to the gas sales and purchases, force majeure may also include events that would effect the ability to deliver or to take delivery of gas

- Security is as important for the provider as it is for the user

- As noted in the chapter 2 discussion of producers, industrial consumers, and LDCs, any organization owing hard assets will require that it have the ability to control its system. Providers own hardware and software, so they will retain control over the system

- Providers will require that users allow them access to software and hardware in a timely manner, and to otherwise cooperate with the provider so that the provider can take immediate action when necessary

- For any user misusing the system, providers will retain the right to remove the system and terminate the contract

The major performance disputes that may occur between a service provider and a user of the system include:

- system malfunctions or glitches
- unauthorized use of a password
- simple mistakes that may result in sizeable loss of value to the user
- inadequate back-up of information
- inadequate alternate sources of energy to remain on-line through interruptions in electric or telecommunications services

---

1. Promoting Wholesale Competition Through Open Access Non-discriminatory Transmission Services by Public Utilities; Recovery of Stranded Costs by Public Utilities and Transmitting Utilities Order No. 888 75 FERC 61,080 (April 24, 1996).

2. Standard & Poor's, a division of the McGraw-Hill Companies.

Chapter 9 New Issues: Over the Counter (OTC) Derivatives Documentation and Electronic Contracting

3. Moody's Investor Service.

4. Contact the International Swaps & Derivatives Association, 600 Fifth Avenue, 27th Floor, Rockefeller Center, New York, NY 10020; ISDA's web site: http/www.isda.org., for information on ordering.

5. ISDA Agreement, Section 3 (a).

6. "[A] contingent property interest that rises upon creation, but is not fully realized unless there is a default on the secured obligation and the secured party enforces its property interest to satisfy the obligation." Russell A. Hakes, *The ABC's of the UCC Article 9: Secured Transactions,* Section of Business Law, the American Bar Association, Chicago, 1996.

7. See the discussion of representations and remedies for misrepresentation in chapter 4.

8. The party providing credit enhancement under the ISDA Agreement.

9. A third party that must be named in the schedule. A specified entity may cause default or termination of the ISDA Agreement even though it is not a signatory party if specified entity defaults on specified indebtedness (indebtedness in respect of borrowed money), a specified transaction (other OTC derivatives transactions), goes bankrupt, or experiences a credit event after a merger that materially reduces its overall creditworthiness.

10. Usually a guaranty provided by the credit support provider or the credit support annex or other security agreement.

11. Any OTC derivative transaction.

12. The parties will elect "market quotation" or "loss" (a general indemnity approach) as the calculation basis.

13. ISDA Agreement Section 11(b).

14. ISDA Agreement Section 11(c).

15. New York General Obligation Law, Section 5-1401.

16. "qualified financial contract" means an agreement in which each party thereto is other than a natural person and which is: (a) for the purchase and sale of foreign

exchange, foreign currency, bullion, coin or precious metals on a forward, spot, next-day value or other basis; (b) a contract (other than a contract for the purchase and sale of a commodity for future deliver on, or subject to the rules of, a contract market or board of trade) for the purchase, sale or transfer of any commodity or any similar good, article, service, right, or interest which is presently or in the future becomes the subject of dealing in the forward contract trade, or any product or byproduct thereof, with a maturity date more than two days after the date the contract is entered into; (c) for the purchase and sale of currency, or interbank deposits denominated in U.S. dollars; (d) for a currency option, currency swap or cross-currency rate swap; (e) for a commodity swap or a commodity option (other than an option contract traded on, or subject to the rules of a contract market or board of trade); (f) for a rate swap, basis swap, forward rate transaction, or an interest rate option; (g) for a security-index swap or option or a security (or securities) price swap or option; (h) an agreement which involves any other similar transaction relating to a price or index including, without limitation, any transaction or agreement involving any combination of the foregoing, any cap, floor, collar or similar transaction with respect to a rate, commodity price, commodity index, security (or securities) price, security-index or other price index); or (i) an option with respect to any of the foregoing. New York General Obligation Law, Section 7-501 (10)(b).

17. This guide is available from the International Swaps & Derivatives Association for a small fee.

18. Compare the right of set-off in settlement payments to *netting*. Netting is a vehicle used by many parties to multiple transactions where either party could be the buyer or the seller under a relevant transaction. Netting is generally associated with routine business practices as an administrative aide.

19. The party to an ISDA Agreement that is not directly effected by the early termination event.

20. Order No. 687-K, (April 2, 1999).

21. The term algorithm (pronounced "al-go-rith-um") is a procedure or formula for solving a problem. The word derives from the name of the Arab mathematician, Al-Khowarizmi (825 AD). A computer program can be viewed as an elaborate algorithm. In mathematics and computer science, an algorithm usually means a small procedure that solves a recurrent problem. (http://www.whatis.com).

22. Restatement (second) of Contracts §5 comments a through c (1981) (terms may become part of contract with or without actual knowledge); Id. §§ 19(2) and 19(3)

Chapter 9 *New Issues: Over the Counter (OTC) Derivatives Documentation and Electronic Contracting*

(conduct may manifest assent even if party engaging in conduct does not assent in fact, if party had reason to know of effect that conduct would have).

23. U.C.C. § 2-207. *Dorton v. Collins & Aikman Corp.*, 453 F.2d 1161 (6th Cir. 1972) (The seller sent an acceptance which included an arbitration clause. Seller later tried to invoke arbitration when a dispute arose. The court held that the arbitration clause fell under U.C.C. Section 2-207(2) and the materiality of change was a question for the trier of fact.)

24. *American Multimedia, Inc. v. Dalton Packaging, Inc.*, 143 Misc.2d 295, N.Y.S.2d 410 (Sup. Ct. 1989) (The buyer faxed only the front side of a form purchase order to seller. At trial, the seller was found to be bound to the clauses contained on the backside of the form as well.)

25. *Tow v. Miners Memorial Hosp. Assn., Inc.*, 305 F.2d 73 (4th Cir. 1921); *Braude & McDonnell v. Isadore Cohen Co.*, 87 W. Va. 763, 106 S.E. 52 (1921).

26. *Bazak International Corp. v. Mast Industries, Inc.*, 73 N.Y.2d 113, 538 N.Y.S.2d 503, 535 N.E.2d 633 (1989) (the court assumed without question that a printout from a fax machine were writings under U.C.C. §2-201.)

27. *Ellis Canning v. Bernstein*, 348 F.Supp. 1212 (D. Colo. 1972) [The court accepted a tape recording in satisfaction of U.C.C. §8-319 writing requirements by finding that the recording had been "reduced to tangible form" within the meaning of U.C.C. §1-201(46)].

28. American National Standards Institute headquartered in New York City. The ANSI web site is http://www.ansi.org.

29. The standard trading partner agreement used by interstate pipelines may be viewed at Internet web sites of the interstate pipeline companies.

Chapter 10: Practical Tips for Gas Contractors

Okay, admit it. Most of you who need some quick answers have probably turned to the last chapter first, and hopefully the tips included in this text will give you the quick answer you are seeking. Most of the information provided in the first nine chapters of this guide has dealt with substantive knowledge rather than useful skills. In applying substantive knowledge, gas contractors should also be aware that drafting, negotiating, and reviewing contracts require a specialized set of skills that can be learned and honed through practice. This chapter will focus on those skills and will include a contractor's checklist for negotiations, and helpful hints for negotiating, drafting, and reviewing contracts. At the end of this chapter, you will find a bibliography of helpful references, including books, magazines, and Internet web site addresses.

Contract Checklist

In this checklist, presume that your organization is Party A to the contract.

Header

1. Full business name of Party B
   Secretary of State (Corporations Division) check required for Party B

2. Legal entity of Party B:
   Corporation
   Partnership
   d/b/a (doing business as)
   If d/b/a, signatory party
   Other
   Sole proprietor

3. Location of Party B:
   Origination
   Business Location
   Secretary of State check

4. Statement of Agreement

Recitals

If you are using the other party's contract:
Are recitals necessary: Yes _____ No _____

Purpose and procedures

Buy/Sell
Buy
Sell
Exchange
Other
Gas
Other Commodities

Definitions

Check each definition for:

words not used in ordinarily understood sense
technical words
ambiguous or vague words
if a definition required by context

Performance obligation

Seller's: to sell _____ deliver _____ schedule _____
Buyer's: to purchase _____ receive _____ schedule _____
Any obligations of only one party
    Party A:
    Party B:

Transportation, nominations, and imbalances

Seller responsible before delivery
Buyer responsible after delivery
Cooperation required
Timely information exchange required
    Time deadlines
Responsibility for imbalance charges
    Seller
    Buyer
    Party that caused
Payment by reimbursement?
Proof/documentation for imbalances

Quality and measurement

Quality
    Seller's transporter
    Buyer's transporter
    Specific gas quality specifications in the contract
    Other
    Limited remedies for failure of gas quality
Measurement
    Seller

  Buyer
  Seller's transporter
  Buyer's transporter
  Other

Remedies/failure to perform

 Seller's failure
  Must buyer cover?
  Time as of which damage computations are made
   Failure
   Failure discovered
   Buyer covers
  May buyer choose its remedy?
  May buyer cover with alternate fuel?
 Buyer's failure
  Must seller find an alternate market?
  Time as of which damage computations are made
   Failure
   Failure discovered
   Seller remedies
  May seller choose its remedy?
 Liquidated damages
 Administrative fee
  Administrative fee due even if failure causes no harm?

Taxes

 Complete definition of taxes
  Fees
  Fines
  Penalties and interest
  Other
 Before delivery
  Based upon time
  Based upon location

Based upon product
Based upon privilege
Taxes on sale
Taxes on seller
At delivery
    Seller responsibility
    Buyer responsibility
After delivery
    Based upon location
    Based upon sale
    Based upon privilege
    Tax on buyer
    Tax on product
New taxes
    Seller responsibility
    Buyer responsibility
    Shared responsibility
    Renegotiations
    Contract termination
    Include new and increased taxes?

Billing, payment, and audit

Seller's invoice responsibility
    Deadline
    Invoice on actuals _____ or transporter's statement _____
        Buyer adjusts invoice
        Seller adjusts on following month's invoice
    Other
Buyer's payment responsibility
    Negotiated Payment Date
    Tied to invoice receipt
    Payment extension for late invoice
Late payments
    Interest on Buyer's late payment _____ on Seller's late payment _____

Interest on disputed amounts
   Is interest waived if dispute resolved in disputing party's favor?
   When is interest payment due?
Audit
   Seller's right
   Buyer's right
   Who pays associated costs?
   Notice/permission required?
   Time of day limitations?
   Other limitations on audit
   Audit right expiration:
      When do rights expire?
      Limited to the ability to audit?

Title, warranty, and indemnity

Seller's express warranties
   Title
   Right to convey
   Other
Disclaimer on implied warranties
   Merchantability
   Fitness for Particular Purpose
   Other
   Conspicuous print?
Buyer's warranties
Seller indemnifies buyer for title claims
   Formal claims
   Frivolous claims
   May buyer suspend payments?
   Limitation on right to suspend
Seller indemnifies buyer for injury/damages
Buyer indemnifies seller for injury/damages
Whose counsel represents the indemnified party in legal proceedings?
Exclusion for negligence

Exclusion for gross negligence/willful misconduct

Notices

Address for formal contract notices
    Party A:
    Party B:
Address for payment
    Party A:
    Party B:
Effective when sent
Effective when received
    Time presumptions if no proof

Financial responsibility

Creditworthiness requirements of Party A:
    Continuing during term of contract:
    Credit enhancements required;
        Guaranty
        Letter of Credit
        Prepayment
        Security Interest
Creditworthiness requirements of Party B:
    Continuing during term of contract:
    Credit enhancements required;
        Guaranty
        Letter of Credit
        Prepayment
        Security Interest
Right to demand adequate assurance of performance
    Seller: Yes _____ No _____
    Buyer: Yes _____ No _____
Limitations on right to demand assurances
Limitations on assurances that may be given
Deadline to provide adequate assurance
Remedy for failure to provide assurances

Force majeure
Definition elements
    Occurrence or event: Yes _____ No _____
    Not anticipated when deal was made Yes _____ No _____
    Couldn't prevent or overcome Yes _____ No _____
    Actually prevented performance Yes _____ No _____
    Partial force majeure allowed Yes _____ No _____
    Result of valid force majeure
        Excuse of performance
        Suspension of performance
        Obligation to make up performance Yes ___ No ___
    List of included events
    List of excluded events
    Requirement of written notice Yes _____ No _____
    Separate strike/lockout provision Yes _____ No _____
    Obligation to cure
        Goal to renew performance
        Rights of party not experiencing force majeure
            Terminate contract
            Renegotiate contract term
            Renegotiate contract price
            Others

Term
Limited Yes _____ No _____
    If evergreen
        Automatically continued
        Require notice to continue
        Require notice to terminate
        Type of notice required to terminate
        Advance notice period

Default and early termination/cancellation
Events of default listed

Separate cure periods listed
Timeline for default
Procedures for default
Calculations
    Detailed
        "Market Quotation"
        "Loss"
    Left to discretion of non-defaulting party
Result of default
    Cancellation
    Termination
    Right to suspend payments
    Right to suspend performance
    Notice requirements
    Must non-defaulting party choose among listed remedies?

Miscellaneous

Headers for convenience

Assignment
    Allowed
        Consent required
        Consent not required
    Not allowed without consent
        Prior
        Written
    Other limitations

Confidentiality
    Contract terms
    Contract existence
    Exclusion for transportation
    Exclusion for hedging
    Exclusion for other purposes

    Notice given to the other party prior to disclosure
Third Party Beneficiary

Waiver
    What may be waived?

Amendments in writing

Amendments signed by both parties

Integration Clause

Governing Law
    Connection to transaction or parties? Yes _____ No _____

Venue
    Exclusive
    Non-exclusive
    No statement regarding venue

Arbitration
            Procedures detailed
                AAA rules
                Other procedural rules
            Rules of evidence requirement
            Rules for arbitrator(s) selection
            Arbitrator(s)' eligibility requirements
                Attorney
                Knowledge in the industry
                Neutrality
            Time limits on process
            Limits on arbitrator's decision
            Decision appealable?

    Mediation

Procedures in contract
Qualifications of mediator
Knowledge of the industry
Attorney
Mediation training
State-sanctioned procedures
Valid Laws

Severability

Authority

Additional representations

Contract signatures

Party A:
Company officer _____
If no officer, has formal delegation of authority been granted?
Title

Party B:
Company Officer
Title

That is a basic contract checklist that a gas contractor new to the industry or new to contracting may wish to expand upon to incorporate any clauses or issues of significance to the company. Every organization that has developed a house contract has included the elements described above, and some have gone even further. These are core elements, and not necessarily all of the elements that would be in a gas contract for any given organization. The list is offered for gas contractors who are trying to establish a method for reviewing contracts to ensure that all necessary contract elements are actually in the contract being reviewed.

Contract Negotiations

To some gas contractors actual contract negotiation is the most fun part of their job—to others, it is a dreaded task—one to be left until the last possible minute. Why is this so? It may lie in the difference between introverted and extroverted personalities, or it may be because many gas contractors have limited authority to make decisions, or it may be possible that some folks just don't have adequate knowledge of the subject matter to make the best and strongest argument for a particular position. No one wants to look foolish when negotiating with another party, so whatever the reason may be, here are some hints you may find helpful if you need that extra push to put you over the top in your negotiating strategies.

It is the responsibility of every contract negotiator to successfully represent the interests of his or her employers or clients to the other negotiating party, who is trying to do exactly the same thing on behalf of his or her employer or client. How can both parties achieve the results desired by their organizations without giving away too much in the process? When gas contractors representing producers negotiate with marketers, industrial consumers, or LDC representatives, there is an instinctive knowledge that parts of the contract will not be agreeable as presented. When you are on one side of such a negotiation, how will you negotiate to the best advantage of your employer or client? No two contractors negotiate the same. We're all unique, and you may encounter every attitude from aggressive rudeness to passivity that leads to a refusal to return phone calls. Some negotiators quite frankly represent the interests of their organization better than others. Some interesting posturing may take place during contract negotiations, not to mention varying personality types. This section is designed to give you some hints on how you can use your own negotiation style, whatever that may be, to achieve desired goals.

The 10 negotiation postures

While many variations on this theme exist, in nearly 20 years of negotiations, first as a lobbyist then later as attorney for a gas marketing organization, I've discovered about 10 different negotiation approaches or tactics—some adapt quite well to the entire negotiation process and others hinder the process. In a nutshell, the 10 approaches are these:

Chapter 10 Practical Tips for Gas Contractors

1. Trust me: This negotiator refuses to listen to your opinion because he or she "knows best." Whenever your counterparty in negotiations adopts this attitude, the best way to handle it may be to ask for documentation for your files

2. Emotion: The emotion being expressed may be anger, impatience, self-pity, or some positive emotion like gratitude. Try not to let the other party's emotions (whether real or a negotiation ploy) deter you from your stated goal. Sometimes the emotional negotiator will only be happy when the other party is as miserable as he or she, so the best advice when dealing with emotions is to remain calm at all costs. Another counter to a negative emotion is to ask if he or she would prefer negotiating on some other day

3. Deadlines: We all have them, but often you will encounter a negotiating counterparty who will give you "faux deadlines." It's a good idea to keep negotiations running smoothly along, but don't be tempted to accept less than you would otherwise just to complete negotiations by the deadline imposed by your counterparty. At the same time, it is most useful to establish your own timeline for negotiation, and try to stick to that timeline

4. Let's trade: If your counterparty offers to give up one of its original changes in exchange for your giving up something you have requested, be aware the other party may have thrown several "loss leaders" into the negotiation. In other words, he or she didn't care about that clause anyway. If you make trades, and every good negotiation contains these, be certain you understand the value of your trades. Know up front what you are willing to give up as well

5. Blame it on "legal:" If your counterparty refuses to accept your requests because "legal won't let me do that," there are ways

to overcome that argument. If you have in-house legal counsel, discuss with him or her whether it would be possible to refer some of these issues to the legal department. Sometimes, just offering to have your counsel review the changes will provoke a more open-minded negotiation

6. We won't: Some organizations simply refuse to accept changes to a contract, or refuse to discuss matters with you. There are few effective ways to counter this—particularly if you need them more than they need you. However, one way to address this situation is by probing into the reasons why language cannot be changed. Sometimes a compromise may be reached by changing other provisions of the contract to achieve the desired goal, while still leaving the original clause intact

7. We can't: If your counterparty says that his or her company just cannot comply with changes you have recommended, the same technique used in paragraph 6 should work. Sometimes, companies really don't have the processes or automation in place to comply with contract provisions. In that event, try to accommodate their needs without compromising yours

8. Intimidation: This is a favorite technique in one of two situations: (a) unequal bargaining power; or (b) unequal experience in the industry. If your negotiating counterparty senses that he or she has more power than you do, or more knowledge of the industry, an opportunity for intimidation is knocking on the door. One way to counteract this tactic is by asking questions. If he or she knows so much, benefit from that knowledge. If you stick with your questioning in a negotiation, you'll be surprised at the other party's willingness to work with you rather than against you

9. One last thing: You've nearly finished your negotiations—or perhaps you thought you were already finished—and your counterparty comes up with something new. Either he or she forgot to mention it earlier, management overrode a decision, or legal wouldn't allow a change. If you are faced with this situation, and the "one last thing" is an important issue, it would be appropriate to find a counterbalance in some issue the other party has either insisted on or refused to accept from you. It's usually not proper to re-open negotiations, but in this instance, it is usually acceptable

10. Open to comments and suggestions: This type of negotiation posture is probably the most amenable to reaching agreement on contract clauses in the least amount of time. Spending an hour with your negotiating counterparty to resolve any differences is, in the long run, much more to your advantage than adopting any of the nine previous attitudes. While you may have recognized some of your own negotiation strategies in types one through nine, I hope you have identified yourself as predominantly this type of negotiator. Some may think that assuming an open position will only weaken any leverage you may have during negotiations. It doesn't—instead, your negotiation counterparty will think much better of you, and you'll probably achieve more in negotiations if you are pleasant and reasonable, and if you offer to work together to reach consensus

Your success as a negotiator

Good negotiators know the subject being negotiated and probably won't rely on delay or intimidation tactics in a contract negotiation. The two additional elements essential to successful contract negotiations are:

- knowledge
- understanding your role in the negotiation process

A Practical Guide to Gas Contracting

First, let's discuss the impact that basic knowledge will have.

Knowledge. Assume that representatives of WIDGET and PUMP are negotiating a standard gas contract. PUMP's sales representative has been in the gas business since the mid 1980s and has been involved in gas contracting for several years. WIDGET, with a low staff budget, has allocated one third of the time of one employee to buy the natural gas needed to fuel its operations. At the outset, it is obvious that PUMP's negotiator has the advantage since it has knowledge of the industry and contracting experience as well. But WIDGET's part-time gas buyer has no gas industry knowledge and no gas contracting experience. Given the example above, absent some positive intervention by fate, WIDGET can only adequately address the relevant contract issues if he or she does any or all of the following:

- *Learn about natural gas transportation.* The best way to do this is to become familiar with the relevant transporter(s)' tariffs. Of particular importance to WIDGET (if it has a transportation agreement) are penalties that its transporter may assess against it for failure to comply with tariff provisions.

- *Learn about the source of the gas.* For example, if the gas is produced in Kansas (or any other mid-continent state), PUMP may have force majeure situations in both winter (freezing of wells and lines of pipe) and summer (tornadoes). If offshore gas is being purchased, hurricanes and hurricane warnings may result in force majeure. If the transporter delivering gas to WIDGET's facilities is constrained on a regular basis at either receipt points or delivery points, the transporter may curtail gas

- *Learn all you can about PUMP.* What are PUMP's estimated gas reserves during the contract term? How numerous are PUMP's gas supply contracts? Is PUMP a reliable gas supplier—should additional deliverability language be added to the contract?

- *Identify the needs and goals of the facility.* Learn as much as possible about the gas burning process. Is the measuring equipment state-of-the-art?

Chapter 10 *Practical Tips for Gas Contractors*

What happens to equipment if gas is not available? (If demand exceeds supply, all transporters have curtailment rules. For LDC's those curtailment rules direct that curtailments affect non-core customers first)

Your role in the negotiation process. Once you have identified the various needs and goals of your organization, the next step involves the level of decision-making power given to gas contractors. Some organizations require all contracts to be reviewed or negotiated by an attorney—others will hire paralegals for contracting purposes—some move employees from other positions in the company to handle contracting. But some natural gas organizations place the contract negotiation, drafting, and revision responsibility on the shoulders of employees who have little or no knowledge of contracting or the gas industry. Sometimes, gas contractors (particularly those who fall into the last category) aren't given much authority to make decisions. One reality of gas contracting is that negotiating counterparties frequently must consult with someone else in the company before any issue is resolved. Even though the reality exists, it is of little comfort to the gas contractor who can't get a contract signed because one negotiator is waiting for an answer from someone "higher up" who has authority to make decisions.

The requirement for management approval is especially important for long-term, high-risk contract negotiations but not necessarily as important for routine gas contracts with relatively low risk. To raise the level of gas contractor competence that would ultimately facilitate faster and more knowledgeable contract negotiations, gas contractors should be given the following information:

- Issues on which the gas contractor may make a decision
- Issues that require higher-level approval
- Contract language that is always required
- Contract language that is never allowed
- Contract clauses that may be included or excluded during negotiations
- The client or employer's bottom line on important issues—how far can you go?
- Appropriate internal channels for resolution of contract issues
    Legal
    Credit
    Tax
    Transportation issues

Gas contractors should know if limits apply to any negotiation matters. Some organizations have an informal policy and others have detailed the responsibilities of gas contractors through job descriptions or other formal means. Without some idea of negotiating goals of the company, gas contractors are often left in the position of starting a race with no clue as to the whereabouts of the finish line. Hopefully, these few negotiation hints will assist you after you have completed your assessment of the counterparty and its gas contracting needs.

Contract Drafting

New contracts are constructed to meet the needs of one particular transaction, the complexity of which is not accommodated by using a standard base gas contract. If you have written a new contract, do you remember the guiding principle behind all your drafting efforts? While the direction given to you may have been to make the contract more user-friendly or to make the contract operate well under a single transaction, that charge should not have been your guiding principle. You must know *how* to write a contract.

Other times, contracts are re-written because the old contract is too long, burdensome, or out-of-date. Many gas contractors have received requests from their trading personnel to make a currently used form contract shorter, or more streamlined to ease the negotiation backlog. If that is the case, your single assignment as a contract drafter might be to determine which words, phrases, and clauses can be stricken from the current form contract being used by your organization.

You may indeed have produced a contract that was the shortest in the industry, with the most resourceful language, and with a new sleek look, but without understanding some very basic principals of legal writing, you might not have achieved your intended goal. Here are 10 tips for your writing skill set:

1. Don't use outdated legal shorthand terms or "legalese" in your contracts. Gas contracts should be easy to read and understandable. What does the following contract clause say to you?

Chapter 10 Practical Tips for Gas Contractors

"Subject to the terms set forth herein, in lieu of paying the applicable commodity charge hereunder, for volumes of gas delivered to buyer and the point(s) of delivery, buyer may lock the price (such price being the "Locked Price") or establish an alternate commodity charge pricing mechanism for all or any portion of the gas to be purchased during any month of the contract for which natural gas futures contracts are then being traded on the New York Mercantile Exchange ("NYMEX") by means of the procedure set forth below:"

Could the clause be changed to reduce the legalese as follows?

"Instead of paying the commodity charge, buyer may lock in a price or establish an alternate pricing mechanism based upon future prices published by the NYMEX at any point prior to settlement of the relevant futures contract, according to the following procedures:"

2. Avoid long sentences. As a general rule, a sentence of more than 30 words is too long and should be divided into smaller segments. You may have noticed that whenever you read anything, you tend to take breaths only at the end of sentences. Some contract clauses leave the reader gasping for air at the end. Were you able to read the original contract clause used as an example in the previous tip comfortably?

3. Proofread the entire contract when negotiated changes have been inserted. Sometimes, instead of composing an entirely new clause to fill a particular need, contract drafters will simply "lift" a particularly good phrase, clause, or sentence from another document. The borrowed passage may contain references that are inappropriate in your contract. The most common elements are: (a) references to "contract"

when the document you are negotiating is called an "agreement;" and (b) internal references to "articles," when the document you are negotiating separates contracts into "sections;" etc. Using individual parts of other contracts is a perfectly acceptable way to construct contract language that works for both parties. Just be sure that you proofread the final version for consistency. To avoid a natural tendency of writers to miss small but important errors when proofing their own work, someone else should proof the contract as well

4. Before you begin writing a gas contract (whether you're starting from scratch or compiling good contract clauses drafted by others) make sure you have a good understanding of the laws that will govern that contract. In the U.S., Article 2 of the UCC governs the sale of goods (with Articles 1, 5, and 9 having relevance as well). In Canada, the Sale of Goods Act in the relevant province or territory governs sales of goods. For international transactions between residents of contracting states, the CISG applies. All of these laws, but particularly the UCC, act as safety nets. When a contract is silent on a particular issue later subject to dispute, the law will step in to solve the dispute. You need to know the difference between UCC provisions that will assist your organization and those that should be excerpted in the contract with wording properly drafted to overcome a deficiency in the UCC. A good example of this practice at work is found in Article 2-609, the *adequate assurance of performance* clause. According to the Code, the party providing the assurance has a reasonable time, not to exceed 30 days, in which to satisfy the demand. Most gas contracts will shorten the response deadline

5. When one word will do, don't use several. For example, many contract clauses begin with the words "in the event." A preferable way to address the same concept is to simply say, "if." Another common contract phrase is "is inclusive of." A preferable alternative is to use the word "includes"

6. Avoid use of "and/or" at all costs. Is this combination of an inclusive and an exclusive word meant to describe a situation that really means *either* or is it being used to describe a situation where the actor (if it takes action at all) must do *both*? This is one case where it's better to use a few more words to describe exactly what is intended

7. Understand the difference between the word *which* and the word *that*. *Which* is used to introduce nonessential information and is preceded by a comma. *That* is used to introduce essential information and is not preceded by a comma

8. Be consistent in form. Words used in past tense, present tense, or future tense should relate to the time-specificity of the particular sentence or clause. Remember that poetry is poetry and contracts are contracts. Writing style for contracts should reflect the underlying purpose of contracts: to control behavior

9. Use the active voice. A simple obligation sentence is: "Party A shall sell and deliver the gas." This is far preferable to the passive voice, which reads: "The gas will be sold and delivered by Party A." Using the passive voice can lead to verbosity and misunderstandings because simply changing the voice from active to passive may blur the meaning of the sentence. Whenever you are reviewing a contract developed by the other party, look for sentences that use the passive voice. If you find any, change them to the active voice and see if the meaning is still the same

10. Use legal terms correctly
     "Is entitled to" = a right
     "Shall" = a duty
     "May" = discretionary action or activity
     "Will" = a term of agreement
     "Must" = creation of a condition precedent
     "Is not entitled to" = negation of a right
     "Is not required to" = negation of a duty

Multitudes of additional rules are available for any contract drafter wishing to hone his or her writing skills. Using the wrong word to describe a duty or inaccurate internal references have caused may a headache for gas contractors.

REFERENCE SUGGESTIONS — SALE OF GOODS LAW

Publication of The West Group, 610 Opperman Drive, St. Paul, MN 55164
Uniform Commercial Code: Sales (Fourth Edition). James J. White, Robert S. Summers, West Publishing Company, St. Paul, Minnesota (1998)

Publications of the American Bar Association, Business Law Section, 750 North Lakeshore Drive, Chicago, Ill. 60611 http://www.abanet.org
The ABCs of the UCC, Article 1: General Provisions, by Fred H. Miller and Kimberly J. Cilke

The ABCs of the UCC Article 2: Sales, by Henry D. Gabriel and Linda J. Rusch

The ABCs of the UCC, Article 9: Secured Transactions, by Russell A. Hakes

The New Article 9, by Corinne Cooper, editor; Steven O. Weise and Edwin Smith, contributing authors

The Portable UCC, Second Edition, Edited by Corinne Cooper
The Portable Bankruptcy Code and Rules, 1998 Edition, edited by Sally M. Henry Meyerson

Internet web sites

FedLaw: http://www.fedlaw.gsa.gov/
FindLaw: http://www.findlaw.com/
FindLaw Canada: http://www.findlaw.com/search/countries/ca.html.
Federal Code of Regulations:
http://www.access.gpo.gov/nara/cfr/index.html.
CISG: http://www.cisg.law.pace.edu/cisg/text/treaty.html.
Text of the UCC: http://www.law.cornell.edu/ucc/ucc.table.html.
Text of Revised Article 2:
http://www.law.upenn.edu/library/ulc/ulc.htm#ucc2.

Electronic Communications Issues

21st Century Money, Banking & Commerce, by Thomas P. Vartanian, Robert H. Ledig and Lynn Bruneau

Web-Linking Agreements: Contracting Strategies and Model Provisions, by the Subcommittee on Interactive Services, Committee on the Law of Commerce in Cyberspace, ABA Section of Business Law

The Commercial Use of Electronic Data Interchange: A Report and Model Trading Partner Agreement, by the Electronic Messaging Services Task Force, UCC Committee, ABA Section of Business Law

Model Funds Transfer Services Agreement and Commentary, by the Working Group on Electronic Financial Services, UCC Committee, ABA Section of Business Law

OVER-THE-COUNTER DERIVATIVES

Court cases

Ukrainian Credit Union Limited v. Nesbitt, Burns Limited, Zanewycz and Pilot, Proceedings commenced November 8, 1995 in the Ontario Court (General Division) – statements of claim and defense illustrate various theories of liability most likely to be raised in Canadian litigation against a financial advisor for losses incurred in derivatives trading

P.T. Adimitra Rayapratama v. Bankers Trust Co., 2 Comm Fut. L. Rep. (CCH) ¶ 26,508 (S.D.N.Y. Aug. 16, 1995) - Choice of law and forum issues

Brane v. Roth, 590 N.E. 2d 587 (Ind. Ct. App. 1992) - Duty of care and loyalty owed by directors to shareholders

Bank Atlantic v. Blythe Eastman Paine Webber Inc., 12 F.3d 1045 (11th Cir. 1994) aff'g 127 F.R.D. 224 (S.D. Fla. 1989) – Duties of advisors, brokers and dealers
Bank Brussels Lambert S.A. v. Intermetals Corp., 779 F. Supp. 741 (S.D.N.Y. 1991) – Implied Duties of principals to derivative transactions

Salomon Forex Inc. v. Tauber, 795 F. Supp. 768 (E.D. Va. 1992) aff'd, 8 F.3d 966 (4th Cir. 1993) cert. Denied, 114 S. Ct. 1540 – State gaming and bucket ship violation

Intershoe v. Bankers Trust Co., 571 N.E. 2d 641 (1991) – Application of parol evidence rule to fx transactions
Hazell v. Hammersmith & Fulham London Borough Council [1993] 2 W.L.R. 17, [1992] Q.B. 697 (Div'l Ct. 1989) – Power and authority to enter into derivatives transactions

Lehman Bros. Commercial Corp. v. Minmetals International Non-Ferrous Metals Trading Co., Fed. Sec. L. Rep (CCH) ¶ 99,001 (S.D.N.Y. 1995)

Chapter 10 *Practical Tips for Gas Contractors*

- Statutes of Fraud for credit support

Daiwa America Corp. v. Rowayton Capital Management, Inc., No. 118148/95 (N.Y. Sup. Ct. filed July 24, 1995) – Statute of Frauds

State of West Virginia v. Morgan Stanley & Co., 1995 W. Va. LENS 94 No. Civ 89-C-3700 June 5, 1995) – Case settled (suitability for a municipality)

Drage v. Procter & Gamble, Action No. A9401998, April, 1994, Court of Common Please, Hamilton County, Ohio – shareholder derivative action (led to Procter & Gamble v. Bankers Trust)

Internet resources

Commodity Futures Trading Commission links page: http://www.cftc.gov/links – provides information on consumer and governmental sites as well as organizations, associations, and domestic exchanges http://risk.icfi.ch/00013028.htm – Derivatives Dictionary

http://www.numa.com/links/regulat.htm - Derivatives Internet Links for Market

Regulators and Associations
  Moody's Investor Services – http://www.moodys.com
  Standard & Poor's – http://www.stockinfo.standardpoor.com
  Dun & Bradstreet – http://www.dnb.com

Printed references

Commodity Trading Manual – prepared by the Market Development Department of the Chicago Board of Trade

Futures & Derivatives Law Report: The Journal on the Law of Investment & Risk Management Products, Glasser Legalworks, 150 Clove Road, Little Falls, JH 07424 (legalwks@aol.com)

DOE Energy Policy Act: OTC Derivative Study, Petroleum Industry Research Associates, Inc., 122 East 42nd Street, Suite 516, New York, NY 10168-0012 (1993)

A Guide to International Financial Derivatives, Francis D. Feeney, Quorum Books, New York/Westport, Connecticut (1991)

Keys to Understanding Securities, Anita Jones Lee, Barron's Educational Series, New York (1989)

Trading Natural Gas: A Nontechnical Guide, Fletcher J. Sturm, PennWell Publishing Company, Oklahoma (1997)

Securities Analysis: A Personal Seminar, New York Institute of Finance New York (1989)

Futures: A Personal Seminar, New York Institute of Finance, New York (1994)

Documentation for Derivatives: Annotated Sample Agreements and Conformations for Swaps and Other Over the Counter Transactions (Third Edition), Anthony C. Gooch and Linda B. Klein, Euromoney Publications (Great Britain) (1993)

Conclusion

This guide was written to provide basic information on gas contracts. Some of the topics covered don't necessarily lend themselves to simplicity, but hopefully, this guide has been a pleasant learning experience for you. In the first chapter, I noted this guide wasn't intended to be a treatise, but rather a first step for gas contractors and others in the industry who want an overall view of gas contracting. Always remember that the particular facts of each issue you encounter in your contract negotiations, drafting, and review is unique and as such must be treated with great care and deliberation before making final decisions. Good luck in your work and happy contracting!

Glossary of Terms

A

ACA UNIT CHARGE - Represents a fee (Annual Charge Adjustment) charged by all interstate pipelines and then paid to the FERC by the pipeline for its allocated share of FERC operating expenses.

ACQUIRING SHIPPER – in the context of capacity release, the shipper that is acquiring firm released capacity from a releasing shipper.

ACTUALS - the as-measured quantity of gas for which a transporter bills the shipper.

AFFILIATED MARKETER - a marketing company that buys and resells gas and is owned by an interstate pipeline or by a corporation that owns an interstate pipeline subsidiary.

AGGREGATOR - a company that consolidates a number of suppliers and/or a number of markets into an aggregate group. A company that aggregates supply or market has greatly increased the number of alternatives to consider when pairing gas supply and market.

AGREEMENT – (1) the bargain made by to parties entering into a contract; (2) the document which describes the terms and conditions under which parties to the relevant transaction have agreed to sell or transport natural gas; this term also refers to the understanding between the parties; also may be called a contract and often the two words are used interchangeably.

B

BACKHAUL - the delivery by a pipeline of gas upstream from its point of receipt. If a downstream producer sells gas to an upstream end-user, an interstate pipeline, whose system flows from south to north, would pick up the end-user's gas from the downstream producer. Equivalent quantities from other sources would then be delivered to the upstream end-user. While all of the gas physically flows downstream, the effect of the transaction is to bring the downstream producer's gas back upstream.

BALANCING - the process of adjusting physical gas takes or deliveries to match what should have been taken or delivered. This is accomplished by making deliveries of gas into or withdrawals of gas from a pipeline and may be done daily, monthly or seasonally, depending on whether the pipeline is an interstate system regulated by the FERC, or an intrastate system.

BALANCING PENALTY - scheduling and imbalance charges, penalties, fees or cashouts contained in a transporter's tariff, which may be assessed against a shipper for failure to satisfy the transporter's balance/nomination/scheduling requirements.

BASELOAD - historically, the part of a gas buyer's gas needs that are either essential or relatively predictable. An LDC, for example, will procure baseload gas for gas needs it can ascertain in advance. Through time and use, this term has become more difficult to quantify, and, indeed the definition may differ from company to company or from region to region.

BASIS – the difference between the price of a single commodity at different times (cash price and futures price), the difference in price at different locations (delivery points) or a patterned relationship between prices of two different commodities. Cash minus Futures equals Basis.

BCF - one billion cubic feet.

BEST BID – in capacity release, the shipper placing the most desirable (highest) bid for released firm capacity; this may also refer to the accepted response to a Request for Proposal.

BEST EFFORTS - a term originally used to describe the nature of a delivery and receipt obligation. This term is susceptible to several differing interpretations. Best efforts could involve doing all that is possible to fulfill one's contractual obligation (similar to a firm obligation) and it could mean that gas flow is interruptible except for economic reasons. Because of the vagueness inherent in this definition and the fact that there is no such thing as "industry standard", best efforts is becoming a definition of the past and is usually not used now to describe delivery and receipt obligations in gas sale contracts.

BID WEEK - the period of time (currently up to 7 trading days) near the end of the month where cash deals are registered for the ensuing month's natural gas flows; coincides with the expiration of the spot futures contract. Originally, bid week was the last week of the month during which first-of-the-month transportation nominations were due.

BLANKET CERTIFICATE - a FERC program under Order No. 436 and its progeny, which allows interstate pipelines to apply for a one-time certificate of "public convenience and necessity" that authorizes transportation of gas for other interstate pipelines, end-users, marketers, brokers and others, on a self-implementing basis. Were it not for the blanket certificate, each individual transportation service agreement would require the approval of FERC. Since Order No. 636 was issued, all interstate transporters were also issued blanket sales certificates.

BROKER – (1) an individual or company that acts as agent for a company which wants to buy or sell a natural gas service. For a commission, a broker will find a compatible party and facilitate a contractual relationship. The major difference between a broker and a marketer is that the broker does not take title to the gas. (2) a person paid a fee or commission for executing buy or sell orders of a customer. In commodity futures trading, the term may refer to (a) *Floor Broker*—a person who actually executes orders on the trading floor of an exchange; (b) *Account Executive, Associated Person, Registered Commodity Representative,* or *Customer's Man*—the person who deals with customers in the offices of futures commission merchants; and (c) the *Futures Commission Merchant.*

BTU (British Thermal Unit) - the amount of heat required to raise one pound of water one degree Fahrenheit. An MMBtu is one million BTU's. This is the measurement of quantity in natural gas contracts.

BUNDLED SERVICE - a way of pricing service where the customer pays one rate for the entire transaction, rather than paying different rates for the various components (transportation, gathering, etc.).

BURNER TIP - the point at which gas is ignited; the end of the gas delivery system. This is typically used in reference to the point at which an industrial, commercial or residential customer takes gas from its utility.

BUTANE (C_4H_{10}) - a natural gas liquid commonly removed from natural gas by processing; it is a widely used household fuel as well as in the manufacture of chemicals and synthetic rubber.

BUTYLENE (C_4H_8) a natural gas liquid. The main use for this hydrocarbon is blending in high-octane gasoline.

BUYER - the party which purchases gas from a supplier. A buyer has the responsibility to pay for gas pursuant to the gas sale agreement.

BY-PASS - the method by which a customer gains access to direct deliveries of gas from a supplier other than its traditional supplier, the pipeline. LDC's may by-pass their pipeline supplier(s) by building interconnects with other pipelines; industrial customers may by-pass their local distributor by tapping directly into an interstate or intrastate pipeline.

C

CALL OPTION - gives the holder the right to "call" (or buy) a specified quantity of a commodity at an agreed price during the life of the option. Also called a cap.

CAPACITY RELEASE - a program adopted by FERC Order No. 636 that provides a procedure for firm transportation shippers to release all or a portion of their firm capacity for a designated period of time through posting of and bidding for the released capacity on the pipeline's electronic bulletin board.

CASH MARKET (OR PHYSICAL MARKET) - the market for a cash commodity where the actual physical product is traded.

CASHOUT - the method utilized by pipelines to eliminate shipper imbalances between receipts and deliveries, usually on a monthly basis. Under a cashout provision, the transporter will pay a shipper for all overages at a predetermined rate and the shipper will be obligated to pay the transporter for all underages at a different predetermined rate.

CITY GATE - the physical location where a local distribution company measures and receives gas from an upstream pipeline company.

COGENERATION - a process by which the reject heat of one process becomes a source of energy for a subsequent conversion process. Typically, natural gas will be used to fuel a cogeneration facility, which produces both steam and electricity.

COMMODITY – any goods that are fungible and may be traded on an organized exchange.

COMMODITY CHARGE - a charges for sales or transportation service based on the volume actually transported ($/Dth.) or the quantity actually purchased ($/MMBtu).

COMMODITY SWAPS - a swap in which one of the payment streams for a commodity is fixed, and the other is floating. Commodity swaps enable producers, end-users and other participants in the industry to hedge commodity price risk. They are usually settled in cash; however, physical delivery also occurs in some instances.

COMMODITY OPTIONS - a contract providing the purchaser the right, but not the obligation, to buy or sell a given quantity of a commodity at a strike price—on or before a given date.

COMPRESSION - the activity of applying pressure to a substance which results in decreased volume of the substance inversely proportionate to the increase in the pressure.

COMPRESSOR STATION - a facility that compresses, or raises the pressure of natural gas in order to allow the gas to be injected into a higher-pressure line. Gas might be compressed to move it from a gathering line into an interstate transmission line and compressor stations are located incrementally along transmission lines.

CONFIRMED NOMINATION - the verification by a pipeline company that a change in the shipper's transportation requirement is matched with the required changes for supply and/or market.

CONTRACT – (1) the total legal relationship between two parties that have reached an agreement; (2) the name of the document setting forth the agreement.

CUBIC FOOT - the most common unit of measurement of gas volume. The amount of gas required to fill a volume of one cubic foot under standard conditions of temperature, pressure and water vapor. The usual measurement is in thousand cubic feet (Mcf) and one Mcf roughly equals one MMBtu (assuming 1,000 Btu's/cf).

CURTAILMENT - a reduction in service or sales lower than the contracted level.

CURTAILMENT POLICY - a policy for reducing natural gas service to designated customers when there is insufficient supply or capacity to meet demand, identified usually with firm transportation service.

D

DAY TRADE – (1) the purchase and sale of a futures contract on the same day. (2) the purchase or sale of natural gas on a daily basis.

DEDICATION OF RESERVES - a contractual commitment made by a producer to sell all of the gas from a given well or gas field to a particular buyer.

DEHYDRATOR, NATURAL GAS - a piece of equipment through which gas is run to remove entrained water. Gas must be relatively free of water vapor to meet pipeline specifications; it may not contain more than 7 pounds of water vapor per million standard cubic feet.

DEKATHERM (Dth)- a metric unit of heat measurement which equals 10 therms. This is a unit of heat equal to 1 MMBtu and the approximate heat content of 1 Mcf.

DELIVERY POINT - the physical point at which gas is delivered by the seller to the buyer. These "points" are more difficult to ascertain when pooled gas is sold or purchased, and may be called "Title Transfer" points.

DEMAND CHARGE - a fixed portion of the price for a sale or purchase of gas; the fixed price paid for firm transportation services. The demand charge is paid whether or not subject gas is sold or transported.

DERIVATIVES – non-regulated securities whose value depends on the values of other more basic assets or an index.

DOWNSTREAM - a word that indicates the direction of flow of gas from supply to market. Since much of the US's natural gas production is located in the south and southwest regions of the country; the downstream flow for interstate gas is generally to the north and northeast.

DOWNSTREAM PIPELINE - a pipeline that receives gas from one or more other pipelines.

DRY BASIS - the heat content of gas is tested on a dry basis when only the actual water vapor content of the sample is included in the test. The test is on a wet basis if water is added before testing until the sample is saturated with water vapor. Wet basis testing results in lower heat content per volume.

DRY GAS – (1) natural gas that contains less than 7 pounds of water vapor per 1,000 Mcf.; (2) natural gas that is measured on a dry basis.

E

ELECTRONIC BULLETIN BOARD (EBB) - the electronic communication system used by pipelines for distribution of information to users.

ELECTRONIC CONTRACT - the exchange of messages between buyers and sellers, structured according to a prearranged format so that the contents are machine-processible and automatically give rise to contractual obligations.

ELECTRONIC DATA INTERCHANGE (EDI) - computer-to-computer exchange of information through a series of standard electronic file formats.

END-USER - a consumer that uses natural gas.

ETHANE - C2H6 - one of the hydrocarbon components of natural gas.

EVERGREEN CLAUSE - a contract provision that extends the term of the contract beyond the primary expiration date. This is usually a month-to-month or year-to-year extension and can be triggered either by affirmative action by one of the parties or by inaction of both of the parties, depending on the contract language; also called a "Rollover" Clause.

EXCHANGE – (1) a type of transportation service or gas gale in which gas is received from one party in exchange for gas delivered to the other party. An exchange usually involves simultaneous deliveries at two different delivery points, not necessarily on the same pipeline; (2) an organized market for trading securities or commodities.

EXCHANGE OF FUTURES FOR PHYSICAL (EFP) - a simultaneous trade between two traders wherein one trader buys the physical and sells the futures contract, while the other trader makes an equal and opposite transaction (sells physical and buys futures). Prices for EFP transactions are mutually agreed upon by the two parties involved. The buyer of an EFP buys the physical commodity and sells futures contracts. The seller of an EFP sells the physical commodity and buys futures contracts.

F

FEDERAL ENERGY REGULATORY COMMISSION (FERC) - the federal agency, established in 1977, that generally has jurisdiction over all interstate gas pipelines, wholesale electric rates, hydroelectric licensing and oil pipeline rates.

FEEDER LINE - a pipeline, a gathering line tied into a trunk line.

FEEDSTOCK - The supply natural gas liquids or natural gas to a refinery or petrochemical plant or the supply of some refined fraction of intermediate petrochemical to some other process.

FIELD - the area encompassing a group of producing oil and gas wells; a pool; a roughly contiguous grouping of wells in an identified area.

FIRM DELIVERY - a term used in sales of gas to denote an absolute obligation on the seller to deliver gas and the same obligation on the buyer to accept delivery of an agreed-to quantity of product.

FIRM STORAGE SERVICE. Pipeline storage service that is not subject to a prior claim by another customer or class or service. All interstate transportation providers are required to offer this service to their customers.

FIRM TRANSPORTATION SERVICE - transportation service that is not subject to a prior claim by another customer or class of service. It is offered to customers (shippers) under tariff authority and associated contracts. Primary firm transportation has a higher priority than secondary firm or interruptible transportation. This level of service requires payment of both a commodity charge (based on the volume/quantity actually transported) and a reservation charge to reserve pipeline capacity.

FIRST PURCHASER - literally, the first purchaser of natural gas after it has been severed from the ground. This term has significance, most notably because of tax burdens that may, by law of the state of extraction, be the responsibility of this person or firm.

FIRST SALE - wellhead and other gas sales defined as "first sales" in the Natural Gas Policy Act; loosely, any sale other than a sale by a pipeline or LDC.

FORCE MAJEURE - from the French, now commonly known in natural gas sales transactions as an unforeseen occurrence beyond the control of the parties to a contract which partially or entirely prevents the performance of one or both parties. Gas sale contracts usually contain a force majeure clause that may list specific instances of force majeure, as well as specific exclusions from force majeure. Transportation service agreements also contain force majeure clauses.

FRACTIONATION – the process of removing liquid hydrocarbons from the gas stream and also separating specific liquid hydrocarbons from each other.

FUEL - gas consumed by a gathering system, processing plant or pipeline as fuel to run the facility. Gathering, processing and transportation agreements often include provisions allowing the operator to retain gas for fuel without paying for it. Fuel costs may or may not be included in the cost of transportation in gas sale contracts.

FUTURES CONTRACT - an exchange-traded contract generally calling for delivery of a specified amount of a particular grade of commodity or financial instrument at a fixed date in the future.

G

GAS – a fluid that has neither independent shape nor volume but tends to expand indefinitely. In the industry, natural gas is frequently referred to simply as Gas.

GAS MEASUREMENT - this generic term refers to a standardized calculation of volumes of natural gas by the use of conversion factors of standard pressure and temperature. The standard pressure is 14.73 pounds per square inch; the standard temperature is 60 degrees Fahrenheit.

GATHERING - the process of collecting gas from several wells prior to delivery to a pipeline. The FERC has jurisdiction over interstate transportation of natural gas, but not over gathering, which is controlled by the states.

GATHERING LINES - non-FERC regulated pipelines that connect wells to a FERC- regulated transmission line. The state that has jurisdiction over a particular gathering line regulates the gathering function.

GATHERING SYSTEM - a system of pipelines and compressor stations used to bring gas from the wellhead to a major pipeline system.

GISB (GAS INDUSTRY STANDARDS BOARD) - a natural gas industry organization whose function is to provide a framework and forum for developing and maintaining standards throughout the gas industry. The five membership categories are (i) producers; (ii) pipelines; (iii) marketers, gatherers, processors, storage operators and associated services; (iv) LDCs; and (v) end-users.

H

HEAT CONTENT - the amount of heat released when gas is burned. This is usually expressed as Btu's per cubic foot, MMBtu's per thousand cubic feet or Dekatherms per thousand cubic feet.

HEDGE - hedge is used to reduce risk by taking a position that offsets existing or anticipated exposure to a change in market rates or prices.

HENRY HUB - a pipeline interchange near Erath, LA, where eight interstate and three intrastate pipelines interconnect. The point of exchange for natural gas futures contracts.

HINSHAW PIPELINE - an interstate pipeline that is not under FERC's jurisdiction because (i) all gas transported and sold by the pipeline in an individual state is ultimately consumed within that state; and (ii) the facilities and rates of the pipeline are regulated by the state. A Hinshaw pipeline may receive a certificate authorizing it to transport natural gas out of the state in which it is located, without giving up its status as a Hinshaw pipeline

HUB OR HUB SERVICES - facilities or services offered by an operator at a point or points where several pipelines or storage facilities connect. Such services include transportation, long term storage; parking (short-term storage), balancing and title transfers.

HYDROCARBONS - organic compounds made up of carbon and hydrogen atoms. For the purpose of a sale of natural gas, these may be identified as the liquid and liquefiable components other than methane contained in natural gas, including but not limited to ethane, propane, normal butane, iso-butane and natural gasoline.

I

INCIDENTAL DAMAGES - costs generically identified as costs incurred to procure an alternate supply of (or market for) gas when the other party has failed to (i) purchase/sell the physical product, or (ii) assume a futures position. The costs giving rise to these damages may, depending on the individual contract, include the cost of the gas itself as well as additional gathering/transportation costs and, in the case of an EFP—if the contract so allows—the cost to liquidate open futures contracts.

INDEPENDENT - Term generally applied to a nonintegrated natural gas company, usually active in only one or two sectors of the industry. An independent marketer buys product from major or independent producers and resells it under its own brand name. There are also independents that are active either in production exclusively and are not controlled by integrated natural gas companies.

INDEX PRICE - a price usually obtained from an industry publication which is intended to represent an average price of gas delivered to a specific point on the pipeline at or during a specified period of time. Examples of indices utilized in natural gas sales are: *"Inside FERC's Gas Market Report," "Gas Daily," "Natural Gas Intelligence" and "Btu Daily."*

INTEGRATION - A term which describes the degree in and to which one given company participates in all phases of the natural gas industry.

INTERPROVINCIAL PIPELINE – In Canada, a pipeline that crosses provincial boundaries and is regulated by the NEB.

INTERRUPTIBLE - while it may be specifically defined otherwise in a particular gas contract, generally this term defines a delivery obligation under a gas sale contract whereby either party may interrupt the flow of gas at any time for any reason or for no reason at all. Sometimes prior notice of the interruption is required in a contract—other times no prior notice is required.

INTERRUPTIBLE SERVICE - natural gas transportation service that receives the lowest priority. Availability of interruptible space is determined by the amount of firm space not being utilized during a specific period of time. A pipeline may, on short notice, interrupt such service for any reason listed in its FERC-approved transportation tariff. Only a commodity charge for gas actually transported is required by the shipper.

INTERSTATE COMMERCE - any commerce that crosses state lines. The FERC regulates all pipelines engaged in interstate commerce.

INTERSTATE PIPELINE - a company that transports natural gas in interstate commerce or makes sales in interstate commerce of natural gas for resale for ultimate public consumption

INTERVENOR - a person, institution or organization admitted as a participant to a FERC proceeding. As an intervenor, such party receives the various pleadings and motions filed in that particular proceeding and is bound by the rules of the FERC relative to that proceeding which affect interventions.

INTRAPROVINCIAL PIPELINE – in Canada, a pipeline system located entirely in one province and is subject to regulation of the province.

INTRASTATE GAS - gas which is not in interstate commerce, and is produced and consumed in one state.

INTRASTATE PIPELINE - a company engaged in natural as transportation (not including gathering) which is not subject to the jurisdiction of FERC under the Natural Gas Act and which is not a Hinshaw pipeline. This type of company is subject to state jurisdiction and regulation.

J

JOINT OPERATING AGREEMENT - agreements between the various interest owners of a well or wells and the operator of the physical facilities which set forth the various rights and responsibilities of the parties.

K

KANSAS CITY BOARD OF TRADE (KCBT) – an exchange that once traded a natural gas futures contract for Waha delivery, then converted that unsuccessful contract to a natural gas basis contract.

L

LAST TRADING DAY - the final day on which futures contracts may be traded. Any contracts left open following this session must be settled by delivery.

LEAN GAS - natural gas that contains little or no liquefiable hydrocarbons..

LINE-PACK - the amount of natural gas packed into a pipeline. Line-pack can be increased by putting gas in faster than it is taken out, thus increasing line pressure. Reversing the process can decrease it. Since a pipeline company may increase line-pack in advance of projected temporary increases in demand, this is a limited form of storage.

LIQUID CONTENT - the amount of natural gas liquids in a natural gas source or sample.

LIQUIDS, NATURAL GAS - the liquid hydrocarbon mixtures that are gases at reservoir temperatures and pressures, but which can be recovered through the processes of condensation or absorption. Propane, butane, other liquefied petroleum gases and natural gasoline are types of natural gas liquids.

LOCAL DISTRIBUTION COMPANY (LDC)- a local utility that sells natural gas to consumers within the city or region that the utility serves. Generally, LDC's are regulated at the state level.

M

MAKE-UP GAS - natural gas taken under a make-up provision of a contract that allows the purchaser to receive supplies that were previously paid for, but not taken. This term may also refer to any instance where natural gas should have been delivered or received in a prior period, and is now being "made-up."

MAINLINE - also called "Transmission Line" or "Trunk Line," this is a large-diameter pipe into which smaller lines connect. These lines usually operate under high pressure and traverse a long distance.

MAJOR - A term broadly applied to those national or multinational energy companies which by virtue of size, age, and/or degree of integration are among the prominent companies in the energy industry.

MARKET AREA - the area along a pipeline system where natural gas is consumed by more than a single end user. Several (to many) market areas may be located on a single interstate pipeline system.

MARKET CENTERS - areas, also identified as "Hubs" where natural gas purchases and sales occur at the intersection of different pipeline systems. Under FERC Order 636, pipelines may not inhibit the development of market centers.

MARKET OUT - a clause in a natural gas sale contract which allows the buyer to lower the purchase price or terminate service in response to market conditions. This type of clause is relatively rare in short term sales to other than end-user LDC's, and may often be expressed as an event of force majeure in lieu of a specific market out clause.

MARKET-BASED RATES – a type of rate determination whereby the market sets rates. If a transportation provider and a shipper agree to a market-based rate for transportation, the rate would be based upon evidence of the market rates at relevant receipt and delivery points.

MARKETER - a company involved in the business of purchasing and reselling natural gas. The company may be independent or affiliated with transmission and production companies as well as LDC's.

MAXIMUM DAILY QUANTITY (MDQ) - the number of units designated as a maximum that may be sold or transported on a daily basis.

MAXIMUM/MINIMUM RATES - the most and the least that an interstate pipeline may charge for self-implementing transportation service.

MERCAPTANS - sulfur compounds often present in refined products, which impart an undesirable odor.

METER - a device for measuring natural gas volume. The gas being measured usually passes through the meter. Typical meters are orifice meters and rotary displacement meters.

METHANE (CH4) - the principal hydrocarbon component of natural gas.

N

NATURAL GAS - a naturally occurring mixture of hydrocarbons or of hydrocarbons and non-combustible gases, in a gaseous state, consisting essentially of methane; also referred to as "Gas."

NATURAL GAS ACT (NGA) - a federal statute of 1938 giving the Federal Power Commission (predecessor of the FERC) jurisdiction over transportation and sales of natural gas for resale in interstate commerce.

NATURAL GAS POLICY ACT OF 1978 (NGPA) - a federal law adopted in phased-in deregulation for gas discovered after 1977. Among other provisions, the NGPA also established incentive prices for certain types of natural gas in order to encourage increased exploration and production.

NATURAL GAS WELLHEAD DECONTROL ACT OF 1989 - a federal law which removed NGPA price controls and NGA jurisdiction from all wellhead and other "first sales."

NOMINATIONS - requests by shippers for transportation under a service agreement.

NOTICE OF PROPOSED RULEMAKING (NOPR) - a document issues by the FERC that describes proposed rules and asks for comments by interested parties.

NO-NOTICE SERVICE - a type of service required by Order No. 636 whereby a shipper may vary its takes from the pipeline within its MDQ without having to make prior nominations for such changes ands without penalty.

NEW YORK MERCANTILE EXCHANGE (NYMEX) – the first U.S. exchange to trade natural gas futures contracts.

O

OFFSHORE GAS - natural gas found underneath coastal bodies of water outside the required 3 mile jurisdictional limit.

ONSHORE GAS - natural gas found underground. Gas which is produced in waters offshore, but inside the 3-mile limit, are also referred to as onshore gas.

OPEN ACCESS - fully equal access to transportation services offered by interstate pipelines.

OPERATIONAL BALANCING AGREEMENT (OBA) - an agreement between the operator of an interconnection and the downstream pipeline whereby predetermined allocation methodologies for receipt and delivery and other matters pertaining to gas balancing are agreed to.

OPERATIONAL FLOW ORDER (OFO) - orders issued by a pipeline to protect the operational integrity of the pipeline itself. Under an OFO, a shipper may have service restricted or be required to perform some affirmative activity.

ORDER NO. 436 - the FERC order issued in 1985 which offered participating pipelines a blanket transportation certificate authorizing self-implementing transportation service to all end users on a non-discriminatory basis. This was essentially the order that "opened up" the natural gas industry to a new type of participant—the marketer.

ORDER NO. 497 - a 1988 rule of the FERC which adopted standards of conduct and reporting requirements to prevent pipelines from discriminating in favor of their own marketing affiliates.

ORDER NO. 500 - the FERC adopted this rule in 1987 to modify Order 436, primarily intended to address the impact of open access on pipelines regarding their take-or-pay obligations.

ORDER NO. 636 - an April 1992 FERC order, which required, among other things, pipelines to unbundle their transportation services from their sales of gas. Pipelines were also required to shift to a straight fixed variable (SFV) rate design. Pipelines in compliance with Order 636 were allowed to sell gas at negotiated rates and to abandon sales upon termination of the contract.

ORIFICE METER - a measuring instrument that records the flow rate of natural gas, enabling the volume of gas delivered or produced to be computed through measurement of the pressure differential across a plate that has a precisely cut hole in its center.

OTC - Over the Counter: A financial product, not listed on an organized exchange. There is no open outcry in the OTC market. Trades are negotiated between parties on an individual basis. Generally, no standardized rules govern OTC trading.

OUTER CONTINENTAL SHELF - the seabed lying more than 3 geographical miles from the ordinary low water mark off the U.S. coast, extending seaward to the outer edge of the continental shelf. This land is under federal jurisdiction, and not that of the state from which the seabed extends.

P

PAPER POOLING - the aggregation, for accounting purposes, of natural gas from a number of receipt points which serve a number of contracts. This is distinguished from a physical pooling point.

PEAK DAY - the day when the highest delivery requirement during a 24-hour period (8 a.m. to 8 a.m.) is placed upon a pipeline system by customer usage.

PEAK DEMAND - the highest demand that a natural gas supplier with flexible demand must meet. For many pipelines and LDC's, peak demand occurs during the coldest part of the winter.

PEAK SHAVING - the practice of meeting some portion of peak demand by using a substitute for normal gas supply which costs more than normal supply when needed, but which costs little or nothing when not needed. An LDC supplementing the natural gas available to it with gasified propane to meet peak demand is using a peak shaving technique.

PIPELINE - refers both to the pipe and other equipment used to transport natural gas and to the company that owns a pipeline.

PIPELINE DAY - the 24-hour period specified by each pipeline company for the operation of its pipeline system, for interstate pipelines now 9 a.m. Central Time until 9 a.m. Central time the next day.

PIPELINE INTERCONNECT - a point at which facilities of 2 or more pipelines interconnect.

POOLING POINT - a point, which may be either actual or theoretical, at which gas is aggregated from a number of receipt points and made available to serve more than one contract without tying a specific receipt point to a specific contract.

PREARRANGED RELEASE - a form of capacity reallocation whereby a firm shipper may agree with another party to a capacity release transaction. In some cases, a prearranged release must be open for bidding prior to taking effect. In all cases, a prearranged release must be posted on the pipeline's EBB for informational purposes.

PRICE REDETERMINATION CLAUSE - a natural gas sale contract provision that allows the adjustment or renegotiation of the price of gas at a designated time.

PROCESSING PLANT - a plant that removes or reduces impurities in wellhead gas to assure that the gas will be of pipeline quality. Also a plant from which liquefiable hydrocarbons such as propane, butane, ethane or natural gasoline are extracted from the gas stream.

PRODUCER - the working interest owner of natural gas.

PRODUCTION AREA - the area along a pipeline system where gas is produced and delivered to the pipeline.

PROPANE (C_3H_8) - a hydrocarbon substance, the primary function of which is residential and commercial heating and cooling.

PROPYLENE (C_3H_6) - a hydrocarbon substance used primarily for residential and commercial heating and cooling as well as a transportation fuel and petrochemical feedstock.

PUT - an option to sell a specific futures contract at a specified price on or before a certain date. Also called a floor.

Q

QUALITY – a reference to pipeline standards regarding heat content, dirt, water and other impurities that must be limited in the gas stream.

QUANTITY – as used in the natural gas industry, a measurement of the heat content of gas. Quantity is typically indicated in MMBtu's (per Mcf) or in Dekatherms (per Mcf).

R

RATE SCHEDULE - a document filed by a pipeline or other regulated entity with the FERC or other regulatory agency to set forth the charges and other terms and conditions of gas sales and transportation.

RATE ZONES - segments along a pipeline that reflect variations in costs of service; usually costs increase as the distance of haul increases.

RAW GAS - natural gas straight from the well before the extraction of impurities and liquefied hydrocarbons.

RECEIPT POINT - a custody transfer point through which natural gas flows from a wellhead or other point on an upstream pipeline system into another system for transportation. A receiving system's receipt point is the delivering system's delivery point.

RELEASE CAPACITY - firm transportation capacity released under a capacity reallocation program.

RICH GAS - natural gas containing significant amounts of liquefiable hydrocarbons.

ROYALTY - a share of production of oil and natural gas properties, which may or may not bear a share of production expenses, depending on the terms of the lease.

ROYALTY OWNER - the person who owns a royalty interest in production.

S

SALE FOR RESALE - the purchase of natural gas with the intent to resell it, as defined by Sec. 1(c) of the NGA.

SCHEDULING – the act of nominating gas between seller and buyer, notifying necessary transporters of receipt and delivery points and quantities at each point, and final confirmation(s) from transporters as to the quantity of gas that will flow under each shipper's transportation service agreement.

SCHEDULING PENALTY - a penalty imposed by pipelines for the difference between the amount of natural gas scheduled to flow and the amount of gas that actually flowed.

SEASONAL RATES - the rates charged by an electric or gas utility for providing service to consumers at different seasons of the year, taking into account demand, based on weather and other factors.

SELF-IMPLEMENTING TRANSPORTATION - transportation that may begin without prior FERC approval; blanket certificate transportation under Order No. 436 is self-implementing.

SELLER - the party that sells natural gas to a buyer. A seller has the responsibility to deliver gas pursuant to the gas sale agreement.

SERVICE AREA - the territory in which an LDC sells and distributes natural gas to end-users.

SETTLEMENT PRICE - the official closing price of the day for each futures contract, established by the exchange as a benchmark for settling margin accounts and determining invoice price for delivery on that day.

SEVERANCE TAX - a tax levied by some states on each barrel of oil or each thousand cubic feet of gas produced.

SHIPPER - the party who contracts with a pipeline for transportation service. A shipper has the obligation to confirm that the volume of gas delivered to the transporter is consistent with nominations. The shipper is obligated to confirm that differences between the volume delivered to the pipeline and the volume delivered by the pipeline back to the shipper is brought into balance as quickly as possible.

SHRINKAGE - the amount by which natural gas volume is reduced as a result of processing, gathering or transportation. Shrinkage rates are often agreed to in advance rather than actually measured.

SHUT-IN WELL - a well at which production has been temporarily halted, usually in order to clean or repair the well, but also because of a lack of market demand or unfavorable prices.

SOUR GAS - natural gas that contains objectionable amounts of sulfur and sulfur compounds..

SPOT - term which describes a one-time open market cash transaction, where a commodity is purchased "on the spot" at current market rates. Spot transactions are in contrast to a term sale, which specifies a steady supply of product over a period of time. Also, "spot month" refers to the nearest futures delivery month.

SPOT MARKET - a generic reference to short-term purchases and sales of natural gas. Spot market prices are reported by pipelines for various places or areas by index publications..

Glossary of Terms

STORAGE - a type of service offered by pipelines and other owners of storage facilities (usually a salt dome cavern or depleted gas reserves) whereby the gas may be injected into the facility and kept until a later date when the owner of the gas may withdraw the stored gas.

STORAGE GAS - natural gas moved from its original location in the ground which is or has been stored in a storage field.

STORAGE FIELD - one or more underground storage reservoirs with associated pipelines, wells, compressors, measurement facilities and other necessary facilities operated as a separate unit.

STRAIGHT FIXED VARIABLE (SFV) RATES – the FERC's preferred way for transportation service providers to determine the rates they charge shippers. The SFV structure places all fixed costs in the demand component of rates and all variable costs in the commodity component.

STRIKE PRICE - the price at which an option may be exercised. Also called an exercise price.

SWAP - a financial transaction in which two counterparties agree to exchange streams of payments over time according to a predetermined rule.

SWEET GAS - natural gas free of significant amounts of hydrogen sulfide (H_2S) when produced; also, gas that has been processed to remove impurities.

SWING - a term used to designate a delivery or receipt obligation that is based upon the ability of the seller to quickly stop or start deliveries up to a maximum quantity. Generally, no more than 24 hours prior notice is required in order to make the change; sometimes no notice is required. Typically, swing sales or purchases are started or stopped in response to changes in market prices.

T

TAILGATE - the outlet of a natural gas processing plant where the residue gas is discharged after processing is completed.

TAKE OR PAY CLAUSE - a provision in natural gas sale contracts which requires a purchaser to take, or if not taken, to pay for a certain quantity of gas. These clauses, popular until passage of Order No. 436 in 1985, are no longer commonly utilized.

TARIFF - a document filed by a regulated natural gas company with the FERC or state regulatory body, if an intrastate pipeline or LDC, which sets forth the prices and conditions of its service to various classes of customers.

TERM – (1) the period of time, expressed in years, months, seasons, days, portions of days, etc. during which a gas sale contract is in effect; (2) a shortened reference to a long-term contract.

TITLE TRANSFER – a sale without physical gas movement.

TRANSPORTATION - the movement of natural gas after gathering and before distribution. This is usually characterized by high pressure, high volume pipelines.

TRANSPORTATION COSTS - the costs associated with the movement of natural gas from one point to another. In addition to the transmission rate, these costs may include gathering rates, fuel and applicable surcharges as set out in the FERC-approved tariff of the transporting pipeline.

TRANSPORTATION SERVICE AGREEMENT - the contract entered into between a pipeline and shipper for transportation service. The agreement sets forth the terms and conditions for the transportation of the product. Terms of a FERC-approved service agreement may not (except in extremely rate circumstances and under non-discriminatory conditions) be changed by parties to an individual agreement.

TRANSPORTATION IMBALANCE - the difference in the amount of natural gas delivered to a customer under a transportation agreement versus the amount of gas actually received by the pipeline for delivery to that customer.

TRANSPORTER - a pipeline that moves natural gas owned by others through its facilities from one point to another, for a fee.

TRIGGER PRICE - a pricing option in some gas sale contracts which allows one party to "pull the trigger" to price the gas at exchange-related pricing at the time of their choice.

TRADING PARTNER AGREEMENT - the master agreement entered into by two parties to facilitate EDI transactions. This agreement sets forth the rights and responsibilities of both parties and allows for valid contracting solely through computer communications.

U

UNASSOCIATED GAS - natural gas that is not in contact with crude oil in the formation and may be produced independently. This product may also be called non-associated gas, dry gas and gas cap gas.

UNIFORM COMMERCIAL CODE (UCC) - a statute in every state but Louisiana which sets the legal rules for natural gas sales. Many states have varied their law from the "uniform" code, but the framework is basically the same in every state that has adopted the UCC.

UPSTREAM - a term that designates all points through which natural gas has already passed in its physical flow downstream; the further upstream a point is, the closer that point is to the supply of gas.

UPSTREAM PIPELINE - a pipeline that delivers natural gas into another pipeline. The delivering pipeline is the upstream pipeline. The receiving pipeline is the downstream pipeline.

V

VOLUME – a unit of measurement that measures the space in cubic units. Transportation service providers measure the flow of gas through their pipelines as volume, as opposed to a quantity measurement, which is consider the heat value or MMBtu value of the gas occupying a specific volumetric unit.

W

WARRANTY COMMITMENT - a provision in a gas sale contract where the seller promises, or warrants, to deliver a specific quantity of gas rather than agreeing to sell whatever is produces from some described source of production. The seller has an absolute obligation to deliver the warranty commitment quantity of gas.

WELLHEAD - the physical point at which natural gas leaves the ground and enters a gathering line or other pipeline system.

WELLHEAD PRICE - the price paid for gas at the well site.

Z

ZONE – (1) a way pipeline companies differentiate between different areas on their pipeline systems. Each pipeline system has a field zone and a market zone. Gas may be delivered into the pipeline's field zone (usually) and taken out of the system in market zones; (2) transportation rates on a single transportation system may vary from zone to zone if the pipeline uses a zone-based rate structure.

ZONE RATE – a transportation service rate whereby costs for transportation will be set based upon the zone or number of zones through which gas travels under a particular transaction.

Index

A

ACA unit charge, 403
Acceptance (contract), 110-111, 154, 219, 365:
    goods, 154
Access/accessibility, 46, 96, 99-100
Acquiring shipper, 403
Acronyms, vii-viii
Actions for breach of contract, 160-161:
    remedies of seller, 160;
    remedies for buyer, 160-161
Actuals, 403
Administrative/overhead fees, 281
Affiliated marketer, 403
Aggregate gas, 4
Aggregator, 403-404
Agreement (definition), 404
Alberta Sale of Goods Act, 153-161:

formation of contract, 153-156;
effects of contract, 156-157;
performance of contract, 157-158;
rights of unpaid seller against goods, 158-159;
actions for breach of contract, 160-161;
supplementary, 161
Alternate dispute resolution, 245, 316-320
Amendments in writing, 328-330, 386
Annual charge adjustment, 403
Anticipatory repudiation, 145, 228, 230
Applying practices, 201-204, 224-226
Arbitration, 317-318, 386-387
Articles of incorporation, 190-191
Asset, 33
Assignment rights, 295
Assignment/transfer (governing law/venue), 323-325
Assurance of performance, 144, 296-297
Audit, 285-286, 381-382: rights, 285-286
Audit rights, 285-286
Authority (contract), 193-195
Automatic fax confirmations, 197

B

Backhaul, 404
Balancing, 268-269, 404:
  penalty, 404
Bankruptcy, 286, 293-300, 350-351:
  trustee, 294-295
Bankruptcy trustee, 294-295
Base agreement, 217
Base contract, 29, 256-286:
  title, 256;
  heading, 256-257;
  definitions, 257-261;
  purposes, 261-262;

procedures, 262-263;
confirmations, 263-264;
transactions, 264-265;
term, 265-266;
obligations, 266-268;
transportation/scheduling/balancing, 268-269;
communications, 269-271;
imbalance charges, 271-272;
gas measurement/quality, 272-275;
delivery failure, 275-281;
billing and payment, 282-286
Baseload, 69-70, 89, 259, 261, 404
Basis (definition), 405
Battle of the forms, 365
Bcf (definition), 405
Best bid, 405
Best efforts, 405
Bid week, 24, 405
Billing and payment, 282-286, 381-382:
  invoice, 282;
  payment, 282-284;
  interest, 284-285;
  audit rights, 285-286
Blanket sales certificate, 104
Blanket transportation certificate, 46, 248, 405
Breach of contract/agreement, 143-146, 160-161, 170-175, 232-234, 291-300, 346-347:
  assurance of performance, 144;
  repudiation, 145;
  substituted performance, 145-146;
  presupposed condition, 145-146;
  actions for, 160-161
Breach of warranty, 232-234
Broker, 406
Btu, 260, 273-274, 406:

definition, 406
Bundled service, 10, 46, 406
Burner tip, 406
Business interests, 54-55:
   disparate needs, 55;
   common contract terms, 55
Business records, 199, 202-204:
   authentication, 199
Butane, 406
Butylene, 406
Buyer, 83-86, 174-175, 231-232, 406:
   deliverability, 84-85;
   title, 85-86;
   credit of seller, 86;
   breach of contract, 174-175;
   repudiation, 231-232;
   definition, 406
Buyer's breach of contract, 174-175
Buyer's repudiation, 231-232
Bypass, 67, 407

C

Calculations, 354-355
California Public Utility Commission, 17
Call/cap (option), 38, 407
Canada, 17-19, 105-107, 153-161, 164-165:
   regulations, 17-19;
   transportation laws/regulations, 105-107;
   tariffs, 105-106;
   tolls, 106;
   natural gas exports, 106-107;
   sale of goods, 153-161
Cancellation, 128-129, 298-300, 351-358, 384-385:
   clauses, 351-358;

right to terminate, 352-353;
effect of designation, 353-354;
calculations, 354-355;
payments, 355-358
Capacity allocation, 96, 101-103:
reallocation, 101-103
Capacity reallocation, 101-103
Capacity release, 101-103, 407
Cash market, 407
Cash prices, 23
Cash-future prices relationship, 26-27
Cashout, 407
Character of transaction, 247, 253-254
Chinese wall, 61
Choice of law, 321
City gate, 4, 88, 407
Clarity of law, 112-114
Cogeneration, 407
Comision de Regularia Energia, 14, 19-20, 107
Commerce rules, 115-116
Commercial laws, 111-118:
  clarity of law, 112-114;
  widget rules, 114-115;
  commerce rules, 115-116;
  gas contract issues, 115-116;
  natural gas, 116-118
Commercial standards, 126-127
Commoditization, 50
Commodity, 25, 104, 408
Commodity charge, 408
Commodity Futures Trading Commission, 25
Commodity options, 408
Commodity rate, 104
Commodity swaps, 408

Common contract terms, 55
Common interest (sellers/buyers), 79-81:
  seller, 80
Common law, 147
Communications, 9, 22-23, 41, 269-271
Compression, 408
Compressor station, 408
Condition of goods, 73
Conditions (contract), 155-156
Confidentiality, 325-326, 385-386
Confirm deadline, 260
Confirmations, 197, 217-225, 263-264:
  controlling confirmation, 219-224
Confirmed nomination, 408
Conflict of terms, 3267-327
Consideration (contract), 110-111
Consumers, 120-122
Consumption tax, 311
Contract, 118-120, 127-128, 377-387, 408:
  sales affected, 118-120;
  checklist, 377-387;
  definition, 408
Contract checklist, 377-387
Contract demand conversion, 97
Contract description, 110-111:
  offer, 110;
  acceptance, 110-111;
  consideration, 111
Contract drafting, 394-398
Contract duration, 214-215
Contract formation, 131, 188-197, 365:
  enforceability, 189-192;
  authority, 193-195;
  in writing, 195-196;

rules of evidence role, 196-197
Contract fundamentals, 241-287, 289-338:
   short-term contracts, 244-246;
   long-term contracts, 246-254;
   standard gas contract elements, 254-255;
   base contract (gas sales/purchases), 256-286;
   financial responsibility, 289-338;
   governing law and venue, 321-335
Contract issues, 8-14, 43-45, 115-116, 188-189, 339-376:
   over-the-counter derivatives, 341-360;
   electronic contracting, 361-372
Contract language, 52, 55-63, 75, 77-79, 109-110:
   control, 56-57;
   force majeure clause (producer), 57-60;
   flexibility, 60-63;
   understanding issues, 78;
   understanding language, 78;
   goals, 79;
   contract writing ability, 79
Contract negotiations, 388-393
Contract performance, 209-226:
   incomplete terms, 210;
   UCC law, 210-212;
   parol evidence rule, 212;
   course of performance, 212-213;
   course of dealing, 213;
   usage of trade, 213-214;
   gap-filling powers, 214;
   delivery point, 214;
   contract duration, 214-215;
   payment terms, 215;
   price, 215-217;
   quantity, 217;
   confirmations, 217-224;

controlling confirmation, 219-224;
applying practices, 224-226;
deal variety/number, 224-225;
performance without contract, 226
Contract problems, 21, 186-188, 201-204, 224-226
Contract subject to conditions, 155-156
Contract writing ability, 79
Contracting essentials, 185-240:
routine practices, 186-188;
contract problems, 186-188;
contract issues, 188-189;
contract formation, 189-197;
records, 197-209;
contract performance, 209-226;
non-performance remedies, 226-236
Contracting states, 161
Control (contract language), 56-57
Control (organization), 56-57
Controlling confirmation, 219-224
Convention on Contracts for the International Sale of Goods (UN), 112, 161-177:
sphere of application, 162-165;
general provisions, 165-167;
formation of contract, 168-170;
sale of goods, 170-175;
passing of risk, 175-177
Convergence, 27
Corporations (enforceability), 190-191
Cost of living index, 27
Course of dealing, 212-213
Course of performance, 212-213
Court deliberations, 210-212, 400-401: cases, 400-401
Cover (delivery failure), 230, 281
Credit of seller, 86
Credit support default, 347-348

Credit support documents, 345
Creditors, 140-142, 295-296
Creditworthiness, 89, 251-252, 341
Critical issues, 8-14, 43-45:
  regulatory changes, 9-14;
  regulators, 14;
  gas contracting, 43-45
Cubic foot (definition), 409
Current prices, 23
Curtailment, 97, 409
Curtailment policy, 97, 40

D

Damages, 228, 234-235, 279-280:
  incidental, 234;
  consequential, 234-235
Day trade, 186-187, 201-204, 409
Deal variety/number, 224-225
Dedication of reserves, 71, 73, 217, 409
Default, 291-293, 346-360, 384-385:
  failure to pay, 346-347;
  breach of agreement, 346-347;
  credit support default, 347-348;
  misrepresentation, 348-349;
  specified transaction, 349;
  bankruptcy, 350-351;
  early termination/cancellation clauses, 351-358
Definitions, vii-viii, 125-129, 244-245, 247, 257-261, 378-379, 403-430:
  base contract, 257-261;
  technical words, 258;
  words not used ordinarily, 258-261;
  vague/ambiguous words, 260-261
Dehydrator (natural gas), 409
Dekatherm (definition), 409

Deliverability, 77, 83-86, 171-173, 247, 251:
  buyer, 83-86
Delivery failure, 275-281:
  word usage, 277;
  imbalance charges, 277-278;
  liability for damage, 279;
  limitation of remedy, 279;
  damages, 279-280;
  market price indicator, 280;
  cover, 281;
  administrative/overhead fees, 281
Delivery point, 214, 409
Delivery rules, 158
Delivery to carrier, 158
Demand charge, 410
Derivatives, 33, 410
Designation effect, 353-354
development (contract), 245-246
Development (gas industry), 3-8
Disclaimers, 81
Discovery, 237
Disparate needs, 55
Displacement, 117
Downstream, 410
Downstream pipeline, 410
Drilling (gas), 3-4
Dry basis, 410
Dry gas, 410
Dual obligations, 132
Dun and Bradstreet number, 257, 286
Duties (seller/buyer), 157-158

E

Early termination/cancellation, 298-300, 341, 346-360, 384-385:

Index

failure to pay, 346-347;
breach of agreement, 346-347;
credit support default, 347-348;
misrepresentation, 348-349;
specified transaction, 349;
bankruptcy, 350-351;
contract clauses, 351-358
Effects of contract, 156-157:
transfer of property, 156;
time of transfer, 156-157;
risk transferred with property, 157;
transfer of title, 157
Electronic bulletin board, 22, 410
Electronic communications, 399
Electronic contract, 410
Electronic contracting, 9, 22-23, 361-372, 410:
internet commerce, 361-363;
legal issues, 363-366;
electronic data interchange, 368-368;
international rules, 368-372
Electronic data interchange, 22, 41, 366-368, 410:
trading partner agreements, 367
Electronic delivery mechanism, 41
Electronic energy management, 225-226
End user, 411
Energy Information Administration, 205, 237
Energy marketing, 9, 20-22
Energy Policy Act of 1992, 14-15
Energy Risk Management Association, 340-342
Energy Risk Management Association agreement, 341-342
Enforceability, 188-192, 201:
contract formation, 189-192;
legal capacity, 189;
power, 189-190;

441

corporations, 190-191;
partnerships, 191-192;
government entities, 192
Entire agreement, 147, 327-328
Ethane, 411
Evergreen clause, 411
Evidence/records, 364
Exchange (definition), 411
Exchange of futures for physical, 411
Exclusive remedy, 277
Excuse, 143-146:
assurance of performance, 144;
repudiation, 145;
substituted performance, 145-146;
presupposed condition, 145-146
Executory contract, 294-295
Expenses, 358
Exploration (gas), 3-4
Exports (Canada), 106-107
Extrinsic evidence, 238

F

Failure (seller/buyer), 145-146, 228, 232, 235-236, 296, 346-347, 380:
presupposed condition, 145-146;
seller, 228, 232;
buyer, 228, 235;
non-delivery, 232;
payment, 235-236, 296, 346-347;
delivery, 235
Failure of presupposed condition, 145-146
Failure to pay, 235-236, 296, 346-347
Failure to take delivery, 235
Fax confirmations, 197
Federal Energy Regulatory Commission. SEE FERC.

Index

Federal Power Commission, 92
Feeder line, 411
Feedstock, 411-412
FERC, 11, 14-17, 40-42, 91, 95-105, 411:
    orders, 40-42, 95-105;
    Order No. 436/500, 95-98;
    Order No. 636, 99-105
FERC orders, 40-42, 95-105
Field (definition), 412
Field zone, 108, 118
Financial responsibility, 289-338, 383:
    default, 291-293;
    bankruptcy, 293-300;
    bankruptcy trustee, 294-295;
    creditors, 295-296;
    failure to pay, 296;
    assurance of performance, 296-297;
    other breaches of contract, 297;
    misrepresentation, 297-298;
    cancellation/early termination, 298-300;
    force majeure, 301-308;
    taxes, 308-311;
    warranties and indemnities, 312-315;
    notices, 315-316;
    miscellaneous, 316;
    alternate dispute resolution, 316-320;
    governing law and venue, 321-335
Firm delivery, 412
Firm offers, 126
Firm storage service, 412
Firm transportation service, 412
First purchaser, 412
First sale, 412
Fitness for purpose, 136-138, 156

Flexibility (organization), 60-63
Force majeure, 57-60, 63-76, 87, 301-308, 384, 412-413:
   producer, 57-60;
   marketer, 63-76;
   LDC, 72-75;
   industrial consumer, 75-76
Force majeure clause (marketer), 63-76:
   reliability, 67-69;
   LDC gas supply load profile, 69-76;
   baseload, 69-70;
   intermediate load, 70;
   peak load, 70-71, 73;
   force majeure clause (LDC), 72-75;
   force majeure clause (industrial consumer), 75-76
Form (contract), 129-132
Formation (contract), 129-132, 153-156, 168-170:
   sale, 154;
   statute of frauds, 154;
   acceptance of goods, 154;
   price, 154;
   conditions, 155;
   warranties, 155;
   contract subject to conditions, 155;
   implied conditions/warranties, 155;
   sale by description, 155;
   implied conditions (quality), 156;
   CISG, 168-170
Fractionation, 413
Fuel (definition), 413
Full output contract, 217
Full requirements contract, 217
Future prices, 23, 25
Future (contracts), 246, 248-254:
   short-term, 246;
   long-term, 248-254

Futures contracts, 9, 23-32, 249, 413:
  gas pricing basics, 23-30;
  markets, 27-28;
  success elements, 30-31;
  gas delivery, 31;
  price risk transfer, 31-32

G

Gap-filling powers, 133, 214, 245
Gas (definition), 260, 413
Gas balancing services, 49
Gas bubble, 11
Gas contract issues, 115-116
Gas contracting, 43-45
Gas day basis, 56
Gas delivery, 31
Gas distribution, 3-4, 92
Gas gathering, 117, 119
Gas industry characteristics, 3-8
Gas Industry Standards Board, 9, 40-42, 50, 53, 55, 86-87, 414:
  contract, 50, 53, 55, 86-87
Gas inventory charges, 98
Gas law (U.S.), 92-95:
  Natural Gas Act of 1938, 92-93;
  Natural Gas Policy Act of 1978, 94-95;
  Natural Gas Wellhead Decontrol Act of 1989, 95
Gas marketing, 3-4, 48, 52-, 60-75
Gas measurement/quality, 272-275, 413:
  measurement, 272-274;
  quality, 272-275
Gas pricing basics, 23-30:
  current prices, 23;
  price indicators, 24;
  bid week, 24;
  monthly pricing, 24-25;

445

future gas prices, 25;
hedging, 25-26;
cash-future prices relationship, 26-27;
parallelism, 27;
convergence, 27;
futures contract markets, 27-28;
futures contracts, 28-30
Gas processing, 3-4
Gas producer, 59-60
Gas producing, 3-6, 52, 56-60
Gas sale laws (U.S.), 109-150:
   contract description, 110-111;
   commercial laws, 111-118;
   sales affected, 118-122;
   Uniform Commercial Code, 122-146
Gas sales/purchases (contract), 57, 74-75, 79-81, 85, 256-286
Gas storage, 3-4
Gas supply contract, 85
Gas transmission/transportation, 3-5, 392
Gas EDI, 42, 50
Gathering, 413
Gathering agreement, 117, 119
Gathering lines, 413
Gathering system, 414
General construction, 125-129, 132-140:
   letter of credit, 134;
   warranties (seller), 134-135;
   title warranty, 135-136;
   merchantability, 136-138;
   fitness for purpose, 136-138;
   warranty disclaimers, 138;
   warranties to third parties, 138-140
General contract provisions, 122-123, 165-167, 170-171:
   UCC, 122-123;

Index

CISG, 165-167, 170-171
General obligation, 132-140:
    letter of credit, 134;
    warranties (seller), 134-135;
    title warranty, 135-136;
    merchantability, 136-138;
    fitness for purpose, 136-138;
    warranty disclaimers, 138;
    warranties to third parties, 138-140
Glossary, 403-430
Goals (contract language), 79
Good faith clause, 165-166
Good faith purchasers, 140-142
Government entities (enforceability), 192
Governing law and venue, 321-335, 386:
    law, 321-322, 386:
    venue, 322-335;
    assignment or transfer, 323-325;
    confidentiality, 325-326;
    conflict of terms, 326-327;
    entire agreement, 327-328;
    amendments in writing, 328-330;
    representations of parties, 330-331;
    severability, 331-332;
    third-party beneficiaries, 332-333;
    waiver, 333-334;
    joint efforts of parties, 334;
    signatures, 334-335
Gray market, 102
Gross receipts tax, 83, 311

H

Header, 378
Heading (contract), 256-257

Heat content, 414
Hedge (definition), 414
Hedging, 9, 25-26, 32-35, 37-38, 247, 252-253, 414:
   futures contracts, 32-35;
   over-the-counter derivatives, 33;
   swaps, 34-35;
   example, 37-38
Hedging example (consumer), 37-38:
   options, 38
Henry Hub (Louisiana), 100, 414
Hinshaw pipeline, 92, 414
History (gas industry), 3-8
Homogenous commodity, 30
Hub/hub services, 414
Hydrocarbons, 415

I

Imbalance charges, 261, 271-272, 277-278
Imbalances, 379
Implied conditions/warranties, 136-137, 155-156:
   warranties, 136-137;
   quality, 156
In the money, 39-40
Incidental damage, 415
Incomplete terms, 210
Incumbency certificates, 194
Indemnities/indemnification, 81-82, 88, 312-315, 382-383
Independent (definition), 415
Index price, 33, 415
Industrial consumer, 52
Industry standards, 53-54
Inside FERC's Gas Market Report, 25
Installment deliveries, 158
Insurable interest, 142-143

Integration, 147, 415
Intention of offeror, 168-170
Interest, 284-285
Intermediate load, 70
International rules, 368-372:
   electronic contracting, 368-372;
   trading partner agreements, 368-370;
   third-party systems, 370-372
International Swaps and Derivatives Association agreement, 340-344
Internet, 361-363, 399, 401:
   commerce, 361-363
Internet commerce, 361-363:
   privacy, 362-363
Interprovincial pipeline, 415
Interruptible (definition), 416
Interruptible service, 416
Interruptible supply, 83-84
Interruptible transportation, 96
Interstate commerce, 14, 416
Interstate pipeline, 108, 416
Intervenor, 416
Intraprovincial pipeline, 416
Intrastate gas, 416
Intrastate pipeline, 16-17, 92, 108, 416
Invoice, 282

J

Joint efforts of parties, 334
Joint operating agreement, 118, 417
Jurisdiction, 358-360

K

Kansas City Board of Trade, 29, 417

L

Last trading day, 417
LDC, 4, 10-14, 16-17, 20-21, 42, 44-45, 52-53, 60, 67-76, 91, 97, 101, 120-122, 241, 248, 418:
   gas supply load profile, 69-76
LDC gas supply load profile, 69-76:
   baseload, 69-70;
   intermediate load, 70;
   peak load, 70-71, 73;
   force majeure clause (LDC), 72-75
Legal capacity, 189
Legal clarity, 112-114
Lean gas, 417
Legal issues, 363-366:
   electronic contracting, 363-366;
   evidence/records, 364;
   contract formation, 365;
   parol evidence rule, 365-366;
   statute of frauds, 366
Legal obligation, 127
Letter of credit, 123-124, 134:
   UCC, 123-124
Letter of intent, 208, 237
Levels of service, 96
Liability (damage/remedy), 80-81, 279:
   limiting, 80-81
Lien rights, 159
Limited warranties, 81-83:
   responsibility, 81;
   indemnification, 81-82;
   taxes, 82-83
Line pack, 417
Liquefied natural gas, 4
Liquid content, 417

Liquidation, 341
Litigation location, 199-200
Local distribution company. SEE LDC.
Long-term contracts, 246-254:
 definition, 247;
 traditional contracts, 248;
 future, 248-254;
 price risk, 249-251;
 seller's deliverability, 251;
 creditworthiness, 251-252;
 mergers/acquisitions, 252;
 hedging activities, 252-253;
 nature of business, 253;
 character of transaction, 253-254

M

Mainline, 418
Major (definition), 418
Make-up gas, 418
Manufactured gas, 7
Market area, 418
Market centers, 99-100, 418
Market out, 418
Market price, 279-280:
 indicator, 280
Market price indicator, 280
Market quotation/loss, 357-358
Market zone, 108
Market-based rates, 419
Marketer, 60-75, 419
Master contract, 217
Material adverse change, 340-341
Material changes, 221
Materiality, 221

Maximum daily quantity, 419
Maximum/minimum rates, 419
Measurement, 379-380
Mediation, 320, 386-387
Mercaptans, 419
Merchant, 125-126
Merchantability, 136-138, 156
Mergers/acquisitions, 247, 252
Meter (definition), 419
Methane, 45-46, 419
Mexico, 19-20, 107, 161:
   regulations, 19-20
Midstream marketing, 120
Minerals Management Service, 59, 88
Misrepresentation, 297-298, 348-349
Monthly pricing, 24-25
Mutual assent, 110

N

National Energy Board, 14, 17-19, 105-107
National Energy Board Act, 18
National Conference of Commissioners on Uniform State Laws, 113
Natural gas, 116-118, 419
Natural Gas Act of 1938, 10, 14, 92-93, 248, 419
Natural gas industry, 3-8
Natural gas liquids, 417
Natural Gas Policy Act of 1978, 11, 14, 94-95, 420
Natural Gas Wellhead Decontrol Act of 1989, 14, 95, 420
Nature of business, 247, 253
Negligence, 314-315
Negotiation, 131-132, 208, 388-394:
   contract, 131-132, 208, 388-393;
   postures/strategies, 388-391;
   knowledge, 392-393;

role, 393-394
Negotiation postures/strategies, 388-391
Negotiations (contract), 131-132, 208, 388-393
New gas, 11
New York Cotton Exchange, 28
New York Mercantile Exchange, 25, 28, 30-31, 420
No-notice service, 420
Nominations, 147, 379, 420
Non-delivery, 232
Non-material changes, 221
Non-performance remedies, 229-236
North American Free Trade Agreement, 14
Notes/memoranda, 208-209
Notice of Proposed Rulemaking, 13, 420
Notices, 315-316, 383
Notional amount, 34
Notional quantity, 34

O

Obligations, 171-175, 266-268, 379:
   seller, 171-173;
   buyer, 173-175
Obligations of buyer (CISG), 173-175:
   buyer's breach, 174-175
Obligations of seller (CISG), 171-173:
   seller's breach, 172-173
Offer (contract), 110, 219-221, 365
Offshore gas, 420
Old gas, 11
Onshore gas, 420
Open access, 46, 420
Operational balancing agreement, 421
Operational flow order, 421
Option exercise, 39-40

Order No. 436, 11-12, 15, 20, 43, 60-61, 95-98, 421:
    nondiscriminatory access, 96;
    unbundled rates, 96;
    interruptible transportation, 96;
    capacity allocation, 96;
    curtailment policy, 97;
    stranded costs, 97;
    contract demand conversion, 97;
    take-or-pay credits, 97-98;
    gas inventory charges, 98
    passthrough of settlement costs, 98
Order No. 497, 12, 421
Order No. 500, 421
Order No. 587, 41-42
Order No. 636, 13-16, 99-105, 421:
    unbundling sales service, 99;
    compatibility of access, 99;
    market centers, 99-100;
    open access storage, 100;
    upstream capacity, 100-101;
    capacity reallocation, 101-103;
    capacity release, 101-103;
    straight fixed-variable rate, 103-104;
    blanket sales certificates, 104;
    receipt and delivery point, 104-105
Organizational needs, 55-63:
    control, 56-57;
    force majeure clause (producer), 57-60;
    flexibility, 60-63
Orifice meter, 421
Origination group, 187, 204-205
Outer continental shelf, 14, 59, 87, 422
Outer Continental Shelf Lands Act, 14
Over-the-counter, 422
Over-the-counter derivatives, 341-360, 400-402, 422:

Index

documentation, 341-360;
court cases, 400-401;
internet, 401;
publications, 401-402;
definition, 422
Over-the-counter derivatives documentation, 341-360:
Energy Risk Management Association agreement, 341-342;
International Swaps and Derivatives Association agreement, 341-344;
representations, 344-346;
default/early termination, 346-360;
expenses, 358;
jurisdiction, 358-360;
set-off, 360

P

Paper pooling, 422
Parallelism, 27
Parol evidence rule, 212, 365-366
Parties of contract, 321-322
Partnership agreement, 191-192
Partnership (enforceability), 191-192
Passing of risk (CISG), 175-177: obligations, 175-177
Passthrough (settlement costs), 98
Payment, 80, 215, 282-284, 355-358, 381-382:
sellers, 80;
terms, 215;
default, 355-358;
early termination, 355-358
Peak day, 422
Peak demand, 422
Peak load, 70-71, 73
Peak shaving, 422
Performance, 84, 110, 142-146, 157-158, 218, 226, 379:
guaranty, 84;

455

    insurable interest, 142-143;
    risk of loss, 142-143;
    duties (seller/buyer), 157-158;
    rules on delivery, 158;
    installment deliveries, 158;
    delivery to carrier, 158;
    without contract, 218, 226
Performance guaranty, 84
Performance without contract, 218, 226
Petroleos Mexicanos SA (PEMEX), 19
Phillips case, 10
Physical gas industry, 4-5
Pipeline, 3-5, 422-423
Pipeline day, 422
Pipeline interconnect, 423
Pipeline pressure, 117
Pipeline quality (gas), 4
Plain vanilla swap, 34
Players (gas industry), 51-90:
    industry standards, 53-54;
    business interests, 54-55;
    organizational needs, 55-63;
    contract language, 55-63, 75, 77-79;
    force majeure clause (marketer), 63-76;
    common interest (sellers/buyers), 79-81;
    limited warranties, 81-83;
    buyers, 83-86
Pooling point, 423
Pooling service, 287
Power (enforceability), 189-190
Practical tips (contractors), 377-402:
    contract checklist, 377-387;
    contract negotiations, 388-393;
    contract drafting, 394-398;
    references/resources, 398-402;

sale of goods law, 398-399;
over-the-counter derivatives, 400-402
Pre-contract writings, 204-205
Prearranged release, 423
Presupposed condition, 145-146
Price, 24, 26, 31-32, 35-37, 94, 154, 215-217, 247, 249-251, 279, 423:
  indicators, 24;
  risk, 26, 31-32, 249-251;
  volatility, 30-31;
  establishment, 35-37;
  pricing categories, 94;
  redetermination clause, 423
Price establishment, 35-37
Price indicators, 24
Price redetermination clause, 423
Price risk, 26, 31-32, 249-251:
  transfer, 31-32
Price risk transfer, 31-32
Price volatility, 30-31
Pricing categories, 94
Prime rate, 285
Privacy (internet), 362-363
Privity of contract, 137, 139, 268
Problem cause (records), 202-204
Problem nature, 202-204
Problem practices, 186-188, 201-204, 224-226
Problem resolution, 203-204
Procedural unconscionability, 133
Procedures, 262-263, 378: contract, 262-263
Processing agreement, 119
Processing plant, 423
Producer (definition), 423
Production area, 423
Propane, 423

Property interest, 373
Property transfer, 156-157
Propylene, 423
Provider issues, 371-372
Public Service Commission, 16-17
Public Utility Commission, 14, 16-17
Publications (resources), 398-399, 401-402
Purpose, 262-262, 378:
   contract, 261-262
Put/floor (option), 38, 424

Q

Quality (definition), 424
Quality control, 379-380
Quantity (contract performance), 217
Quantity (definition), 424

R

Rate schedule, 424
Rate zones, 424
Raw gas, 424
Readjustment (contract), 129-132
Reasonableness/prudency review, 68-69, 249
Receipt/delivery point, 104-105, 108, 424:
   designation, 108
Recitals, 378
Records, 193-195, 197-209:
   automatic fax confirmation, 197;
   tape recordings, 198;
   business records, 199;
   litigation location, 199-200;
   enforceability, 201;
   applying practices, 201-204;
   problem, 202-204;

pre-contract writings, 204-205;
request for proposal, 205-207;
letters of intent, 208;
negotiations, 208;
notes/memoranda, 208-209
Records problem, 202-204:
nature of, 202-204;
resolution, 203-204
Redact, 236
References/resources, 398-402:
sale of goods law, 398-399;
over-the-counter derivatives, 400-402
Regulations, 5, 14-20:
U.S., 14-17;
Canada, 17-19, 105-107;
Mexico, 19-20, 107
Regulators, 9, 14
Regulatory changes, 9-14
Release capacity, 424
Reliability, 67-69, 363
Remedies, 146, 160-161, 226-236, 380:
seller, 160;
buyer, 160-161;
non-performance, 226-236
Remedies for non-performance, 226-236:
failure (seller/buyer), 228, 232, 235;
repudiation, 228-232;
breach of warranty, 232-234;
damages, 234-235;
failure to take delivery, 235;
failure to pay, 235-236
Replevin, 231, 239
Representations, 330-331, 344-346, 387
Repudiation, 143-146, 178, 228-232:
assurance of performance, 144;

anticipatory, 145, 230;
substituted performance, 145-146;
presupposed condition, 145-146;
seller's, 230-231;
buyer's, 231-232
Request for proposal, 205-207
Resale (buyer/seller), 159
Responsibility (warranties), 81
Rich gas, 424
Right to terminate, 352-353
Rights of unpaid seller against goods, 158-159:
stoppage of goods in transit, 159;
resale (buyer/seller), 159
Risk of loss, 142-143, 157, 243, 247:
transferred with property, 157
Risk transferred with property, 157
Routine practices, 186-188, 201-204, 224-226
Royalty, 425
Royalty owner, 425
Rules of evidence role, 196-197
Rules on delivery, 158

S

Sale by description, 155
Sale contract, 154
Sale for resale, 425
Sale of goods, 124, 151-184:
Alberta Sale of Goods Act, 153-161;
Canada, 153-161;
Convention on Contracts for the International Sale of Goods (UN), 161-177
Sale of goods (CISG), 170-175:
general provisions, 170-171;
obligations of seller, 171-173;
obligations of buyer, 173-175

Index

Sale of goods (North America), 151-184:
  Alberta Sale of Goods Act, 153-161;
  Convention on Contracts for the International Sale of Goods (UN), 161-177
Sale of goods law, 398-399:
  publications, 398-399;
  internet, 399;
  electronic communications, 399
Sales (UCC), 124-146:
  title/construction/subject matter, 125-129;
  form/formation/readjustment, 129-132;
  general obligation and construction, 132-140;
  title/creditors/good faith purchasers, 140-142;
  performance, 142-143;
  breach/repudiation/excuse, 143-146;
  remedies, 146
Sales affected, 118-122:
  contracts, 118-120;
  midstream marketing, 120;
  consumers, 120-122;
  LDCs, 120-122
Sales tax, 83
Schedule, 259, 267
Scheduling, 268-269, 425
Scheduling penalty, 425
Scope of Agreement, 261
Seasonal rates, 425
Secondary market, 102
Secured transactions (UCC), 124
Security interest, 345
Security of supply, 77
Self-implementing transportation, 12, 47, 425
Seller, 79-81, 172-173, 230-231, 251, 425:
  payment, 80;
  limiting liability, 80-81;

breach of contract, 172-173;
repudiation, 230-231;
deliverability, 251
Seller's breach of contract, 172-173
Seller's deliverability, 251
Seller's repudiation, 230-231
Service area, 425
Service marketing, 44
Set-off, 360
Settlement amount, 293
Settlement price, 426
Severability, 33-332, 387
Severance tax, 82, 426
Shipper, 426
Short-term contracts, 244-246:
    definition, 244-245;
    development, 245-246;
    future, 246
Shrinkage, 426
Shut in well, 11, 426
Signatures, 334-335, 387
Sour gas, 426
Specified transaction, 349
Sphere of application (CISG), 162-165
Spot (definition), 426
Spot market, 245, 427
Standard contract elements, 254-255
Standard in the industry, 54
Standards development, 40-42
Statute of frauds, 154, 195-196, 366
Stoppage of goods in transit, 159
Storage, 100, 427
Storage access, 100
Storage field, 427
Storage gas, 427

Index

Storms, 59
Straight fixed-variable rate, 16, 103-104, 427:
　rate setting, 16
Stranded costs, 97-98
Strike price, 427
Subject matter, 125-129
Substantitive unconscionability, 133
Substituted performance, 145-146
Success elements (futures contracts), 30-31:
　homogenous commodity, 30;
　substantial trading activity, 30;
　volatility, 30-31
Suppliers, 65, 73
Swap, 36-37, 259, 427:
　examples, 36-37
Swap examples, 36-37:
　producer/supplier, 35-36;
　marketer, 36-37
Swaptions, 40
Sweet gas, 427
Swing, 259, 427

T

Tailgate, 428
Take-or-pay clause, 428
Take-or-pay credits, 907-98
Take-or-pay settlement costs, 98
Tape recordings, 198
Tariff, 105-106, 428
Tax exemption certificate, 62, 88
Taxes, 82-83, 308-311, 380-381
Technical language, 258
Term (contract), 265-266, 384
Term (definition), 428
Term contract, 246-254, 260

463

Term sale, 253
Termination/cancellation, 128-129, 351-358:
   right to terminate, 352-353;
   effect of designation, 353-354;
   calculations, 354-355;
   payments, 355-358
Termination vs. cancellation, 128-129
Terminology, vii-viii, 403-430. SEE ALSO Definitions.
Third-party beneficiaries, 332-333
Third-party systems, 370-372
Time of transfer, 156-157
Title, 85-86, 124-129, 135-136, 140-142, 157, 256, 382-383, 428:
   buyer, 85-86;
   transfer, 124-125, 157, 428;
   contract, 125-129, 140-142, 256;
   warranty, 135-136;
   contract, 256
Title transfer, 124-125, 157, 428
Title warranty, 135-136
Tolls, 106
Trading activity, 30
Trading partner agreement, 367-370, 429
Traditional contracts, 248
Transaction, 253-254, 264-265, 349:
   characteristics, 253-254;
   specified, 349
Transfer of property, 156
Transfer of title, 124-125, 157, 428
Transportation, 91-95, 99, 105-107, 268-270, 379, 428-429:
   service agreements, 91, 268-270, 428;
   U.S. gas law, 92-95;
   service provider, 99;
   Canada, 105-107;
   Mexico, 107;
   costs, 428;

imbalance, 429
Transportation costs, 428
Transportation imbalance, 429
Transportation laws/regulations, 91-108:
  gas law (U.S.), 92-95;
  FERC orders, 95-105;
  Canada, 105-107;
  Mexico, 107
Transportation laws/regulations (Canada), 105-107:
  tariffs, 105-106;
  tolls, 106;
  natural gas exports, 106-107
Transportation laws/regulations (Mexico), 107
Transportation laws/regulations (U.S.), 92-105:
  gas law, 92-95;
  FERC orders, 95-105
Transportation service agreements, 91, 268-270, 428
Transportation service provider, 99
Transporter, 429
Trigger price, 429
Trunkline pipelines, 47-48, 117

U

U.S., 14-17, 92-150, 161:
  regulations, 14-17;
  gas law, 92-95;
  Natural Gas Act of 1938, 92-93;
  Natural Gas Policy Act of 1978, 94-95;
  Natural Gas Wellhead Decontrol Act of 1989, 95;
  FERC orders, 95-105;
  gas sale laws, 109-150
U.S. gas sale laws, 109-150:
  contract description, 110-111;
  commercial laws, 111-118;
  sales affected, 118-122;

UCC, 122-146
U.S. regulations, 14-17, 92-105:
  interstate pipelines, 16-17;
  LDCs, 16-17;
  gas law, 92-95;
  FERC orders, 95-105
UCC, 111-113, 122-146, 151-153, 210-212, 429:
  general provisions, 122-123;
  letters of credit, 123-124;
  secured transactions, 124;
  sales, 124-146
Unassociated gas, 429
Unbundled rates, 96
Unbundling sales service, 99
Understanding issues, 78
Understanding language, 78
Uniform Commercial Code. SEE UCC.
Uniform Law Conference of Canada, 152
United Nations Convention on Contracts for the International Sale of Goods, 161-177:
  sphere of application, 162-165;
  general provisions, 165-167;
  formation of contract, 168-170;
  sale of goods, 170-175;
  passing of risk, 175-177
Upstream, 429
Upstream capacity, 100-101
Upstream pipeline, 429
Usage of trade, 212-214
Use tax, 83
User issues, 370-371
Utility tax, 311

V

Venue and governing law, 321-335, 386:

governing law, 321-322;
  legal venue, 322-335, 386
Volatility, 30-31
Volume, 430

W

Waha contract, 29
Waiver, 333-334, 386
Warrant of deliverability, 84
Warranties, 71, 81, 84, 134-135, 138-140, 155, 312-315, 382-383, 430:
  commitment, 71, 430;
  disclaimers, 81, 138;
  deliverability, 84;
  by seller, 134-135;
  to third parties, 138-140
Warranty commitment, 71, 430
Warranty disclaimers, 81, 138
Weather phenomena, 59-60
Wellhead (definition), 430
Wellhead price, 430
Widget rules, 115-116
Word usage, 78, 188, 258-261, 277:
  technical, 258;
  unordinary, 258-261;
  vague/ambiguous, 260-261
Written contract, 195-196

Z

Zone (definition), 430
Zone rate, 430